AF361863

Anticancer Drugs Activity and Underlying Mechanisms

Anticancer Drugs Activity and Underlying Mechanisms

Editor

Domenico Iacopetta

MDPI • Basel • Beijing • Wuhan • Barcelona • Belgrade • Manchester • Tokyo • Cluj • Tianjin

Editor
Domenico Iacopetta
University of Calabria
Italy

Editorial Office
MDPI
St. Alban-Anlage 66
4052 Basel, Switzerland

This is a reprint of articles from the Special Issue published online in the open access journal *Applied Sciences* (ISSN 2076-3417) (available at: https://www.mdpi.com/journal/applsci/special_issues/anticancer_drugs_Activity_and_mechanisms).

For citation purposes, cite each article independently as indicated on the article page online and as indicated below:

LastName, A.A.; LastName, B.B.; LastName, C.C. Article Title. *Journal Name* **Year**, *Volume Number*, Page Range.

ISBN 978-3-0365-2119-0 (Hbk)
ISBN 978-3-0365-2120-6 (PDF)

Cover image courtesy of Domenico Iacopetta.

© 2021 by the authors. Articles in this book are Open Access and distributed under the Creative Commons Attribution (CC BY) license, which allows users to download, copy and build upon published articles, as long as the author and publisher are properly credited, which ensures maximum dissemination and a wider impact of our publications.
The book as a whole is distributed by MDPI under the terms and conditions of the Creative Commons license CC BY-NC-ND.

Contents

About the Editor

Domenico Iacopetta received his degree in Chemistry and Pharmaceutical Technology and, later, his Doctoral degree in "Cellular Biochemistry and drug activity in Oncology", University of Calabria (Italy). He was Postdoctoral Fellow in Biochemistry (Italy) and Postdoctoral Associate in cancer research at the Baylor College of Medicine, Houston (TX, USA). He was Postdoctoral fellow in Medicinal Chemistry at the CERMN (University of Caen, France). He is contract professor the Department of Pharmacy, Health and Nutritional Science, "Dept. of Excellence", at the University of Calabria for the Master's course of level II in "Nutrition and Nutraceutical Supplements". His scientific activity encompasses medicinal chemistry, oncology, biochemistry and molecular biology studies, particularly the synthesis and biological evaluation of compounds interacting with cell targets, amongst them human topoisomerases, actin and tubulin. He is the author of high H-index papers and patents.

Preface to "Anticancer Drugs Activity and Underlying Mechanisms"

This Special Issue of *Applied Sciences* includes some of the latest research and literature studies regarding the synthesis and biological evaluation of new and old drugs, aiming to detangle the complexity of the molecular mechanisms underlying their action in cancer cells. The main goal of this Special Issue is the dissemination of scientific results and inncvative ideas amongst researchers and students, stimulating their interest in drug science and pushing them to contribute to further progress in the field of Medicinal Chemistry and Cancer Biology. Finally, it is a great pleasure to express my sincere gratitude to the Editors and Assistants of Applied Sciences that kindly supported me for this Special Issue and all the authors who contributed.

Domenico Iacopetta

Editor

Editorial

Special Issue on "Anticancer Drugs Activity and Underlying Mechanisms"

Domenico Iacopetta

Department of Pharmacy, Health and Nutritional Sciences, University of Calabria, 87036 Arcavacata di Rende, Italy; domenico.iacopetta@unical.it

check for **updates**

Citation: Iacopetta, D. Special Issue on "Anticancer Drugs Activity and Underlying Mechanisms". *Appl. Sci.* **2021**, *11*, 8169. https://doi.org/10.3390/app11178169

Received: 26 August 2021
Accepted: 1 September 2021
Published: 3 September 2021

Publisher's Note: MDPI stays neutral with regard to jurisdictional claims in published maps and institutional affiliations.

Copyright: © 2021 by the author. Licensee MDPI, Basel, Switzerland. This article is an open access article distributed under the terms and conditions of the Creative Commons Attribution (CC BY) license (https://creativecommons.org/licenses/by/4.0/).

1. Introduction

Cancer is a reputed non-communicable disease, namely a non-transmittable illness affecting humankind, which represents a major public health issue and is one of the leading causes of death worldwide [1]. In 2021, it is predicted that there will be 1,898,160 new cancer cases and 608,570 cancer deaths in the United States [2]; in the European Union and in United Kingdom [3], 1,267,000 and 176,400 cancer deaths are predicted to occur, respectively.

Several factors concur with cancer onset and progression, leading to an uncontrollable cell growth, the type of which has been used to classify the different kinds of cancer [1]. To date, surgery, radiation and chemotherapy are considered to be the main approaches to cancer, but an important contribution has been made by an extensive and careful prevention campaign that has been met with public consent. Nevertheless, efforts made by the scientific community, and the continuous increase of improved tools to prevent/treat this disease, numerous obstacles, including the heterogeneity of cancer types/subtypes, the limited treatment efficacy and the occurring toxicity, together with resistance onset and relapse phenomena, make this fight hard to win completely [4]. The chemotherapeutic approach may reach cancer cells in all body tissues and may hamper both the cell growth at the original site and the possible metastases, but the drawbacks represented by dramatic side effects and cancer cell resistance often arise [5]. The ability of cancer cells to exploit the salvage or compensatory pathways that counteract the efficacy of chemotherapy is only the tip of the iceberg. Indeed, a first approach to drug design was based on obtaining a drug targeting a primary (and single) cell component. However, a drug (or its metabolites) can potentially manifest several "off-target" activities, which can be adverse (the so-called negative side effects), neutral, or, hopefully, beneficial [6]. More recently, a multi-target pharmacological drug approach has been recorded, particularly in terms of drug design, discovery, and repositioning [7,8] as well as for the employment of relevant and effective drug combinations/synergy [9]. Great help comes from nature. Indeed, natural compounds and their derivatives represent a valuable source of compounds with anticancer or preventive properties [10].

In this context, great interest in proposing new compounds, or repurposing the old ones, with anticancer properties has been recorded. Thus, this Special Issue, which includes four research papers, a hypothesis, and five literature reviews, offered the opportunity to approach cancer treatment from different points of view.

2. Contributions

The need of suitable therapies to treat cancer is the "primum movens" of the drug discovery process that, as reported by Cava et al. [11], often starts from academic studies that, hopefully, can be translated "from the bench to the bedside" and offer a larger and alternative arsenal to fight cancer. From this point of view, the combined use of in silico and in vivo studies is essential in medicinal chemistry in order to identify putative targets and to explore the anticancer properties of newly synthesized molecules [12] in a given cell model

1

prior to proceeding towards in vivo studies. Amongst them, the metal complexes, following the successful employment of *Cis*platin in cancer treatment, has attracted the attention of many researchers because of their different chemical and biological properties [13,14].

In this context, two research studies and one review paper published in this Special Issue are representative of this idea. The first, by Skoupilova et al. [15] reported a series of ferrocene derivatives from the general formula [Fe(η^5-C$_5$H$_4$CH$_2$(p-C$_6$H$_4$)CH$_2$(N-het))$_2$], which bear either substituted or unsubstituted saturated five- and six-membered nitrogen-containing heterocycles that are able to inhibit the cervical cancer cell growth, with or without the contemporaneous exposure to ionizing radiation. The authors demonstrated that these complexes possessed higher anticancer activity than *Cis*platin, which was used as reference, and that the exposure to an ionizing radiation increased the anticancer properties, probably because of a radiosensitizing phenomenon that should be further investigated. Cervical cancer cells treated with the lead ferrocene complex underwent intrinsic apoptosis and autophagy with ROS levels increase, as demonstrated by flow cytometry, immunofluorescence, and Western blotting studies. Most importantly, the lead complex exhibited a mild cytotoxic effect on the normal cell lines, which were used as controls, revealing a better cytotoxic profile than *Cis*platin. The second, by Iacopetta et al. [16], describes the study of some N-heterocyclic carbene (NHC)-gold(I) complexes, whose multiple biological activities have already been described [14,17–19], disclosing another important target, namely the intracellular actin, and demonstrating that these complexes hamper actin polymerization by means of docking simulations, immunofluoresce, and direct enzymatic assays. These studies highlight the multi-target potential of NHC-gold(I) complexes, whose anticancer properties come together with negligible cytotoxic effects on the normal cells, paving the way to further modifications with the goal of obtaining new and effective anticancer drugs.

This even more topical field of research has been reviewed by Ielo et al. [20] in detail, who summarized the most salient and up to date papers related to antitumor activity, particularly in terms of breast cancer treatment, drug delivery systems, nanosystems, and complexes based on gold. The authors highlighted how gold-based systems are able to overcome *Cis*platin resistance and its dramatic toxic effects, indicating a desirable therapy personalization that may offer a targeted and more effective treatment and less side effects.

An interesting contribution to the area of drug repurposing against cancer has been made by Barbarossa et al. [21], who reported a library of thalidomide analogs. Thalidomide, a historically well-known drug, has recently been repurposed for its anticancer, antiangiogenic, and immunomodulatory actions, and several analogs with improved efficacy and reduced toxicity have been proposed [22,23]. In this paper, the authors reported the anticancer properties of phthalimide derivatives in a panel of cancer cell lines, mostly against A2058 melanoma cells, individuating a lead compound that is able to block melanoma cell growth by interfering with the tubulin network. Exposure to this compound leads to DNA damage and triggers melanoma cell death by means of the apoptotic mechanism without affecting the growth of normal cell lines.

Again, Catalano et al. [24] reviewed the different properties of diarylureas [8,25], namely ureas bearing two aromatic moieties as substituents, focusing on their role as important pharmacophore in anticancer drugs over the past 10 years. A clinically used member of this class, sorafenib was the lead compound approved from Food and Drug Administration (FDA) and the European Medicinal Agency (EMEA) for the treatment of advanced metastatic hepatocellular carcinoma and advanced renal cell carcinoma, paving the way to other diarylureas derivatives, such as regorafenib, linifanib, tivozanib and ripretinib, which share the ability to inhibit the kinases.

Finally, Iacopetta et al. [26] conducted a literature study reviewing the research conducted on on mono- and bis-Schiff bases within the past few decades, determining several applications and various biological properties [27] and highlighting the compounds with high antitumor properties that fall in the micromolar to nanomolar range. The authors extrapolated upon the literature results, evidencing the versatility of these compounds, both by themselves or in association with metal complexes, indicating a high and broad range anticancer activity with few or no effects on the viability of normal cells. These compounds are able to target different cell components, such as DNA, kinases, redox enzymes, etc.

Nature is an incredible source of drugs [28–30], including interesting bioactive anti-cancer molecules, even though they sometimes exhibit bioavailability issues that can be overcome by the use of proper vehicles or chemical modifications [31–33]. Concerning this, Do et al. [34] reported the anticancer properties of 1-(5,7-dimetoxy-2,2dimetyl-2H-cromen-8-yl)-but-2-en-1-on (malloapelta B, malB), isolated from *Mallotus apelta*, which is able to inhibit the activation of nuclear factor kappa B (NF-kB) and is responsible for down-regulating pivotal genes involved in inflammation. However, this compound possesses unfavorable features, such as low solubility and high toxicity; thus, in order to overcome these pharmaceutical limitations, the authors entrapped the malB into nanoliposomes, fully characterized, and studied their anticancer and antitumor properties against the lung carcinoma in vitro and in vivo, demonstrating improved antitumor activity in vivo with respect to the free malB form.

A review study from Chang et al. [35] reported the recent results determining the role of second mitochondria-derived activators of apoptosis (smac) mimetics, birinapant, LCL161, and GDC-0152, in cancer treatment. These molecules, which have entered in phase 1 and 2 clinical trials, are able to induce the non-canonical NF-kB signaling pathway and downregulate the protein expression inhibitor level of apoptosis proteins (IAPs), leading to cells death by apoptosis, even though other mechanisms are still under investigation. The versatility of the smac mimetics resides, for instance, in their possible combination with other clinically used anticancer strategies ("classic" chemotherapy, radiotherapy, and immune therapy) that, together with their safety and the possibility of being coupled with the nanotechnology, make these molecules very attractive in the fight against cancer.

Finally, an interesting hypothesis has been reported by Metzler et al. [36] in their description of a case of a patient with a low-grade ovarian cancer, in which ibrutinib treatment leads to CA-125 suppression, which is reported for the first time in this study. However, further studies are needed in order to understand the underlying mechanisms.

Funding: This research received no external funding.

Acknowledgments: I would like to heartily thank all the authors who contributed to this Special Issue, the reviewers, and the editorial team of *Applied Sciences*. A particular thank to the Assistant Editor, Sara Zhan. Her professional and constant support contributed to making this Special Issue successful. Last but not least, my deepest gratitude to Professor Sebastiano Andò, who pushed me to improve my scientific skills overseas and to Professor Maria Stefania Sinicropi, who welcomed me into her laboratories, making it possible to complete my scientific research.

Conflicts of Interest: The author declares no conflict of interest.

References

1. Pereyra, C.E.; Dantas, R.F.; Ferreira, S.B.; Gomes, L.P.; Silva, F.P., Jr. The diverse mechanisms and anticancer potential of naphthoquinones. *Cancer Cell Int.* **2019**, *19*, 207. [CrossRef]
2. Siegel, R.L.; Miller, K.D.; Fuchs, H.E.; Jemal, A. Cancer Statistics, 2021. *CA Cancer J. Clin.* **2021**, *71*, 7–33. [CrossRef] [PubMed]
3. Carioli, G.; Malvezzi, M.; Bertuccio, P.; Boffetta, P.; Levi, F.; La Vecchia, C.; Negri, E. European cancer mortality predictions for the year 2021 with focus on pancreatic and female lung cancer. *Ann. Oncol.* **2021**, *32*, 478–487. [CrossRef]
4. Hutchinson, L.; Kirk, R. High drug attrition rates—Where are we going wrong? *Nat. Rev. Clin. Oncol.* **2011**, *8*, 189–190. [CrossRef]
5. Wang, X.; Zhang, H.; Chen, X. Drug resistance and combating drug resistance in cancer. *Cancer Drug Resist.* **2019**, *2*, 141–160. [CrossRef]
6. Overington, J.P.; Al-Lazikani, B.; Hopkins, A.L. How many drug targets are there? *Nat. Rev. Drug Discov.* **2006**, *5*, 993–996. [CrossRef] [PubMed]
7. Puxeddu, M.; Shen, H.; Bai, R.; Coluccia, A.; Bufano, M.; Nalli, M.; Sebastiani, J.; Brancaccio, D.; Da Pozzo, E.; Tremolanti, C.; et al. Discovery of pyrrole derivatives for the treatment of glioblastoma and chronic myeloid leukemia. *Eur. J. Med. Chem.* **2021**, *221*, 113532. [CrossRef]
8. Catalano, A.; Iacopetta, D.; Pellegrino, M.; Aquaro, S.; Franchini, C.; Sinicropi, M.S. Diarylureas: Repositioning from Antitumor to Antimicrobials or Multi-Target Agents against New Pandemics. *Antibiotics* **2021**, *10*, 92. [CrossRef]

9.   Shi, D.; Khan, F.; Abagyan, R. Extended Multitarget Pharmacology of Anticancer Drugs. *J. Chem. Inf. Modeling* **2019**, *59*, 3006–3017. [CrossRef] [PubMed]

10.   Huang, M.; Lu, J.-J.; Ding, J. Natural Products in Cancer Therapy: Past, Present and Future. *Nat. Prod. Bioprospecting* **2021**, *11*, 5–13. [CrossRef]

11.   Cava, C.; Castiglioni, I. Integration of Molecular Docking and In Vitro Studies: A Powerful Approach for Drug Discovery in Breast Cancer. *Appl. Sci.* **2020**, *10*, 6981. [CrossRef]

12.   Iacopetta, D.; Catalano, A.; Ceramella, J.; Barbarossa, A.; Carocci, A.; Fazio, A.; La Torre, C.; Caruso, A.; Ponassi, M.; Rosano, C.; et al. Synthesis, anticancer and antioxidant properties of new indole and pyranoindole derivatives. *Bioor. Chem.* **2020**, *105*, 104440. [CrossRef] [PubMed]

13.   Bruno, G.; Nicolò, F.; Lo Schiavo, S.; Sinicropi, M.S.; Tresoldi, G. Synthesis and spectroscopic properties of di-2-pyridyl sulfide (dps) compounds. Crystal structure of [Ru(dps)2Cl2]. *J. Chem. Soc. Dalton Trans.* **1995**, *1*, 17–24. [CrossRef]

14.   Ceramella, J.; Mariconda, A.; Iacopetta, D.; Saturnino, C.; Barbarossa, A.; Caruso, A.; Rosano, C.; Sinicropi, M.S.; Longo, P. From coins to cancer therapy: Gold, silver and copper complexes targeting human topoisomerases. *Bioor. Med. Chem. Lett.* **2020**, *30*, 126905. [CrossRef] [PubMed]

15.   Skoupilova, H.; Rak, V.; Pinkas, J.; Karban, J.; Hrstka, R. The Cytotoxic Effect of Newly Synthesized Ferrocenes against Cervical Carcinoma Cells Alone and in Combination with Radiotherapy. *Appl. Sci.* **2020**, *10*, 3728. [CrossRef]

16.   Iacopetta, D.; Ceramella, J.; Rosano, C.; Mariconda, A.; Pellegrino, M.; Sirignano, M.; Saturnino, C.; Catalano, A.; Aquaro, S.; Longo, P.; et al. N-Heterocyclic Carbene-Gold(I) Complexes Targeting Actin Polymerization. *Appl. Sci.* **2021**, *11*, 5626. [CrossRef]

17.   Iacopetta, D.; Rosano, C.; Sirignano, M.; Mariconda, A.; Ceramella, J.; Ponassi, M.; Saturnino, C.; Sinicropi, M.S.; Longo, P. Is the Way to Fight Cancer Paved with Gold? Metal-Based Carbene Complexes with Multiple and Fascinating Biological Features. *Pharmaceuticals* **2020**, *13*, 91. [CrossRef]

18.   Iacopetta, D.; Mariconda, A.; Saturnino, C.; Caruso, A.; Palma, G.; Ceramella, J.; Muia, N.; Perri, M.; Sinicropi, M.S.; Caroleo, M.C.; et al. Novel Gold and Silver Carbene Complexes Exert Antitumor Effects Triggering the Reactive Oxygen Species Dependent Intrinsic Apoptotic Pathway. *ChemMedChem* **2017**, *12*, 2054–2065. [CrossRef] [PubMed]

19.   Saturnino, C.; Barone, I.; Iacopetta, D.; Mariconda, A.; Sinicropi, M.S.; Rosano, C.; Campana, A.; Catalano, S.; Longo, P.; Ando, S. N-heterocyclic carbene complexes of silver and gold as novel tools against breast cancer progression. *Future Med. Chem.* **2016**, *8*, 2213–2229. [CrossRef] [PubMed]

20.   Ielo, I.; Iacopetta, D.; Saturnino, C.; Longo, P.; Galletta, M.; Drommi, D.; Rosace, G.; Sinicropi, M.S.; Plutino, M.R. Gold Derivatives Development as Prospective Anticancer Drugs for Breast Cancer Treatment. *Appl. Sci.* **2021**, *11*, 2089. [CrossRef]

21.   Barbarossa, A.; Catalano, A.; Ceramella, J.; Carocci, A.; Iacopetta, D.; Rosano, C.; Franchini, C.; Sinicropi, M.S. Simple Thalidomide Analogs in Melanoma: Synthesis and Biological Activity. *Appl. Sci.* **2021**, *11*, 5823. [CrossRef]

22.   Barbarossa, A.; Iacopetta, D.; Sinicropi, M.S.; Franchini, C.; Carocci, A. Recent Advances in the Development of Thalidomide-Related Compounds as Anticancer Drugs. *Curr. Med. Chem.* **2021**, *28*. [CrossRef]

23.   Iacopetta, D.; Carocci, A.; Sinicropi, M.S.; Catalano, A.; Lentini, G.; Ceramella, J.; Curcio, R.; Caroleo, M.C. Old Drug Scaffold, New Activity: Thalidomide-Correlated Compounds Exert Different Effects on Breast Cancer Cell Growth and Progression. *ChemMedChem* **2017**, *12*, 381–389. [CrossRef]

24.   Catalano, A.; Iacopetta, D.; Sinicropi, M.S.; Franchini, C. Diarylureas as Antitumor Agents. *Appl. Sci.* **2021**, *11*, 374. [CrossRef]

25.   Catalano, A.; Iacopetta, D.; Rosato, A.; Salvagno, L.; Ceramella, J.; Longo, F.; Sinicropi, M.S.; Franchini, C. Searching for Small Molecules as Antibacterials: Non-Cytotoxic Diarylureas Analogues of Triclocarban. *Antibiotics* **2021**, *10*, 204. [CrossRef] [PubMed]

26.   Iacopetta, D.; Ceramella, J.; Catalano, A.; Saturnino, C.; Bonomo, M.G.; Franchini, C.; Sinicropi, M.S. Schiff Bases: Interesting Scaffolds with Promising Antitumoral Properties. *Appl. Sci.* **2021**, *11*, 1877. [CrossRef]

27.   Catalano, A.; Sinicropi, M.S.; Iacopetta, D.; Ceramella, J.; Mariconda, A.; Rosano, C.; Scali, E.; Saturnino, C.; Longo, P. A Review on the Advancements in the Field of Metal Complexes with Schiff Bases as Antiproliferative Agents. *Appl. Sci.* **2021**, *11*, 6027. [CrossRef]

28.   Ceramella, J.; Loizzo, M.R.; Iacopetta, D.; Bonesi, M.; Sicari, V.; Pellicanò, T.M.; Saturnino, C.; Malzert-Fréon, A.; Tundis, R.; Sinicropi, M.S. Anchusa azurea Mill. (Boraginaceae) aerial parts methanol extract interfering with cytoskeleton organization induces programmed cancer cells death. *Food Funct.* **2019**, *10*, 4280–4290. [CrossRef]

29.   Fazio, A.; Iacopetta, D.; La Torre, C.; Ceramella, J.; Muià, N.; Catalano, A.; Carocci, A.; Sinicropi, M.S. Finding solutions for agricultural wastes: Antioxidant and antitumor properties of pomegranate Akko peel extracts and β-glucan recovery. *Food Funct.* **2018**, *9*, 6618–6631. [CrossRef] [PubMed]

30.   Tundis, R.; Iacopetta, D.; Sinicropi, M.S.; Bonesi, M.; Leporini, M.; Passalacqua, N.G.; Ceramella, J.; Menichini, F.; Loizzo, M.R. Assessment of antioxidant, antitumor and pro-apoptotic effects of Salvia fruticosa Mill. subsp. thomasii (Lacaita) Brullo, Guglielmo, Pavone & Terrasi (Lamiaceae). *Food Chem. Toxicol.* **2017**, *106*, 155–164. [PubMed]

31.   Iacopetta, D.; Grande, F.; Caruso, A.; Mordocco, R.A.; Plutino, M.R.; Scrivano, L.; Ceramella, J.; Muià, N.; Saturnino, C.; Puoci, F.; et al. New insights for the use of quercetin analogs in cancer treatment. *Future Med. Chem.* **2017**, *9*, 2011–2028. [CrossRef]

32.   Scrivano, L.; Iacopetta, D.; Sinicropi, M.S.; Saturnino, C.; Longo, P.; Parisi, O.I.; Puoci, F. Synthesis of sericin-based conjugates by click chemistry: Enhancement of sunitinib bioavailability and cell membrane permeation. *Drug Deliv.* **2017**, *24*, 482–490. [CrossRef] [PubMed]

33. Iacopetta, D.; Lappano, R.; Mariconda, A.; Ceramella, J.; Sinicropi, M.S.; Saturnino, C.; Talia, M.; Cirillo, F.; Martinelli, F.; Puoci, F.; et al. Newly Synthesized Imino-Derivatives Analogues of Resveratrcl Exert Inhibitory Effects in Breast Tumor Cells. *Int. J. Mol. Sci.* **2020**, *21*, 7797. [CrossRef] [PubMed]
34. Do, T.T.; Nguyen, T.N.; Do, T.P.; Nguyen, T.C.; Trieu, H.P.; Vu, P.T.T.; Le, T.A.H. Improved Anticancer Activity of the Malloapelta B-Nanoliposomal Complex against Lung Carcinoma. *Appl. Sci.* **2020**, *10*, 8148. [CrossRef]
35. Chang, Y.-C.; Cheung, C.H.A. An Updated Review of Smac Mimetics, LCL161, Birinapant, and GDC-0152 in Cancer Treatment. *Appl. Sci.* **2020**, *11*, 335. [CrossRef]
36. Metzler, J.M.; Fink, D.; Imesch, P. Ibrutinib Could Suppress CA-125 in Ovarian Cancer: A Hypothesis. *Appl. Sci.* **2020**, *11*, 222. [CrossRef]

 *applied sciences*

Article

The Cytotoxic Effect of Newly Synthesized Ferrocenes against Cervical Carcinoma Cells Alone and in Combination with Radiotherapy

Hana Skoupilova [1], Vladimir Rak [2], Jiri Pinkas [3], Jindrich Karban [4] and Roman Hrstka [1,*]

[1] Research Centre for Applied Molecular Oncology, Masaryk Memorial Cancer Institute, Zluty kopec 7, 656 53 Brno, Czech Republic; hana.skoupilova@med.muni.cz

[2] Department of Radiation Oncology, Masaryk Memorial Cancer Institute, Zluty kopec 7, 656 53 Brno, Czech Republic; vladimir.rak@mou.cz

[3] J. Heyrovský Institute of Physical Chemistry, Academy of Sciences of the Czech Republic, v. v. i., Dolejškova 2155/3, 182 23 Prague 8, Czech Republic; jiri.pinkas@jh-inst.cas.cz

[4] Institute of Chemical Process Fundamentals, Academy of Sciences of the Czech Republic, v. v. i., Rozvojová 135, 165 02 Prague 6, Czech Republic; karban@icpf.cas.cz

* Correspondence: hrstka@mou.cz; Tel.: +420-543-133-306

Received: 11 May 2020; Accepted: 26 May 2020; Published: 28 May 2020

Abstract: Cervical cancer is one of the most common types of cancer in women, with approximately 500,000 new cases and 250,000 deaths every year. Radiotherapy combined with chemotherapy represents the treatment of choice for advanced cervical carcinomas. The role of the chemotherapy is to increase the sensitivity of the cancer cells to irradiation. Cisplatin, the most commonly used drug for this purpose, has its limitations. Thus, we used a family of ferrocene derivatives (in addition, one new species was prepared using standard Schlenk techniques) and studied their effects on cervical cancer cells alone and in combination with irradiation. We applied colorimetric assay to determine the cytotoxicity of the compounds; flow cytometry to analyze the production of reactive oxygen species (ROS), cell cycle, and mitochondrial membrane potential (MMP); immunochemistry to study protein expression; and colony forming assay to evaluate changes in radiosensitivity. Treatment with ferrocenes exhibited significant cytotoxicity against cervical cancer cells, associated with increasing ROS production and MMP changes, suggesting the induction of apoptosis. The combined activity of ferrocenes and ionizing radiation highlighted ferrocenes as potential radiosensitizing drugs, while their higher single-agent toxicity in comparison with routinely used cisplatin could also be promising. Our results demonstrate antitumor activity of several tested ferrocenes both alone and in combination with radiotherapy.

Keywords: ferrocenes; chemotherapy; cytotoxic effect; radiotherapy; radiosensitization; irradiation; cell death

---

1. Introduction

Worldwide, cervical carcinoma is the second most common malignancy specific to women. It was estimated that in 2012, more than half a million women were diagnosed with cervical cancer around the world, and approximately quarter of a million died [1,2].

Fortunately, the incidence of cervical cancer is steadily declining in most developed countries [3]. Two main reasons for this trend are effective screening and vaccination against the most common oncogenic human papilloma virus (HPV) strains which cause almost all cervical cancers [4]. Treatment of cervical cancer depends on the stage of disease and ranges from conization (simple removal of abnormal cervical epithelium) or trachelectomy (removal of the whole cervix) to hysterectomy and/or radiotherapy [5]. Radiotherapy is usually used in more advanced cases, either to remove remaining microscopic disease after surgery (adjuvant therapy) or as a main treatment when surgery cannot be performed (curative therapy).

Curative radiotherapy is often combined with chemotherapy to improve the treatment outcome [6]. The most commonly used chemotherapeutic drug in this case is cisplatin. Adding a weekly infusion of cisplatin to radiotherapy improves five-year overall survival by 6% and disease-free survival by 8% [6].

Although platinum substances are widely used, they can cause serious side effects, including renal, neural, and gastrointestinal toxicity, that limit their usefulness [7]. Thus, not only the advantages but also the disadvantages of cisplatin stimulated further research into other types of compounds containing metal in their structure. Besides platinum complexes, species containing iron, titanium, ruthenium, gold, or palladium have been synthesized and tested [8–14]. Among the most intensively studied substances exhibiting radiosensitization effects are ruthenium compounds [15–17] or gold nanoparticles [16], as demonstrated also by in vivo screenings [18–20]. Nevertheless, despite a great deal of research into iron-containing antitumor compounds, there have been only a few investigations into the potential radiosensitizing effects of these compounds in cancer cells [21,22].

This article is focused on the series of ferrocene derivatives of the general formula $[Fe(\eta^5\text{-}C_5H_4CH_2(p\text{-}C_6H_4)CH_2(N\text{-het}))_2]$ bearing either substituted or unsubstituted saturated five- and six-membered nitrogen-containing heterocycles (Figure 1). In vitro cytotoxicity analysis was performed on cell lines derived from cervical cancer. In the case of highly toxic ferrocenes, we focused especially on their mechanisms of action in terms of the disruption of cell metabolism and examination of cellular mechanisms leading to cell death. The combined effect of ferrocenes with ionizing radiation was also determined to test the possible use of these ferrocenes as potential radiosensitizing agents.

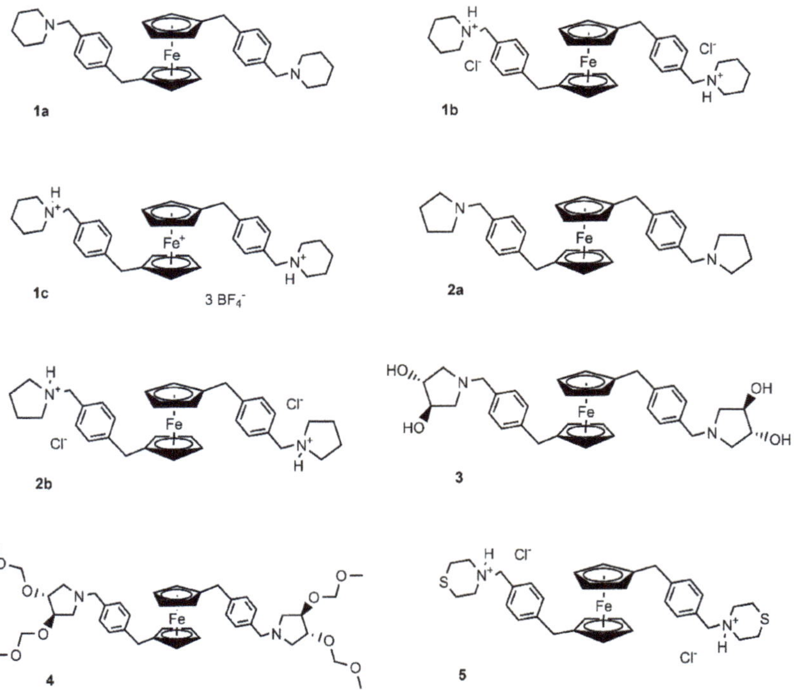

Figure 1. Structures of the tested ferrocenes.

2. Materials and Methods

2.1. Preparation and Characterization of Ferrocenes

Ferrocenes **1a**, **1b**, **2a**, **2b**, **3**, **4**, and **5** were prepared under argon atmosphere using standard Schlenk techniques, as was previously published [23]. Ferrocene **1c** was prepared from **1b** (155 mg, 244 µmol) and AgBF$_4$ (149 mg, 764 µmol) (Scheme 1). The substances were dried under vacuum for 1 h. The vacuum was replaced with argon and dry acetone (10 mL) was added, which caused an immediate color change from yellow to dark green. The mixture was then stirred for 2 days with light exclusion and then filtered. The filtrate was evaporated to dryness to obtain **1c** as a green solid substance with yield 200 mg (99%). M.p. 82 °C. ATR (Si); cm^{-1}: 3151 (sh, vw), 3115 (w), 2961 (w), 2932 (sh, vw), 2870 (vw), 1614 (vw), 1518 (w), 1462 (m), 1417 (m), 1367 (w), 1284 (w), 1226 (vw), 1054 (vs), 1035 (sh, s), 857 (m), 839 (sh, w), 764 (vw), 578 (w), 520 (m), 416 (w). Elemental analysis for **1c**—C$_{36}$H$_{46}$B$_3$F$_{12}$FeN$_2$ calculated C, 52.53; H, 5.63; N, 3.40%, found C, 51.41; H, 5.49; N, 3.35%. Its infrared spectrum (ATR) is characterized by the presence of a very strong and broad band at 1054 cm^{-1}, which could be assigned to stretching B–F vibration of the [BF$_4$]$^-$ anion.

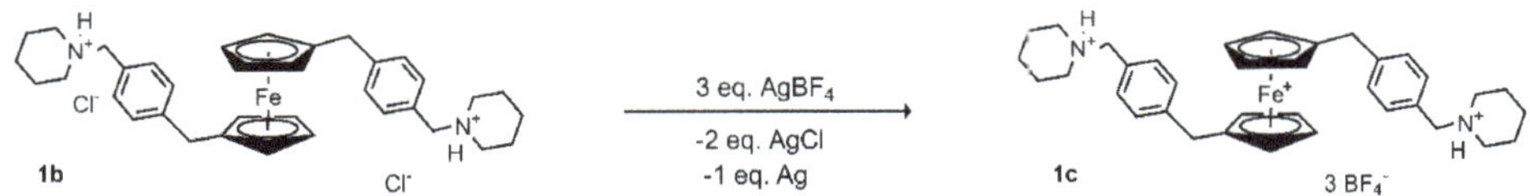

Scheme 1. Preparation of **1c**.

Ferrocene derivatives **1a**, **1b**, **2a**, **2b**, **3**, **4**, and **5** were characterized by elemental analysis, melting point, nuclear magnetic resonance (NMR), electrospray ionization mass spectrometry (ESI-MS), and X-ray diffraction analysis, as reported previously [23,24].

2.2. Cell Lines and Cultivation

Cervical cancer cell lines Ca Ski (ATCC® CRL-1550™), SiHa (ATCC® HTB-35™), and HeLa (ATCC® CCL-2™) were used. As nonmalignant controls we used the hTERT (human telomerase reverse transcriptase) immortalized retinal epithelium cell line RPE-1 (ATCC® CRL-4000™) and immortalized HEK 293 cells (ATCC® CRL-1573™) derived from human embryonic renal epithelium. All these cell lines were obtained from the American Type Culture Collection (ATCC, Manassas, Virginia, USA). Ca Ski and SiHa were maintained in high-glucose RPMI-1640 Medium (Sigma-Aldrich, St. Louis, USA) at 37 °C in a humidified atmosphere with 5% CO$_2$. HeLa and noncancerous cell lines HEK 293 and RPE-1 were maintained in high-glucose Dulbecco's modified Eagle's medium (DMEM) (Sigma-Aldrich, St. Louis, Missouri, USA) under the same conditions. Both media were supplemented with 10% fetal bovine serum (Gibco, Thermo Fisher Scientific, Waltham, Massachusetts, USA), 300 µg/mL L-glutamine (Sigma-Aldrich, St. Louis, Missouri, USA), and 100 µg/mL HyClone Penicillin–Streptomycin 100× solution (BioSera, Nuaille, France). The culture medium was changed during each cell passage. Cells were grown to 60–80% confluence prior to experimental treatments with ferrocenes at concentrations from 1 to 100 µM. Cells were mycoplasma-free throughout the duration of all experiments.

2.3. Cell Viability Assay

Due to their different sizes and growth rates, SiHa and Ca Ski cells were seeded at a density of 10,000 cells; HeLa, 8000 cells; HEK 293, 5000 cells; and RPE-1, 4000 cells per well in 96-well plates. The next day, the cells were exposed to the tested ferrocenes diluted in dimethyl sulfoxide (DMSO; Sigma-Aldrich, St. Louis, Missouri, USA) in concentrations from 0 to 100 µM (each in pentaplicate) for 24 h. Cell viability was measured using colorimetric MTT (3-(4,5-dimethylthiazol-2-yl)-2,5-diphenyltetrazolium bromide) assay as described previously [25]. All experiments were performed in three independent runs (twice, when cytotoxicity was over 100 µM). Data from cytotoxicity assays were measured using a Microplate Reader: Infinite® M1000 PRO (Tecan, Männedorf, Zürich, Switzerland) and analyzed using GraphPad Software (San Diego, California, USA) as IC_{50} values (concentrations of compounds that cause metabolic inhibition of 50% of cells).

2.4. Cell Cycle

The cell cycle was measured by a modified propidium iodide (PI) staining protocol as described previously [26,27]. Cells in 6-well plates were treated with 5 µM ferrocenes or cisplatin (positive control), and half of the wells were irradiated with 4 Gy (discussed in detail in Section 2.11); all samples were then incubated for 24 h. Afterwards, cells were trypsinized, washed with PBS (phosphate-buffered saline), and centrifuged at 1000 rpm for 5 min. Pellets were washed and resuspended in 0.5 mL of PBS. Cells were then fixed in 70% EtOH for at least 4 h at 4 °C. After fixation, the cells were centrifuged at 1000 rpm for 5 min and washed in PBS again. Subsequently, the cells were stained in 1 mL of staining solution: 0.1% Triton X-100, 10 µg/mL PI (both Sigma-Aldrich, St. Louis, USA), and 100 µg/mL DNase-free RNase A (Invitrogen, Carlsbad, California, USA) for 10 min at 37 °C. The DNA content was measured using a flow cytometer (Navios, Beckman Coulter, USA).

2.5. Reactive Oxygen Species (ROS) Production

SiHa or HeLa cells were seeded at 8000 cells per well in dark 96-well plates and incubated under standard conditions for 24 h. The next day, the medium was changed with 100 µL of Hanks' balanced salt solution (HBSS) (0.137 M NaCl, 5.4 mM KCl, 0.25 mM Na_2HPO_4, 0.1 g glucose, 0.44 mM KH_2PO_4, 1.3 mM $CaCl_2$, 1.0 mM $MgSO_4$, 4.2 mM $NaHCO_3$). Cells were incubated for 1 h. The solution was subsequently aspirated, and 100 µL of HBSS with 5 µM concentration of general oxidative stress indicator 5-(and-6)-chloromethyl-2′,7′-dichlorodihydrofluorescein diacetate, acetyl ester (CM-H_2DCFDA; Invitrogen, Carlsbad, California, USA) was added. After 30 min the cells were washed twice and treated with 10 µM ferrocenes or 50 µM H_2O_2, serving as a positive control, or 10 mM N-acetylcysteine (NAC), serving as a negative control. ROS production was measured after 2, 4, and 6 h.

2.6. Mitochondrial Membrane Potential Changes

The changes in mitochondrial membrane potential were measured using a 1,1′,3,3′-Tetraethyl-5,5′,6,6′-tetrachloroimidacarbocyanine iodide dye—mitochondrial membrane potential probe (JC-1; Invitrogen, Carlsbad, California, USA) [28,29]. The cells were harvested with trypsin and seeded at a density of 0.15–0.2 million of cells per well in 12-well plates, then allowed to adhere overnight. The cells were exposed to ferrocenes in 5 µM concentration for 24 h. Since SiHa cells were more resistant, the conditions were modified to 40 µM concentration for 6 h. Treatment with valinomycin (Molecular probes, Eugene, Oregon, USA) in 50 µM concentration for 2 h served as a positive control. Cells were collected with a rubber scraper, washed with PBS, and centrifuged at 1000 rpm for 5 min. Pellets were washed, centrifuged, and resuspended in PBS with JC-1 probe in concentration 5 µg/mL. Mitochondrial potential changes were measured using a BD FACS Aria sorter (BD Biosciences, Franklin Lakes, New Jersey, USA).

2.7. Annexin V–Fluorescein Isothiocyanate (FITC)/PI Binding Assay

HeLa cells were incubated for 24 h at 37 °C in 6-well plates with 10 µM ferrocenes. SiHa cells were more resistant; thus, the conditions were modified to 20 µM concentrations. Cells were harvested by acutase, washed twice with PBS, centrifuged at 1000 rpm for 5 min, and then resuspended in Annexin V Binding buffer (10 mM HEPES/NaOH, pH 7.4; 14 mM NaCl; 2.5 mM $CaCl_2$) at a concentration of 1 million cells/mL. A volume of 100 µL of cell suspension was pipetted into the 1.5 mL tube and mixed with fluorescein isothiocyanate (FITC)-labeled Annexin V (BioLegend, San Diego, USA) and PI solutions. The cells were gently vortexed and incubated for 15 min at 22 °C in the dark. After incubation,

400 µL of Annexin V Binding buffer was added and samples were measured using a flow cytometer (Navios, Beckman Coulter, USA).

2.8. Western Blot Analysis

Hela and SiHa cell lines were treated with 2 µM concentration of selected ferrocenes, and half of them were subsequently irradiated with 4 Gy of ionizing radiation. All samples were then incubated for 24 h. Cells were washed twice with ice-cold PBS, scraped off with a rubber scraper, and then lysed in nonyl phenoxypolyethoxylethanol (NP-40) lysis buffer (150 mM NaCl, 50 mM TrisHCl pH 8.0, 5 mM NaF, 5 mM EDTA, 1% NP-40, 1:100 phosphatase inhibitor cocktail, and 1:100 protease inhibitor cocktail, both cocktails from Sigma-Aldrich, St. Louis, USA). The protein concentrations were measured via Bradford protein assay (Bio-Rad, Hercules, California, USA). A quantity of 20 µg of protein lysate per sample was applied and separated on 10% SDS polyacrylamide gel, then transferred onto nitrocellulose blotting membrane (Pall Life Sciences, New York, USA). The accuracy of sample loading was verified with Ponceau staining. Membranes were blocked in 5% milk with 0.1% Tween 20 in PBS and probed overnight with the following antibodies: anti β-actin monoclonal antibody (Sigma-Aldrich, St. Louis, Missouri, USA) served as a loading control, SQSTM1 p62 antibody (A-6) (Santa Cruz Biotechnology, Dallas, Texas, USA), and LC3B antibody (Novus Biologicals, Littleton, USA). Membranes were washed with PBS containing 0.1% Tween and incubated with secondary IgG antibodies SWAR-Px (Swine Anti-Rabbit Immunoglobulins- horseradish peroxidase, #P0217) and RAM-Px (Rabbit Anti-Mouse Immunoglobulins- horseradish peroxidase, #P0161) (Dako, Glostrup, Denmark) for 1 h. Positive signals were visualized with enhanced chemiluminescence reagent (ECL; Amersham Pharmacia Biotech, UK) using a G:BOX Chemi XX6 System (Syngene, Cambridge, United Kingdom). Ordinarily used chemicals were obtained from Sigma-Aldrich (St. Louis, Missouri, USA).

2.9. Immunofluorescence Staining

Cells were seeded on coverslips in 12-well plates. After 24 h of incubation at 37 °C with 5 µM concentration of ferrocenes **1b**, **2a**, and **3** and with 0.2 µM Bafilomycin A as a positive control, coverslips with adherent cells were washed with PBS solution and fixed with 4% formaldehyde. After permeabilization with 0.2% Triton-X100 (Sigma-Aldrich, St. Louis, Missouri, USA), cells were incubated with primary antibodies recognizing p62 and LC3B, respectively, for 1 h at 37 °C. The cells were then washed and incubated with fluorescent-dye-conjugated secondary antibodies (ab96899 for LC3B and ab96881 for p62; Abcam, Cambridge, United Kingdom) for 1 h at room temperature. In parallel, Hoechst staining was used to visualize nuclei. PBS was used for washing and Vectashield (Cole-Parmer, Vernon Hills, Illinois, USA) for mounting the coverslip. Cells were visualized on a Nuance Multispectral Tissue Imaging System FX (PerkinElmer, Waltham, Massachusetts, USA).

2.10. Colony Forming Assay (CFA)

The radiosensitizing properties of the selected compounds were tested in vitro according to the standardized protocol for colony forming assay [30]. Briefly, cells were trypsinized, centrifuged, and resuspended in fresh medium. The suspensions were then plated at a density of 250 cells per well in 12-well plates and left for 24 h in an incubator to adhere. The next day, selected ferrocenes were added in concentrations of 0.5 and 1.0 µM for HeLa cells and 1.0 and 2.0 µM for the more-resistant SiHa cells. Following 1–2 h of incubation, the cells were irradiated according to the protocol in Section 2.11. The cells were then further incubated under standard conditions for 14 days, and colonies were fixed with crystal violet staining/fixing solution—1% methanol (Penta, Chrudim, Czech Republic), 0.05% crystal violet (Merck Millipore, Burlington, Massachusetts, USA), and 3.7% formaldehyde (Sigma-Aldrich, St. Louis, USA) in PBS and manually counted. Surviving fractions (SF) were calculated by comparing the number of colonies in irradiated and nonirradiated (control) plates using a linear-quadratic model for cell death after irradiation. To obtain the final radiosensitizing effect, dose-modifying factors (DMF) were calculated as the ratios of surviving fractions in cells irradiated with and without tested ferrocenes at selected doses.

2.11. Ionizing Radiation

Orthovoltage X-ray irradiation was performed using an Xstrahl 200 radiotherapy system (Xstrahl, Surrey, England) as a single fraction of 2, 4, or 6 Gy with energy 200 kV with a half-value layer of 1 mm of Cu. The field size was 20×20 cm, the distance from source to irradiated wells was 50 cm, and the dose rate was 0.38 Gy/min. To obtain an adequate dose at a surface, wells were covered with a 1 cm thick bolus material. Control cells were taken from the incubator and handled similarly to treated cells, except during irradiation when they were placed outside the treatment room to avoid any exposure to ionizing radiation. All treatments were done at room temperature.

3. Results

3.1. Cytotoxic Activity

The cytotoxicity of ferrocenes was tested in vitro against cervical cancer cell lines CaSki, SiHa, and HeLa using MTT tests. Since nephrotoxicity remains the main side effect of cisplatin treatment, the toxicity of the selected (most active) complexes against noncancerous human embryonic kidney cells (HEK 293) was evaluated along with retinal epithelial cells (RPE-1). Table 1 shows the IC_{50} values for particular ferrocenes. According to the cytotoxicity results, the three most active substances—**1b**, **2a**, and **3**—were selected and used for further testing.

Table 1. Cytotoxic effects of the studied ferrocenes in micromolar concentrations after 24 h against selected cell lines.

Comp.	Ca Ski	SiHa	HeLa	RPE-1	HEK 293
Cisplatin	35.1 ± 8.7	26.6 ± 4.1	28.2 ± 5.9	46.8 ± 6.0	25.3 ± 8.6
1a	69.4 ± 8.6	17.1 ± 3.4	61.5 ± 2.9	82.8 ± 3.7	>100
1b	6.2 ± 2.1	9.9 ± 2.2	6.4 ± 1.6	5.1 ± 0.8	35.6 ± 6.2
1c	12.5 ± 0.6	18.2 ± 3.1	10.6 ± 2.5	11.0 ± 2.4	16.7 ± 2.0
2a	8.3 ± 2.3	14.1 ± 3.4	7.3 ± 1.2	4.7 ± 0.1	6.2 ± 1.6
2b	15.0 ± 1.0	30.2 ± 2.9	13.0 ± 0.6	8.2 ± 0.4	5.4 ± 0.6
3	7.6 ± 1.3	11.1 ± 1.5	6.5 ± 0.4	8.5 ± 1.1	9.5 ± 1.0
4	>100	>100	>100	>100	>100
5	>100	91.8 ± 9.4	7.7 ± 1.1	7.4 ± 2.0	3.0 ± 0.3

3.2. Effect of Selected Ferrocenes on Cell Cycle

To determine the effect of the tested ferrocenes on the cell cycle, PI assay was used. HeLa cells showed significant changes in cell cycle distribution, especially an increased proportion of cells in S phase and a corresponding decrease in G0/G1 phases (Figure 2). On the other hand, SiHa cells showed no significant changes in cell cycle distribution in response to the tested ferrocenes (data not shown). Importantly, the applied doses of irradiation in our experiment had no significant effect on cell cycle distribution except for increased accumulation of HeLa cells in the G2/M phase in response to **3**.

3.3. Analysis of Cell Death

Since MTT tests clearly confirmed the cytotoxicity of the selected compounds, closer examination of the specific mechanisms and pathways associated with cell death was performed.

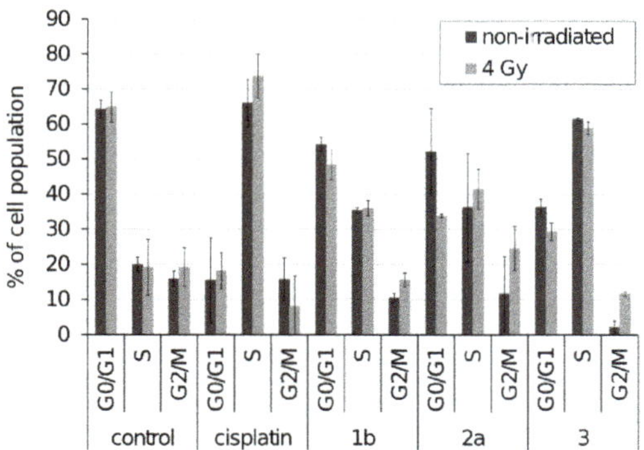

Figure 2. Cell cycle analysis of HeLa cells exposed for 24 h to 5 µM ferrocenes either with or without 4 Gy irradiation.

3.3.1. Effect of Selected Ferrocenes on ROS Production

A cell-membrane-permeable chloromethyl derivative of H_2DCFDA was used to determine the production of reactive oxygen species. An increased amount of ROS was measured after 2 h of treatment with selected ferrocenes in 10 µM concentration. Untreated cells and cells treated with hydrogen peroxide were used as controls. The relative fluorescence values are summarized in Figure 3.

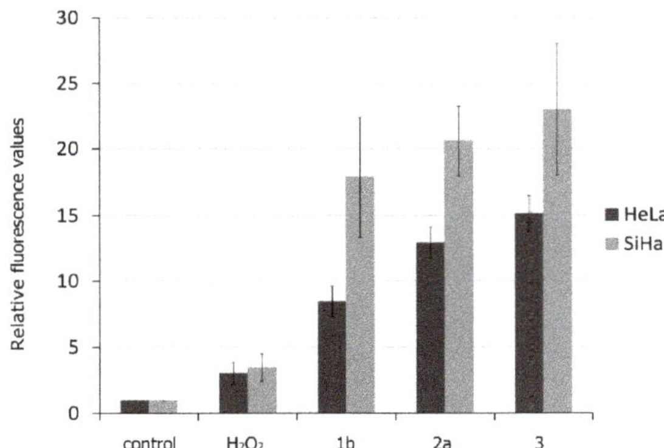

Figure 3. Reactive oxygen species (ROS) production in HeLa and SiHa cells after treatment with the most active ferrocenes. Untreated cells were used as a negative control, H_2O_2 was used as a positive control. Data were normalized to control untreated cells.

3.3.2. Effect of **1b**, **2a**, and **3** on Mitochondrial Membrane Potential

Increased levels of reactive oxygen species can be closely associated with mitochondrial metabolism [31–33]. Thus, the JC-1 probe, routinely used for measuring the state of mitochondrial membranes and their potential in a wide spectrum of cell types, was applied [34–37]. Increased mitochondrial potential in HeLa cells (by about 50% to 100%) was observed after treatment with 5 µM ferrocenes. SiHa cells were resistant under the same conditions, but increasing the concentration to 40 µM caused a similar change in mitochondrial potential (Figure 4).

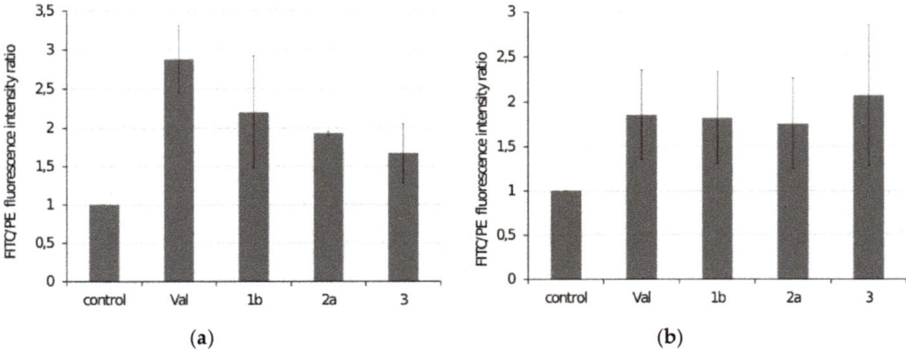

Figure 4. Determination of mitochondrial membrane depolarization in (**a**) HeLa cells exposed to 5 μM ferrocenes for 24 h and (**b**) SiHa cells exposed to 40 μM ferrocenes for 6 h. Valinomycin (Val) was used as a positive control. Fluorescein isothiocyanate (FITC) and Phycoerythrin (PE) fluorescence intensity ratio were determined. Data are reported in relation to untreated cells (control).

3.3.3. Analysis of Apoptosis

Increased production of reactive oxygen radicals along with the disruption of the mitochondrial membrane may indicate the involvement of apoptosis as a possible mode of cell death. Annexin V labeling with FITC was used for cytometric measurement of the translocation of phosphatidylserine to the outer surface of the plasma membrane [38–40]. PI staining of DNA was used to detect the late phase of apoptosis. Treatment with active ferrocenes was associated with the induction of both early and late phases of apoptosis in either cell line. Necrotic cells were not observed in significant amounts (Figure 5).

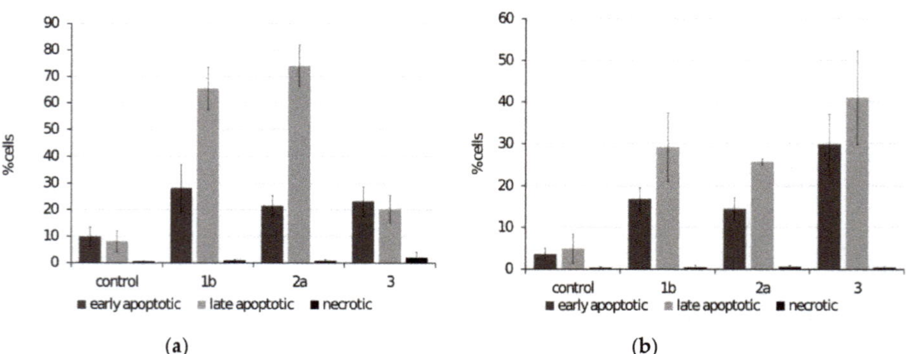

Figure 5. Determination of apoptosis by flow cytometry in (**a**) HeLa cells treated with 10 μM ferrocenes and (**b**) SiHa cells treated with 20 μM ferrocenes.

3.4. Autophagy Detection

The production of ROS is not specific only to apoptotic cells. Autophagy can also be induced by higher levels of reactive oxygen species [41,42]. To elucidate whether autophagy is also elevated in response to treatment with ferrocenes, the levels of autophagy-associated proteins p62 and cleaved LC3B were determined by immunofluorescent staining in HeLa cells. Indeed, increased levels of both p62 and LC3B were observed in these cells exposed to particular ferrocenes (Figure 6). Clear elevation of the LC3B level was also confirmed by Western blot analysis in both cell lines exposed to particular ferrocenes (Figure 7). In parallel, we also analyzed the combined effect of irradiation on LC3B cleavage. Interestingly, a clear decrease in the LC3B level was observed, predominantly in SiHa cells.

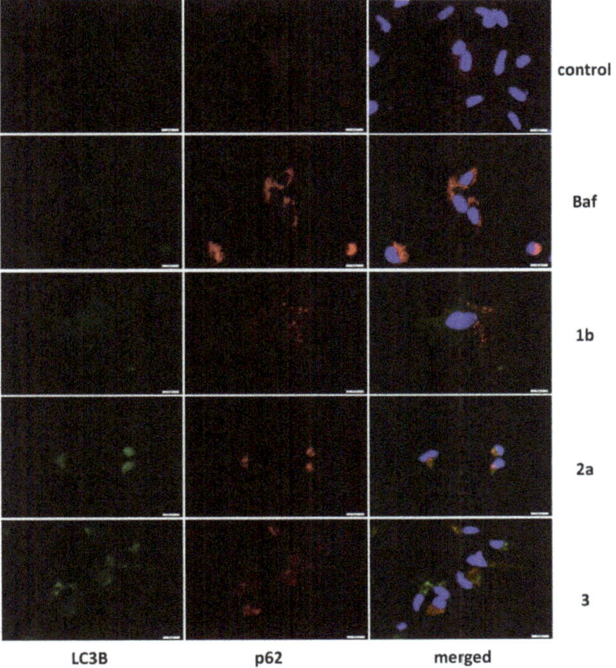

Figure 6. Detection of autophagy-related proteins in HeLa cells exposed to 5 µM ferrocenes for 24 h. Disruptor of autophagic flux Bafilomycin A (Baf) [43] was used as a positive control. Nuclei are stained with DAPI.

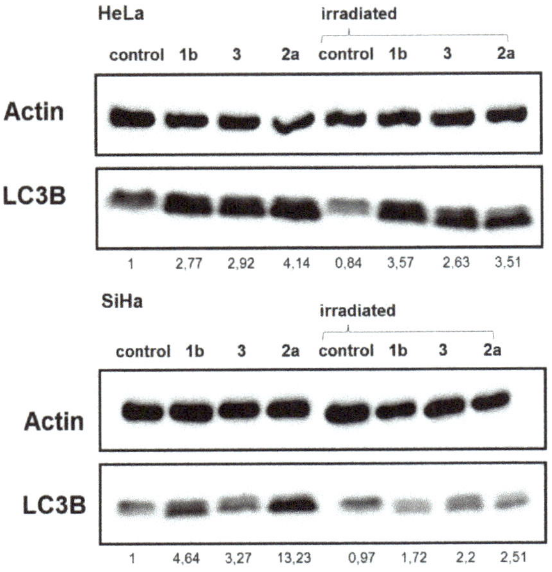

Figure 7. Detection of autophagy-associated protein LC3B in HeLa (upper part) and SiHa (lower part) cells. β-actin served as a loading control. Cells were treated with 2 µM concentrations of selected ferrocenes with or without a 4 Gy dose of irradiation.

3.5. Sensitivity to Ionizing Radiation

Radiotherapy is a standard treatment option in advanced cervical carcinomas. Colony forming assay was used to evaluate the potential radiosensitizing effect of the selected ferrocenes [30].

3.5.1. Determination of the Surviving Fraction

The surviving fraction (SF) calculated from CFA is the ratio of surviving colonies in the irradiated and nonirradiated plates. Figure 8 shows that ferrocene **1b** had a potent radiosensitizing effect on both cell lines. In the SiHa cell line, a similar, albeit smaller, effect was observed for **2a**.

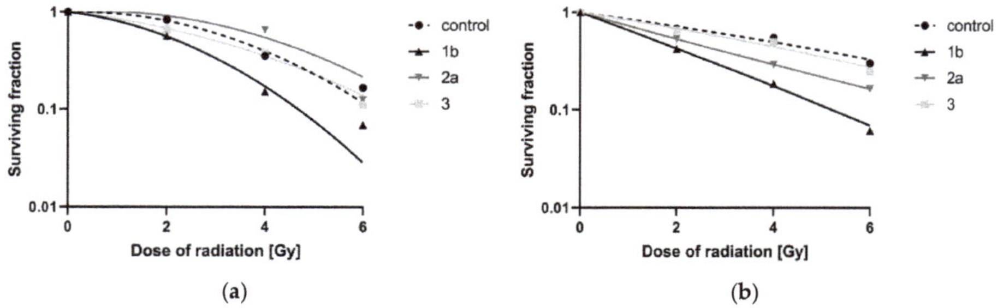

Figure 8. Graphical interpretation of the surviving fraction (SF) for the combination of selected ferrocenes with different doses of irradiation in (**a**) HeLa cells exposed to 1 µM ferrocenes and (**b**) SiHa cells treated with 2 µM ferrocenes.

3.5.2. Evaluation of Dose-Modifying Factors

Dose-modifying factors (DMFs) assess the additive effect of a tested drug when used in combination with ionizing radiation. They are the ratios of doses needed to obtain the same surviving fractions in cells irradiated with and without the tested ferrocenes [44]. Values below 0.8 were considered as showing an antagonistic effect, values between 0.8 and 1.2 as showing no effect, and values above 1.2 as showing a synergistic effect (Table 2).

Table 2. Dose-modifying factors for the selected ferrocenes in HeLa and SiHa cells.

HeLa Cells				SiHa Cells			
Conc.	1b	2a	3	Conc.	1b	2a	3
0.5 µM	0.9 ± 0.08	1.0 ± 0.1	0.9 ± 0.08	1 µM	1.4 ± 0.19	1.3 ± 0.07	0.9 ± 0.08
1 µM	1.6 ± 0.13	0.9 ± 0.2	1.2 ± 0.08	2 µM	1.8 ± 0.5	1.4 ± 0.17	1.2 ± 0.08

4. Discussion

In recent articles, syntheses of some ferrocenes have been published [23,24]. Several of these compounds showed a clear cytotoxic effect that was equal or even greater when compared with cisplatin, which has become a mainstay of cancer therapy. We hypothesize that ferrocenes, similarly to cisplatin, may, in combination with ionizing radiation, show a synergistic cytotoxic effect. This radiosensitization effect could be particularly beneficial in the treatment of chemoresistant malignancies, e.g., cervical tumors that are generally perceived as resistant to cisplatin [45,46].

In previous studies, the mechanisms of transport of substituted ferrocenes were investigated using differential pulse voltammetry and inductively coupled plasma mass spectrometry. We identified membrane transferrin-receptor-mediated endocytosis of transferrin-bound ferrocene as the major mechanism of cellular uptake. Importantly, the rate of ferrocene accumulation in cancer cells is proportional to its cytotoxic effects [24,47]. To assess the exact mode of action of the studied ferrocenes, several methods were used. First, while no changes in cell cycle distribution were observed in SiHa cells, HeLa cells showed increased accumulation of cells in the S phase in response to treatment with all tested ferrocenes.

However, the increase in proportion of cells in the S phase predominantly after treatment with **1b** and **2a** was significantly lower when compared to that in cisplatin-treated cells. This could be of potential interest, since the doses of cisplatin used probably slow down DNA synthesis due to repairing its damage, which is associated with the accumulation of cells in the S phase [48], and cells in S phase are generally considered to be more radioresistant [49]. Thus, even though cisplatin is commonly used with radiotherapy, the effect of ionizing radiation in this combination is probably partly attenuated because cisplatin actually increases the proportion of cells in the S phase [50,51]. Therefore, treatment with ferrocenes resulting in the induction of cell cycle arrest in the S phase to a much lesser extent compared to cisplatin would be more effective, bringing greater benefit to these patients.

Second, a potent increase in reactive oxygen species was observed upon treatment with the selected ferrocenes. This can be explained by the basic chemical structure of the ferrocene core that is known as a catalyst in a Fenton reaction generating both hydroxyl radicals and higher oxidation states of the iron. Thus, in the presence of transition metals, H_2O_2, a product of mitochondrial oxidative respiration, is reduced inside the cells; this generates oxygen radicals responsible for damage to all macromolecules, including DNA, proteins, and membrane phospholipids that are damaged by the peroxidation of unsaturated fatty acids in exposed cells [52–54]. In recent years, several studies have confirmed potent induction of ROS by a range of organometallic complexes, including ferrocenes, in cancer cells [55–62]. These studies imply their possible utilization in cancer research and treatment. Our current results show a large increase in levels of reactive oxygen species after treatment with ferrocenes **1b**, **2a**, and **3**, which indicates their impact on the redox homeostasis in tumor cells.

Third, due to the aforementioned increase in ROS production, the effect of the tested compounds on mitochondrial function was also analyzed. Our experiments showed clear changes in mitochondrial membrane potential. These changes are generally considered to be an early step in apoptosis after treatment with different drugs [63,64]. Depolarization of the mitochondrial membrane leads to the release of cytochrome c from the mitochondria. This process triggers the formation of an apoptosome and the activation of caspases, launching the intrinsic apoptotic pathway [65]. Furthermore, early and late stages of apoptosis were both detected via Annexin V assay and DNA labeling by PI. These findings are in agreement with those of other studies showing that many different ferrocene-containing compounds, e.g., ferrocifens, Pt–ferrocene complexes [66,67], and ferrocenes combined with retinoids, are able to activate apoptotic cell death in cancer cells [68–70].

Elevated levels of ROS can also lead to the induction of autophagy as a defense mechanism that allows cell survival during stress conditions. An increase in autophagy-related proteins was observed, especially for cleaved-form LC3B. These changes were most prominent after treatment with ferrocenes **1b** and **2a**. It is well known that autophagy can play a dual role in response to drug treatment. On one hand, upregulation of autophagy could help in the formation of cancer cells resistant to chemotherapy [71]. On the other hand, excessive elevation of autophagy may assist in the induction of programmed cell death [72]. In our study, combination of ferrocenes with irradiation led to a decrease in the amount of LC3B protein. The decrease was most pronounced in response to ferrocene **2a**. It can therefore be assumed that autophagy induction helps cells to overcome exposure to ferrocenes. In line with these findings, combined treatment with ionizing radiation attenuates autophagy induction and results in significantly higher anti-tumor efficacy. This correlates with the CFA results, which showed decreased colony formation rates after combination of ferrocene **1b** and partially **2a** with radiotherapy when compared to radiotherapy alone. To quantify this interplay, the dose-modifying factor for each ferrocene was calculated. The most pronounced synergistic effect was observed for ferrocene **1b** (1.6× increased effectiveness of radiotherapy in HeLa and 1.8× in SiHa cell lines) and for ferrocene **2a** (1.4× for the SiHa cell line).

5. Conclusions

A series of ferrocenes based on the general formula $[Fe(\eta^5\text{-}C_5H_4CH_2(p\text{-}C_6H_4)CH_2(N\text{-het}))_2]$ bearing either substituted or unsubstituted saturated five- and six-membered nitrogen-containing heterocycles showed higher cytotoxic activity than cisplatin against cervical cancer cell lines. These ferrocenes were able to increase the production of reactive oxygen species and disrupt mitochondrial homeostasis. As a consequence of these processes, the onset of a cell death mechanism via apoptosis and the activation of autophagy were observed. Furthermore, two of these ferrocenes—**1b** and **2a**—showed increased cytotoxic effects when combined with irradiation. However, the precise relationship between ionizing radiation and ferrocene derivatives needs further investigation. On the other hand, to put this problem into perspective, precise knowledge of the interaction between ionizing radiation and cisplatin is also unknown, and this holds true for many other chemotherapeutic drugs as well.

The combination of ferrocene **1b** and, to a lesser degree, **2a** with irradiation resulted in greater efficacy than either treatment alone in cervical cancer cell lines. Thus, our results suggest that the selected ferrocenes could be used in combination with radiotherapy, representing promising candidates for further investigation. Alternatively, the addition of ferrocene compounds to standard chemoradiation with cisplatin could also be considered. However, care needs to be taken to evaluate the potential increase in toxicity, since the studied ferrocenes and cisplatin have partly overlapping effects, mainly in the production of reactive oxygen species.

Author Contributions: Conceptualization, H.S. and V.R.; methodology, H.S.; software, H.S. and V.R.; validation H.S. and V.R.; chemical synthesis, J.P. and J.K., formal analysis, H.S.; investigation, H.S.; writing—original draft preparation, H.S.; writing—review and editing, V.R. and H.R.; visualization, H.S.; supervision, R.H.; project administration, R.H.; funding acquisition, J.P., J.K. and R.H. All authors have read and agreed to the published version of the manuscript.

Funding: This research was funded by GACR 17-05838S, MEYS-NPS I-LO1413 and MH CZ-DRO (MMCI, 00209805).

Acknowledgments: At this point we would like to thank L. Sommerova for help with flow cytometry analyses.

Conflicts of Interest: The authors declare no conflict of interest.

References

1. International Agency for Research on Cancer. Cancer Incidence in Five Continents Time Trends. Available online: http://ci5.iarc.fr/CI5plus/Default.aspx (accessed on 24 June 2018).
2. Chovanec, J.; Náležinská, M. Přehled diagnostiky a léčby karcinomu děložního hrdla. *Onkologie* **2014**, *8*, 269–274.
3. Ferlay, J.; Colombet, M.; Bray, F. Cancer Incidence in Five Continents, CI5plus: IARC CancerBase No. 9. Available online: http://ci5.iarc.fr (accessed on 18 October 2019).
4. Walboomers, J.M.; Jacobs, M.V.; Manos, M.M.; Bosch, F.X.; Kummer, J.A.; Shah, K.V.; Snijders, P.J.; Peto, J.; Meijer, C.J.; Munoz, N. Human papillomavirus is a necessary cause of invasive cervical cancer worldwide. *J. Pathol.* **1999**, *189*, 12–19. [CrossRef]
5. Marth, C.; Landoni, F.; Mahner, S.; McCormack, M.; Gonzalez-Martin, A.; Colombo, N. Cervical cancer: ESMO Clinical Practice Guidelines for diagnosis, treatment and follow-up. *Ann. Oncol. Off. J. Eur. Soc. Med. Oncol.* **2018**, *29*, 262. [CrossRef] [PubMed]
6. Vale, C.; Tierney, J.F.; Stewart, L.A.; Brady, M.; Dinshaw, K.; Jakobsen, A.; Parmar, M.K.; Thomas, G.; Trimble, T.; Alberts, D.S.; et al. Reducing uncertainties about the effects of chemoradiotherapy for cervical cancer: A systematic review and meta-analysis of individual patient data from 18 randomized trials. *J. Clin. Oncol.* **2008**, *26*, 5802–5812.
7. Galanski, M.; Jakupec, M.A.; Keppler, B.K. Update of the preclinical situation of anticancer platinum complexes: Novel design strategies and innovative analytical approaches. *Curr. Med. Chem.* **2005**, *12*, 2075–2094. [CrossRef] [PubMed]
8. Peter, S.; Aderibigbe, B.A. Ferrocene-Based Compounds with Antimalaria/Anticancer Activity. *Molecules* **2019**, *24*, 3604. [CrossRef] [PubMed]
9. Sirignano, E.; Saturnino, C.; Botta, A.; Sinicropi, M.S.; Caruso, A.; Pisano, A.; Lappano, R.; Maggiolini, M.; Longo, P. Synthesis, characterization and cytotoxic activity on breast cancer cells of new half-titanocene derivatives. *Bioorg. Med. Chem. Lett.* **2013**, *23*, 3458–3462. [CrossRef]

10. Saturnino, C.; Sirignano, E.; Botta, A.; Sinicropi, M.S.; Caruso, A.; Pisano, A.; Lappano, R.; Maggiolini, M.; Longo, P. New titanocene derivatives with high antiproliferative activity against breast cancer cells. *Bioorg. Med. Chem. Lett.* **2014**, *24*, 136–140. [CrossRef]

11. Saturnino, C.; Barone, I.; Iacopetta, D.; Mariconda, A.; Sinicropi, M.S.; Rosano, C.; Campana, A.; Catalano, S.; Longo, P.; Ando, S. N-heterocyclic carbene complexes of silver and gold as novel tools against breast cancer progression. *Future Med. Chem.* **2016**, *8*, 2213–2229. [CrossRef]

12. Iacopetta, D.; Mariconda, A.; Saturnino, C.; Caruso, A.; Palma, G.; Ceramella, J.; Muia, N.; Perri, M.; Sinicropi, M.S.; Caroleo, M.C.; et al. Novel Gold and Silver Carbene Complexes Exert Antitumor Effects Triggering the Reactive Oxygen Species Dependent Intrinsic Apoptotic Pathway. *Chem. Med. Chem.* **2017**, *12*, 2054–2065. [CrossRef]

13. Alessio, E.; Messori, L. The Deceptively Similar Ruthenium(III) Drug Candidates KP1019 and NAMI-A Have Different Actions. What Did We Learn in the Past 30 Years? *Met. Life Sci.* **2018**, *18*, 141.

14. Ari, F.; Cevatemre, B.; Armutak, E.I.; Aztopal, N.; Yilmaz, V.T.; Ulukaya, E. Apoptosis-inducing effect of a palladium(II) saccharinate complex of terpyridine on human breast cancer cells in vitro and in vivo. *Bioorg. Med. Chem.* **2014**, *22*, 4948–4954. [CrossRef]

15. Carter, R.; Westhorpe, A.; Romero, M.J.; Habtemariam, A.; Gallevo, C.R.; Bark, Y.; Menezes, N.; Sadler, P.J.; Sharma, R.A. Radiosensitisation of human colorectal cancer cells by ruthenium(II) arene anticancer complexes. *Sci. Rep.* **2016**, *6*, 20596. [CrossRef]

16. Mesbahi, A. A review on gold nanoparticles radiosensitization effect in radiation therapy of cancer. *Rep. Pract. Oncol. Radiother. J. Greatpoland Cancer Cent. Pozn. Pol. Soc. Radiat. Oncol.* **2010**, *15*, 176–180. [CrossRef]

17. Gill, M.R.; Menon, J.U.; Jarman, P.J.; Owen, J.; Skaripa-Koukelli, I.; Able, S.; Thomas, J.A.; Carlisle, R.; Vallis, K.A. (111)In-labelled polymeric nanoparticles incorporating a ruthenium-based radiosensitizer for EGFR-targeted combination therapy in oesophageal cancer cells. *Nanoscale* **2018**, *10*, 10596–10608. [CrossRef]

18. Hainfeld, J.F.; Slatkin, D.N.; Smilowitz, H.M. The use of gold nanoparticles to enhance radiotherapy in mice. *Phys. Med. Biol.* **2004**, *49*, 309–315. [CrossRef]

19. Chang, M.Y.; Shiau, A.L.; Chen, Y.H.; Chang, C.J.; Chen, H.H.; Wu, C.L. Increased apoptotic potential and dose-enhancing effect of gold nanoparticles in combination with single-dose clinical electron beams on tumor-bearing mice. *Cancer Sci.* **2008**, *99*, 1479–1484. [CrossRef]

20. Hainfeld, J.F.; Smilowitz, H.M.; O'Connor, M.J.; Dilmanian, F.A.; Slatkin, D.N. Gold nanoparticle imaging and radiotherapy of brain tumors in mice. *Nanomedicine* **2013**, *8*, 1601–1609. [CrossRef]

21. Teicher, B.A.; Jacobs, J.L.; Cathcart, K.N.; Abrams, M.J.; Vollano, J.F.; Picker, D.H. Some complexes of cobalt(III) and iron(III) are radiosensitizers of hypoxic EMT6 cells. *Radiat. Res.* **1987**, *109*, 36–46. [CrossRef]

22. Joy, A.M.; Goodgame, D.M.; Stratford, I.J. High efficiency of ferricenium salts as radiosensitizers of V79 cells in vitro and the KHT tumor in vivo. *Int. J. Radiat. Oncol. Biol. Phys.* **1989**, *16*, 1053–1056. [CrossRef]

23. Hodík, T.; Lamač, M.; Červenková Šťastná, L.; Cuřínová, P.; Karban, J.; Skoupilová, H.; Hrstka, R.; Císařová, I.; Gyepes, R.; Pinkas, J. Improving cytotoxic properties of ferrocenes by incorporation of saturated N-heterocycles. *J. Organomet. Chem.* **2017**, *846*, 141–151. [CrossRef]

24. Bartosik, M.; Koubkova, L.; Karban, J.; Cervenkova Stastna, L.; Hodik, T.; Lamac, M.; Pinkas, J.; Hrstka, R. Electrochemical analysis of a novel ferrocene derivative as a potential antitumor drug. *Analyst* **2015**, *140*, 5864–5867. [CrossRef]

25. Kvardova, V.; Hrstka, R.; Walerych, D.; Muller, P.; Matoulkova, E.; Hruskova, V.; Stelclova, D.; Sova, P.; Vojtesek, B. The new platinum(IV) derivative LA-12 shows stronger inhibitory effect on Hsp90 function compared to cisplatin. *Mol. Cancer* **2010**, *9*, 147. [CrossRef]

26. Darzynkiewicz, Z.; Halicka, H.D.; Zhao, H. Analysis of cellular DNA content by flow and laser scanning cytometry. *Adv. Exp. Med. Biol.* **2010**, *676*, 137–147.

27. Koubkova, L.; Vyzula, R.; Karban, J.; Pinkas, J.; Ondrouskova, E.; Vojtesek, B.; Hrstka, R. Evaluation of cytotoxic activity of titanocene difluorides and determination of their mechanism of action in ovarian cancer cells. *Investig. New Drugs* **2015**, *33*, 1123–1132. [CrossRef]

28. Matarrese, P.; Tinari, A.; Mormone, E.; Bianco, G.A.; Toscano, M.A.; Ascione, B.; Rabinovich, G.A.; Malorni, W. Galectin-1 sensitizes resting human T lymphocytes to Fas (CD95)-mediated cell death via mitochondrial hyperpolarization, budding, and fission. *J. Biol. Chem.* **2005**, *280*, 6969–6985. [CrossRef]

29. Ruan, Q.; Lesort, M.; MacDonald, M.E.; Johnson, G.V. Striatal cells from mutant huntingtin knock-in mice are selectively vulnerable to mitochondrial complex II inhibitor-induced cell death through a non-apoptotic pathway. *Hum. Mol. Genet.* **2004**, *13*, 669–681. [CrossRef]

30. Franken, N.A.; Rodermond, H.M.; Stap, J.; Haveman, J.; van Bree, C. Clonogenic assay of cells in vitro. *Nat. Protoc.* **2006**, *1*, 2315–2319. [CrossRef]

31. Starkov, A.A. The role of mitochondria in reactive oxygen species metabolism and signaling. *Ann. N. Y. Acad. Sci.* **2008**, *1147*, 37–52. [CrossRef]

32. Marchi, S.; Giorgi, C.; Suski, J.M.; Agnoletto, C.; Bononi, A.; Bonora, M.; De Marchi, E.; Missiroli, S.; Patergnani, S.; Poletti, F.; et al. Mitochondria-ros crosstalk in the control of cell death and aging. *J. Signal Transduct.* **2012**, *2012*, 329635. [CrossRef]

33. Di Meo, S.; Reed, T.T.; Venditti, P.; Victor, V.M. Role of ROS and RNS Sources in Physiological and Pathological Conditions. *Oxid. Med. Cell. Longev.* **2016**, *2016*, 1245049. [CrossRef]

34. Cossarizza, A.; Ceccarelli, D.; Masini, A. Functional heterogeneity of an isolated mitochondrial population revealed by cytofluorometric analysis at the single organelle level. *Exp. Cell Res.* **1996**, *222*, 84–94. [CrossRef]

35. Di Lisa, F.; Blank, P.S.; Colonna, R.; Gambassi, G.; Silverman, H.S.; Stern, M.D.; Hansford, R.G. Mitochondrial membrane potential in single living adult rat cardiac myocytes exposed to anoxia or metabolic inhibition. *J. Physiol.* **1995**, *486*, 1–13. [CrossRef]

36. Sick, T.J.; Perez-Pinzon, M.A. Optical methods for probing mitochondrial function in brain slices. *Methods* **1999**, *18*, 104–108. [CrossRef]

37. White, R.J.; Reynolds, I.J. Mitochondrial depolarization in glutamate-stimulated neurons: An early signal specific to excitotoxin exposure. *J. Neurosci.* **1996**, *16*, 5688–5697. [CrossRef]

38. Sekine, C.; Moriyama, Y.; Koyanagi, A.; Koyama, N.; Ogata, H.; Okumura, K.; Yagita, H. Differential regulation of splenic CD8- dendritic cells and marginal zone B cells by Notch ligands. *Int. Immunol.* **2009**, *21*, 295–301. [CrossRef]

39. Galski, H.; Oved-Gelber, T.; Simanovsky, M.; Lazarovici, P.; Gottesman, M.M.; Nagler, A. P-glycoprotein-dependent resistance of cancer cells toward the extrinsic TRAIL apoptosis signaling pathway. *Biochem. Pharm.* **2013**, *86*, 584–596. [CrossRef]

40. Seguier, S.; Tartour, E.; Guerin, C.; Couty, L.; Lemitre, M.; Lallement, L.; Folliguet, M.; El Naderi, S.; Terme, M.; Badoual, C.; et al. Inhibition of the differentiation of monocyte-derived dendritic cells by human gingival fibroblasts. *PLoS ONE* **2013**, *8*, e70937. [CrossRef]

41. Azad, M.B.; Chen, Y.; Gibson, S.B. Regulation of autophagy by reactive oxygen species (ROS): Implications for cancer progression and treatment. *Antioxid. Redox Signal.* **2009**, *11*, 777–790. [CrossRef]

42. Cordani, M.; Donadelli, M.; Strippoli, R.; Bazhin, A.V.; Sanchez-Alvarez, M. Interplay between ROS and Autophagy in Cancer and Aging: From Molecular Mechanisms to Novel Therapeutic Approaches. *Oxid. Med. Cell. Longev.* **2019**, *2019*, 8794612. [CrossRef]

43. Yamamoto, A.; Tagawa, Y.; Yoshimori, T.; Moriyama, Y.; Masaki, R.; Tashiro, Y. Bafilomycin A1 prevents maturation of autophagic vacuoles by inhibiting fusion between autophagosomes and lysosomes in rat hepatoma cell line, H-4-II-E cells. *Cell Struct. Funct.* **1998**, *23*, 33–42. [CrossRef] [PubMed]

44. Thames, H.D., Jr.; Rasmussen, S.L. A test for dose-modifying factors. *Radiat. Res.* **1978**, *76*, 308–324. [CrossRef] [PubMed]

45. Minagawa, Y.; Kigawa, J.; Itamochi, H.; Kanamori, Y.; Shimada, M.; Takahashi, M.; Terakawa, N. Cisplatin-resistant HeLa cells are resistant to apoptosis via p53-dependent and -independent pathways. *Jpn. J. Cancer Res.* **1999**, *90*, 1373–1379. [CrossRef] [PubMed]

46. Chen, J.; Solomides, C.; Parekh, H.; Simpkins, F.; Simpkins, H. Cisplatin resistance in human cervical, ovarian and lung cancer cells. *Cancer Chemother. Pharm.* **2015**, *75*, 1217–1227. [CrossRef]

47. Skoupilova, H.; Bartosik, M.; Sommerova, L.; Pinkas, J.; Vaculovic, T.; Kanicky, V.; Karban, J.; Hrstka, R. Ferrocenes as new anticancer drug candidates: Determination of the mechanism of action. *Eur. J. Pharm.* **2020**, *867*, 172825. [CrossRef]

48. Velma, V.; Dasari, S.R.; Tchounwou, P.B. Low Doses of Cisplatin Induce Gene Alterations, Cell Cycle Arrest, and Apoptosis in Human Promyelocytic Leukemia Cells. *Biomark. Insights* **2016**, *11*, 113–121. [CrossRef]

49. Pawlik, T.M.; Keyomarsi, K. Role of cell cycle in mediating sensitivity to radiotherapy. *Int. J. Radiat. Oncol. Biol. Phys.* **2004**, *59*, 928–942. [CrossRef] [PubMed]

50. Cruet-Hennequart, S.; Villalan, S.; Kaczmarczyk, A.; O'Meara, E.; Sokol, A.M.; Carty, M.P. Characterization of the effects of cisplatin and carboplatin on cell cycle progression and DNA damage response activation in DNA polymerase eta-deficient human cells. *Cell Cycle* **2009**, *8*, 3039–3050. [CrossRef]

51. Wagner, J.M.; Karnitz, L.M. Cisplatin-induced DNA damage activates replication checkpoint signaling components that differentially affect tumor cell survival. *Mol. Pharm.* **2009**, *76*, 208–214. [CrossRef]

52. Wang, Y.; Yin, W.; Ke, W.; Chen, W.; He, C.; Ge, Z. Multifunctional Polymeric Micelles with Amplified Fenton Reaction for Tumor Ablation. *Biomacromolecules* **2018**, *19*, 1990–1998. [CrossRef]

53. Liu, C.; Chen, W.; Qing, Z.; Zheng, J.; Xiao, Y.; Yang, S.; Wang, L.; Li, Y.; Yang, R. In Vivo Lighted Fluorescence via Fenton Reaction: Approach for Imaging of Hydrogen Peroxide in Living Systems. *Anal. Chem.* **2016**, *88*, 3998–4003. [CrossRef] [PubMed]

54. Halliwell, B.; Gutteridge, J.M. Role of free radicals and catalytic metal ions in human disease: An overview. *Methods Enzymol.* **1990**, *186*, 1–85. [PubMed]

55. Acevedo-Morantes, C.Y.; Meléndez, E.; Singh, S.P.; Ramírez-Vick, J.E. Cytotoxicity and Reactive Oxygen Species Generated by Ferrocenium and Ferrocene on MCF7 and MCF10A Cell Lines. *J. Cancer Sci. Ther.* **2012**, *4*, 271–275.

56. Gasser, G.; Ott, I.; Metzler-Nolte, N. Organometallic anticancer compounds. *J. Med. Chem.* **2011**, *54*, 3–25. [CrossRef]

57. Mooney, A.; Tiedt, R.; Maghoub, T.; O'Donovan, N.; Crown, J.; White, B.; Kenny, P.T. Structure–activity relationship and mode of action of N-(6-ferrocenyl-2-naphthoyl) dipeptide ethyl esters: Novel organometallic anticancer compounds. *J. Med. Chem.* **2012**, *55*, 5455–5466. [CrossRef] [PubMed]

58. Peng, X.; Gandhi, V. ROS-activated anticancer prodrugs: A new strategy for tumor-specific damage. *Ther. Deliv.* **2012**, *3*, 823–833. [CrossRef]

59. Podolski-Renić, A.; Bösze, S.; Dinić, J.; Kocsis, L.; Hudecz, F.; Csámpai, A.; Pešić, M. Ferrocene-Cinchona Hybrids with Triazolyl-chalcone Linker Act as Prooxidants and Sensitize Human Cancer Cell Lines to Paclitaxel. *Metallomics* **2017**, *9*, 1132–1141. [CrossRef]

60. Kovjazin, R.; Eldar, T.; Patya, M.; Vanichkin, A.; Lander, H.M.; Novogrodsky, A. Ferrocene-induced lymphocyte activation and anti-tumor activity is mediated by redox-sensitive signaling. *FASEB J. Off. Publ. Fed. Am. Soc. Exp. Biol.* **2003**, *17*, 467–469.

61. van Staveren, D.R.; Metzler-Nolte, N. Bioorganometallic chemistry of ferrocene. *Chem. Rev.* **2004**, *104*, 5931–5985. [CrossRef]

62. Santos, M.M.; Bastos, P.; Catela, I.; Zalewska, K.; Branco, L.C. Recent Advances of Metallocenes for Medicinal Chemistry. *Mini Rev. Med. Chem.* **2017**, *17*, 771–784. [CrossRef]

63. Satoh, T.; Enokido, Y.; Aoshima, H.; Uchiyama, Y.; Hatanaka, H. Changes in mitochondrial membrane potential during oxidative stress-induced apoptosis in PC12 cells. *J. Neurosci. Res.* **1997**, *50*, 413–420. [CrossRef]

64. Zhang, Y.H.; Wu, Y.L.; Tashiro, S.; Onodera, S.; Ikejima, T. Reactive oxygen species contribute to oridonin-induced apoptosis and autophagy in human cervical carcinoma HeLa cells. *Acta. Pharm. Sin.* **2011**, *32*, 1266–1275. [CrossRef] [PubMed]

65. Elena-Real, C.A.; Diaz-Quintana, A.; Gonzalez-Arzola, K.; Velazquez-Campoy, A.; Orzaez, M.; Lopez-Rivas, A.; Gil-Caballero, S.; De la Rosa, M.A.; Diaz-Moreno, I. Cytochrome c speeds up caspase cascade activation by blocking 14-3-3epsilon-dependent Apaf-1 inhibition. *Cell Death Dis.* **2013**, *9*, 365. [CrossRef]

66. Niioka, T.; Uno, T.; Yasui-Furukori, N.; Takahata, T.; Shimizu, M.; Sugawara, K.; Tateishi, T. Pharmacokinetics of low-dose nedaplatin and validation of AUC prediction in patients with non-small-cell lung carcinoma. *Cancer Chemother. Pharm.* **2007**, *59*, 575–580. [CrossRef]

67. Jaouen, G.; Vessieres, A.; Top, S. Ferrocifen type anti cancer drugs. *Chem. Soc. Rev.* **2015**, *44*, 8802–8817. [CrossRef]

68. Cortes, R.; Tarrado-Castellarnau, M.; Talancon, D.; Lopez, C.; Link, W.; Ruiz, D.; Centelles, J.J.; Quirante, J.; Cascante, M. A novel cyclometallated Pt(II)-ferrocene complex induces nuclear FOXO3a localization and apoptosis and synergizes with cisplatin to inhibit lung cancer cell proliferation. *Metallomics* **2014**, *6*, 622–633. [CrossRef]

69. Wlassoff, W.A.; Albright, C.D.; Sivashinski, M.S.; Ivanova, A.; Appelbaum, J.G.; Salganik, R.I. Hydrogen peroxide overproduced in breast cancer cells can serve as an anticancer prodrug generating apoptosis-stimulating hydroxyl radicals under the effect of tamoxifen-ferrocene conjugate. *J. Pharm. Pharm.* **2007**, *59*, 1549–1553. [CrossRef]

70. Ivanova, D.; Gronemeyer, H.; de Lera, A.R. Design and stereoselective synthesis of retinoids with ferrocene or N-butylcarbazole pharmacophores that induce post-differentiation apoptosis in acute promyelocytic leukemia cells. *Chem. Med. Chem.* **2011**, *6*, 1518–1529. [CrossRef]
71. Guo, J.Y.; Chen, H.Y.; Mathew, R.; Fan, J.; Strohecker, A.M.; Karsli-Uzunbas, G.; Kamphorst, J.J.; Chen, G.; Lemons, J.M.; Karantza, V.; et al. Activated Ras requires autophagy to maintain oxidative metabolism and tumorigenesis. *Genes Dev.* **2011**, *25*, 460–470. [CrossRef]
72. Yonekawa, T.; Thorburn, A. Autophagy and cell death. *Essays Biochem.* **2013**, *55*, 105–117.

© 2020 by the authors. Licensee MDPI, Basel, Switzerland. This article is an open access article distributed under the terms and conditions of the Creative Commons Attribution (CC BY) license (http://creativecommons.org/licenses/by/4.0/).

 *applied sciences*

Review

Integration of Molecular Docking and In Vitro Studies: A Powerful Approach for Drug Discovery in Breast Cancer

Claudia Cava [1],* and Isabella Castiglioni [2]

[1] Institute of Molecular Bioimaging and Physiology, National Research Council (IBFM-CNR), Via F.Cervi 93, 20090 Segrate-Milan, Italy

[2] Department of Physics "Giuseppe Occhialini", University of Milan-Bicocca Piazza dell'Ateneo Nuovo, 20126 Milan, Italy; isabella.castiglioni@unimib.it

* Correspondence: claudia.cava@ibfm.cnr.it

Received: 29 August 2020; Accepted: 2 October 2020; Published: 6 October 2020

Abstract: Molecular docking in the pharmaceutical industry is a powerful in silico approach for discovering novel therapies for unmet medical needs predicting drug–target interactions. It not only provides binding affinity between drugs and targets at the atomic level, but also elucidates the fundamental pharmacological properties of specific drugs. The purpose of this review was to illustrate newer and emergent uses of docking when combined with in vitro techniques for drug discovery in metastatic breast cancer. We grouped the selected articles into five main categories; namely, systematic repositioning of drugs, natural drugs, new synthesized molecules, combinations of drugs, and drug latentiation. We focused on new promising drugs that have a good affinity with their targets, thus inducing a favorable biological response. This review suggests that the integration of molecular docking and in vitro studies can accelerate cancer drug discovery showing a good consistency of the results between the two approaches.

Keywords: molecular docking; in vitro; metastatic breast cancer; drug discovery

---

1. Introduction

Breast cancer (BC) is the most common type of tumor in women, but metastases are the main cause of death. Metastasis is a complex process where cancer cells move into the blood vessels, invade other tissues, and determine a colony in secondary sites. Indeed, BC initiates as a local disease but can spread with metastases to distant sites, such as the lymph nodes and different organs [1]. This process involves the expression of a series of genes that regulate the survival and invasion of cancer cells. Therefore, drugs that modulate the genes/proteins that regulate cancer cell survival, metastasis, apoptosis, and invasion are of great importance as potential drug targets in the drug discovery process [2,3]. However, although the development of new therapies has significantly reduced mortality for metastatic BC, the resistance to anticancer agents can lead to treatment failure [4].

A drug discovery process originates because there is a clinical condition without a suitable therapy. The first step of research often begins in academia, where a hypothesis is generated; for example, the inhibition or induction of a protein or pathway as a therapeutic effect in a disease condition [5]. Indeed, a crucial point of the research process is the selection of a target, which can be a range of biological entities such as proteins, RNA, and genes that can be selected via bioinformatics analyses [6,7]. An optimal target must be accessible to the putative drug molecule and the binding drug–target complex should induce a biological response [5], which can be quantified with in vitro models. The most used in vitro BC models are cell lines, as they share many molecular and genomic features of BC. The binding affinity between the drug and the target can be calculated in silico with molecular docking. Thus, in silico and in vitro screenings may help to quickly identify the toxicity of the tested drugs/molecules, thus avoiding further steps such as in vivo and preclinical studies (in case of unfavorable results from in silico and in vitro methods) [5].

In silico approaches with docking studies require at least two elements: a protein/drug database and a molecular docking algorithm. Protein and drug databases are a collection of the structures of proteins and drugs. The rapidly increasing number of structures has created big data, which offer a wide range of biological and chemical information and are a recent opportunity to develop better knowledge of the relationships between drugs and targets (usually proteins), drugs and diseases, and targets and diseases. However, although the available data are often heterogeneous and incomplete, computational methods can exploit this knowledge to deepen these interactions [8]. Given the cost and time consumption of experimental methods, high-performing computational algorithms for drug discovery processes are needed. The computational technique known as "docking" can predict the binding of drug–target complexes, as well as the conformation of the ligand upon binding to a protein target. The binding free energy of target–drug interactions establishes the affinity of an association and the conditions for forming a complex. Ranked binding free energies are not always precise, but they can be used to select new drugs such as small molecules to be experimentally tested in a virtual screening approach [9–11]. Small molecules are promising new drugs with a low molecular weight, which allows them to penetrate cells easily [12]. In addition, molecular docking can be also used for predicting the effects of a drug; for example, the identification of an undesired interaction between a compound and off-targets. To date, 57,000 abstracts/papers have been published on molecular docking, indicating the importance of this computational method in drug development [13–15].

Despite encouraging results, the real condition of the cellular environment, such as the pH and temperature, cannot be fully replicated in a docking study. Each docking algorithm has its limitations and advantages. Therefore, it has been reported that a binding free energy that integrates the results from different docking algorithms can lead to a higher performance in a virtual screening process [16]. Moreover, molecular docking, being a structure-based method, is limited to receptors and ligands with a known stable structure. Thus, the integration of in vitro and in vivo studies as a validation step of in silico methods is an indispensable part of the drug discovery process. These techniques can study different aspects of potential drugs, such as absorption, regulation of targets, metabolic stability, and toxicity [17].

The goal of this review was to describe recent studies in metastatic BC that used molecular docking and in vitro studies for drug development.

2. Materials and Methods

Papers published on the PubMed platform on the use of in vitro and molecular docking strategies in metastatic BC were included in this review. Papers were included in the review only if: (1) they were published in the last five years (from 2015 to the end of the search on 27 August 2020); (2) they were published in full-text English language in a peer-reviewed journal (excluding short communications and abstracts); (3) they included in vitro and docking analyses in metastatic BC. Papers were excluded if: (1) only one in vitro or docking analysis was performed; (2) they did not provide in vitro studies on BC cell lines.

3. Results

We categorized the selected papers into five main groups based on the characteristics of the drugs and approaches used: systematic repositioning of drugs/molecules, natural molecules, new synthesized molecules, combinations of drugs, and drug latentiation.

3.1. Systematic Repositioning of Drugs/Molecules

Drug discovery is a time-consuming and labor-intensive work process. On average, the development of a new drug takes 10–15 years. Drug repositioning, namely, the use of old drugs for new diseases, is an efficient strategy for its low-cost and riskless characteristics [15]. Several studies have performed systematic approaches, using a combination of in silico and experimental methods, to reposition known drugs/molecules. Table 1 presents the works in the last five years that have implemented in silico and in vitro models for drug discovery using a systematic repositioning of drugs for metastatic BC.

In particular, the study of Rymbai et al. [18] focused on the similarity between the side effects of two different drugs. Indeed, the assumption of the study was that if the two drugs have common side effects, then they can also have common gene targets and clinical indications. In vitro cell line and molecular docking studies have shown that ropinirole shares many side effects with letrozole (used in the management of advanced and metastatic BC) and is efficient in the treatment of breast cancer. An in vitro study of ropinirole on MCF-7 cells by a 3-(4,5-dimethylthiazol-2-yl)-2, 5-diphenyltetrazolium bromide assay (MTT) was performed to test the ability of ropinirole to inhibit cell growth. Molecular docking showed a good interaction with a good binding affinity of −7.8 kcal/mol between ropinirole and aromatase, a well-known target of letrozole. The three-dimensional structure of aromatase (Protein Data Bank (PDB) ID: 3EQM) is available for computational docking analysis in the Protein Data Bank (PDB).

In a recent study, Liu et al. screened more than 1000 known small molecular compounds and focused on 15 compounds [19], studying their effect on breast cancer cell viability and migration with in vitro studies on MCF-7, MDA-MB-231, and BT-474 BC cell lines. Molecular docking was performed to demonstrate the ability of the binding of the 15 compounds with chemokine ligand 18 (CCL18), as previous studies suggested that CCL18 is a chemokine derived from tumor-associated macrophages (TAMs) to induce BC metastasis [20]. Therefore, CCL18 is considered a potential drug target (PDB ID: 4MHE). Narrowing the selection of ligands from more than 1000 compounds to 15 was performed by evaluating the binding energy with CCL18 obtained via molecular docking studies. In this way, 15 compounds were selected as potential drugs targeting CCL18, and the toxicity of these 15 compounds was evaluated by using cell counting kit-8 (CCK-8) assays, which indicated that most of the compounds did not influence cell viability. A total of 6 of the 15 compounds inhibited CCL18-induced cell migration; this anticancer activity was also confirmed by adherence and invasion assays [19].

In vitro, in silico, in vivo, and ex vivo analyses have been performed to evaluate the cytotoxic action of etoposide (ET), doxorubicin (DOX), pifithrin-α (PIF), and dexamethasone (DEX) in triple-negative BC (TNBC) [21]. TNBC is a molecular subtype of BC that is negative for three hormone receptors—namely, estrogen receptors (ERs), progesterone receptors (PRs), and human epidermal receptor 2 (HER2) [21]. ET, a podophyllotoxin derivative, is a chemotherapy medication used against a wide range of cancers (e.g., lung cancer, lymphoma, lung cancer, leukemia, and glioblastoma multiforme), but its efficacy against TNBC is still unknown [22]. DOX, alone or in combination with other drugs, is an effective therapy for numerous cancers, including breast cancer [23]. The antiapoptotic and non-topoisomerase inhibitors PIF and DEX were considered in a previous study as a negative control [21]. In this study, the authors identified tumor necrosis factor-related apoptosis-inducing ligand (TRAIL) and death receptor 5 (DR5) as potential drug targets (PDB ID: 4N90) [21]. Molecular docking and molecular dynamics studies have shown the ability of ET, DOX, PIF, and DEX to stabilize the TRAIL–DR5 complex. In vitro technologies have demonstrated that ET and DOX increase apoptosis, demonstrating their synergistic effect with TRAIL. These results were confirmed with the binding energy for the ternary complexes TRAIL–DR5–ET and TRAIL–DR5–DOX [21].

Thionine (TH), an organic dye, and its derivatives have been proposed as promising drugs for the photodynamic therapy of cancer [24]. Previous studies reported that TH shows possible genotoxic and cytotoxic activity in prokaryotic cells [25]. Since there is a wide range of applications of TH, it is important to evaluate the site-specific interaction of TH with human serum albumin (HSA), the main protein in plasma, which is responsible for the maintenance and regulation of the colloidal osmotic pressure of blood [24]. HSA has several binding sites to which various ligands can bind. The binding of ligands to this protein can influence drug distribution, as HSA plays a crucial role in the transport of endogenous/exogenous compounds. Therefore, HSA is widely used in clinical applications as a drug delivery system. Manivel et al. revealed that TH interacts with the hydrophobic cavity of subdomain IIA of HSA (PDB ID:1AO6) and that the complex shows a good level of cytotoxicity in cancer cells through in vitro studies [24].

Previous studies have shown the anticancer and antimetastatic effects of the microbial polyketide 2,4-diacetylphloroglucinol (DAPG) through the regulation of NF-κB activity [26]. The induction of NF-κB also mediates the expression of other antiapoptotic proteins and protein kinases in cancer cells. However, the mechanism of action of DAPG acting on metastatic proteins such as Matrix metalloproteinase 9 (MMP9), MMP2, NF-κB, and the antiapoptotic Bcl-2 family proteins are not yet known. In a recent study [27], binding energies, computed by molecular docking, revealed that MMP2, MMP9, and NF-κB achieved a higher interaction with DAPG. In vitro studies have confirmed that DAPG compounds are able to inhibit cancer cells in several cancer cell lines, including MDA-MB-231 [27].

Drug repositioning offers new clinical indications for known drugs/molecules using an efficient, low-cost and riskless strategy. However, despite encouraging results, drug repositioning is a complex process that involves different elements, such as the use of big medical data, to develop an appropriate approach for drug repositioning and a new framework for the integration of available resources.

3.2. Natural Molecules

Historically, natural products, which contain a wide range of compounds for drug discovery, have been considered in multiple clinical trials, especially as anticancer and antimicrobial agents [28]. The advantage of natural products with respect to synthetic compounds is that they are also metabolites. Therefore, they are biologically active and can also be substrates for transporter systems [28]. Table 2 shows the works in the last five years that have implemented in silico and in vitro models for drug discovery using natural ligands as drugs for metastatic BC.

Hypercholesterolemia has been reported to play a role in the progression of BC and resistance to hormonal therapy. High low-density lipoprotein (LDL) levels, primarily caused by familial hypercholesterolemia, are also one of the risk factors of the initiation and promotion of BC [29]. Proprotein convertase subtilisin/kexin type-9 (PCSK9) binds to LDL receptors (avoiding binding with LDL) and regulates the cholesterol metabolism, targeting the receptor for lysosomal degradation and thus leading to the degradation of LDL. Pseurotin A (PS) is a microbial secondary metabolite originally isolated from the fungal culture of *Pseudeurotium ovalis* in 1976 [30]. PS shows different biological activities, including the inhibition of the fungal chitin synthase [31], the activation of cell differentiation [32], and apomorphine antagonist activity [33]. Abdelwahed et al. showed that PS reduces PCSK9 secretion, suggesting its potential as a drug [34]. In particular, they performed predictive molecular modeling to evaluate the binding of PS to PCSK9 (PDB ID: 4NE9, 4NMX, and 3GCW). Docking analysis has revealed that PS is able to successfully bind to the PCSK9 domain, thereby disrupting PCSK9–LDL receptor interactions. Furthermore, this finding has been validated in vitro by surface plasmon resonance, confirming the capacity of PS to interfere with the PCSK9–LDL complex at their binding interface.

Table 1. Characteristics of the studies reported herein categorized as "systematic repositioning of drugs" used in metastatic breast cancer. The table reports the drug, its target with Protein Data Bank (PDB) ID, the in silico/in vitro/in vivo methods used to test the drug, clinical trials, the original use, and the reference.

Drug	Target	In Silico	In Vitro	In Vivo	Clinical Trials	Original Use	Ref.
Ropinirole	aromatase enzyme (PDB [1] ID: 3EQM)	Docking studies	MTT [2] assay	-	-	antiparkinsonism	[19]
15 small molecular compounds	CCL18 (PDB ID: 4MHE)	Docking studies	Cell viability, Boyden chamber, adherence assay	Tumor xenografts	-	CCL18 antagonist	[20]
Topoisomerase inhibitor etoposide (ET) and doxorubicin (DOX)	TRAIL-DR5 (PDB ID: 4N90)	Docking, mutational and dynamics studies	MTT assay, FACS [3]	Tumor xenografts	NCT00004 906	against a wide range of cancers	[21]
Thionine	human serum albumin (HSA) (PDB ID:1AO6)	Dockingstudies	MTT assay and Fluorescence microscopic	-	-	against bacteria, viruses and yeasts	[24]
2,4–diacetylphloroglucinol	Bcl-2 (PDB ID: 4AQ3), Bcl-xL (PDB ID: 2YQ6), Bcl-w (PDB ID: 2Y6W), MMP2 (PDB ID: 1HOV), MMP9 (PDB ID: 1GKC), NF-jB p65 (PDB ID: 1VKX)	Docking studies	MTT and invasion assay	-	-	antimicrobial, antiviral, and anticancer	[27]

[1] PDB: Protein Data Bank, [2] MTT: 3-[4, 5-dimethylthiazol-2-yl]-2, 5-diphenyltetrazolium bromide, [3] FACS: Fluorescence-activated cell sorting.

In an another study, the interactions of atranorin (ATR), a metabolite of many lichens, with the target proteins that are overexpressed in BC, such as AKT, BCL-2, BAX, BCL-W, and BCL-XL, were studied with docking and in vitro analyses [35]. Commonly, ATR is metabolite present in numerous lichens such as *Stereocaulon cacspitorim*, *Everniastrum vexans*, *Parmatrema* species, and others [35]. The complex formed by ATR and AKT shows a better binding energy, but interacting residues show minor affinities to inhibiting the overexpression of BC biomarkers. Interestingly, previous studies have reported the activation of AKT in drug resistance [36] and in vitro studies have been performed on MDA-MB-231 and MCF-7. The inhibitory activity of ATR has been tested with MTT assays, showing a downregulation of oncoproteins. Furthermore, gene expression analysis of the ATR–AKT model has shown the induction of apoptosis in BC cell lines [35].

Another natural compound, eugenol, a phenylpropanoid obtainable from honey and some essential oils with antioxidant and anticancer properties, has been tested with in silico and in vitro approaches [37]. In vivo and in vitro studies have demonstrated that eugenol promotes the inhibition of β-catenin, a biomarker associated with the progression of cancer and the development of lymph node metastasis. β-catenin accumulation in the nucleus is evident in BC because of aberrant wnt signaling. Western blot analysis has confirmed a significant modification in the expression level of total β-catenin in in vivo and in vitro models. In addition, eugenol demonstrates a downregulation of the expression of cancer stem cell biomarkers. Docking studies have revealed many binding sites of the complex and the data support a good interaction between the ligand and the β-catenin protein (PDB ID: 3BCT) [37].

Pharmacological research in recent years has also proposed the plant *Astragalus membranaceus* (AM) as a natural product for cancer treatment. A recent study using a multidisciplinary approach constituted by gene expression analysis, pharmacokinetic screening, biological network analysis, and in vitro approaches investigated the possible and novel mechanism of AM in TNBC [38]. All of the ingredients of AM were collected for a total of 87 compounds and 16 active components and *Astragalus polysaccharides* (APS) was proposed as a potential compound against BC. Indeed, docking analysis showed good results between APS and the proteins AKT, BCL2, and PIK3CG. Indeed, in vitro experiments confirmed that the compound can inhibit migration and invasion and can induce apoptosis [38].

Natural compounds can be also administrated to avoid the side effects of drugs. In a recent study, drug design methods were performed to investigate plant-derived inhibitors against sirtuin (SIRT) proteins [39]. As SIRT, which comprises seven human isoforms, is associated with the metastatic and oncogenic progression of advanced BC, its inhibition is a promising approach against tumorigenesis. A previous study considered 21 plant-derived inhibitors as ligands and the seven human isoforms of SIRT as targets [39], and molecular docking with the binding energies of the ligand–receptor complexes showed that sulforaphane, kaempferol, and apigenin can achieve the highest binding energies against SIRT1, 3, and 6, respectively. To validate these in silico results, they explored the role of these potential small molecules against BC cell lines on cellular viability using MTT assays [39].

Noscapine (NOS) is a phthalide isoquinoline alkaloid derived from the *Papaver somniferum* plant. As several studies have reported the ability of NOS to inhibit the growth of tumor cells and to activate apoptosis, there are ongoing phase I/II clinical trials for cancer management, but not for BC [40]. Maurya et al. [41] studied the interaction between NOS and carrier protein HSA using different techniques, including in vitro approaches and computational methods. The cytotoxicity findings by MTT assay indicated that NOS has good potential in cancer. Meanwhile, the in silico results showed that the main binding site for NOS was site I (subdomain II A) of HSA [41].

Shikonin (SK), a phytochemical derived from the medicinal plant *Lithospermum erythrorhizon*, has been demonstrated to induce tumor immunogenicity [42]. However, the molecular mechanisms of action and the pharmacological processes are still unknown. First, in a recent study [43] the authors performed a computation prediction analysis, applying molecular docking and a virtual screening system. They screened 27,317 human protein structures deposited in the Protein Data Bank in order to search for the molecular targets of SK. Heterogeneous nuclear ribonucleoprotein A1 (hnRNPA1) obtained the highest binding energy with SK (−15.3 kcal/mol), and this binding leads to the suppression of post-transcriptional mRNA processing in vitro [43]. In addition, the SK– hnRNPA1 complex inhibits the splicing and suppresses the nuclear export activities of specific inflammation-associated genes, resulting in a reduction in acute cytokine storms [43].

Matrix metalloproteinase 9 (MMP9) and matrix metalloproteinase 2 (MMP2) regulate the tumor microenvironment and tumor metastasis. Therefore, the inhibition of MMP9 and MMP2 could reduce invasion and tumor metastasis [44]. Currently, traditional Chinese medicines (TCMs) are proposed in the prevention and treatment of several diseases, including cancer. In a recent study, plantamajoside (PMS), a Chinese herbal medicine, was proposed as an inhibitor of MMP9 and MMP2. Indeed, molecular docking and in vitro analyses confirmed the good interaction of the molecules with the proteins and the inhibition of the proliferation, migration, and invasion of BC cell lines [44]. Specifically, in vitro studies on the MDA-MB-231 and 4T1 BC cell lines have demonstrated that PMS reduces the activity of MMP9 and MMP2. In addition, after treatment with PMS, BC cell lines show a decrease in cell proliferation.

The aryl hydrocarbon receptor (Ahr), a helix–loop–helix transcription factor, is a promising regulator of the invasiveness and metastasis of BC cells. It is regulated by a wide variety of natural molecules such as flavonoids—among which, flavipin is the least studied. Hanieh et al. [45] applied in silico and in vitro methods to investigate the relationships between flavipin and Ahr. Docking analysis revealed eight hydrogen bonds involving the Phe115, Leu116, and Ala119 residues of the Ahr molecule. The results of the in vitro analysis showed that flavipin has inhibitory effects on the migration of MDA-MB-231 and T47D cells [45].

The studies reported above show that the growing application of in silico and in vitro techniques may contribute to the recovery of interest in natural products for drug discovery.

Table 2. Characteristics of the studies reported herein categorized as "natural drugs" used in metastatic breast cancer. The table reports the drug, its target with PDB ID, the in silico/in vitro/in vivo methods used to test the drug, clinical trials, the mechanism of action, and the reference.

Drug	Target	In Silico	In Vitro	In Vivo	Clinical Trials	Mechanism of Action	Ref.
Pseurotin A	PCSK9 (PDB[1] ID: 4NE9, 4NMX, and 3GCW)	Docking studies	MTT[2] assay	Tumor xenografts	-	cholesterol metabolism	[34]
Atranorin	AKT, BCL-2, BAX, BCL-W and BCL-XL (PDB ID: NA)	Docking studies	MTT assay	-	-	apoptosis	[35]
Eugenol	β-catenin (PDB ID: 3BCT)	Docking studies	MTT assay	Tumor xenografts	-	Cancer Stem Cell	[37]
Astragalus membranaceus	AKT (PDB ID:3QKK), BCL2 (PDB ID: 4AQ3), and PIK3CG (PDB ID: CHX)	Differential expression analysis Docking, dynamics studies	CCK-8[3], Chamber, FITC[4] assay	-	NCT03314805, NCT03634150	apoptosis	[38]
21 plant-derived inhibitors	human sirtuin (SIRTs 1-7). SIRT1 (PDB ID: 4I5I), SIRT2 (PDB ID: IJ8F), SIRT3 (PDB ID: 5D7N), SIRT5 (PDB ID: 2B4Y), SIRT6 (PDB ID: 3K35) and SIRT7 (PDB ID: 5IQZ)	Docking and dynamics studies	MTT, trypan blue, sirtuin, Anchorage-dependent clonogenic assay	-	-	sirtuin inhibitors	[39]
Noscapine	human serum albumin (HSA) (PDB ID: 1AO6)	Docking and dynamics studies	MTT assay	-	-	inhibition of cell growth	[41]
Shikonin	27,317 human protein structures	Docking studies	calorimetry analysis and electrophoretic mobility shift assay	Tumor xenografts	-	suppression of post-transcriptional mRNA processing	[42]
Plantamajoside	Matrix metalloproteinase 2 and 9 (PDB ID: NA)	Docking studies	CCK-8, chamber wound assay	Tumor allografts	-	inhibition of cell growth	[44]
Flavipin	Aryl hydrocarbon receptor (Ahr) (PDB ID: 4M4X)	Docking studies	CCK-8 and Boyden chamber	-	-	cancer cell motility	[45]

[1] PDB: Protein Data Bank, [2] MTT: 3-[4, 5-dimethylthiazol-2-yl]-2, 5-diphenyltetrazolium bromide, [3] CCK-8 cell counting kit-8, [4] FITC Annexin-V fluorescein isothiocyanate.

3.3. New Synthesized Molecules

Since there is not yet a definite solution for the treatment of cancer, there is a clear need to research new molecules with anticancer properties. The combination of in vitro techniques and molecular docking can also be used also to test new synthesized molecules. Table 3 shows the works in the last five years that have implemented in silico and in vitro models for drug discovery using new synthesized molecules as drugs for metastatic BC.

For example, Nashaat et al. [46] synthesized a new series of compounds; i.e., new benzimidazole derivatives against BC-targeting peptidylprolyl cis-trans isomerase NIMA-interacting 1 (PIN1). PIN1 proteins play a role in cell cycle regulation, and their overexpression is correlated with human cancer [46]. It has also been described that PIN1 increases DNA binding to estrogen receptors [47]. Anticancer effects have been tested in vitro with MTT assays using the MCF-7 cell line: Several synthesized compounds have shown a very strong efficacy against this BC cell line. Furthermore, the interactions of three synthesized compounds with PIN1 crystal structures (PDB: 4TYO) have been validated using molecular docking, and the formed complexes show some essential interactions for PIN1 inhibition [46]. Specifically, the interaction with Lys63 is crucial for PIN1 inhibition.

In another study, Vaz et al. [48] synthesized the new hybrid dihydroquinoline derivative (M-CNP) compound. They designed this hybrid compound composed of sulfonamide, quinoline, and chalcone. The idea behind the study was that molecules containing sulfonamide, quinoline, and chalcone could originate new hybrid architectures with novel anticancer properties. The aim is that compounds derived from more molecules can enhance the pharmacological effectiveness. Anticancer activity of the novel compound was demonstrated by regulating aldehyde dehydrogenase 1 family member A1 (ALDH1A1), which, by converting retinal (retinaldehyde) to retinoic acid (RA), plays a role in the differentiation of cells and signaling events. Interestingly, an in vivo test on a metastatic BC cell lines (i.e., MDA-MB-231) has confirmed the promising role of ALDH1A1. Indeed, cytotoxicity assays have demonstrated the specific activity of the compounds against tumor cells. Furthermore, molecular docking analyses have demonstrated that the M-CNP derivative plays a role as an anticancer drug because of its good affinity with ALDH1A1 [48].

Previous studies have shown that compounds can bind to DNA with covalent and non-covalent interactions; for example, cisplatin, a well-known anticancer drug, interacts with DNA via a covalent link. However, covalent interactions can induce serious side effects, highlighting the importance of non-covalent interactions between DNA and drugs [49]. Among metal complexes, nickel aroylhydrazone Schiff base complexes show non-covalent interactions. Following this assumption, Li et al. [49] synthesized two nickel-derived complexes, and molecular docking revealed that both compounds could bind to DNA through the interaction of the phenyl rings with the double helix. In addition, the association between DNA and bovine serum albumin (BSA) is able to modify the secondary structure of BSA. The anticancer activity of individual complexes was also evaluated with in vitro cytotoxicity assays on the A549, MCF-7, and L-02 cell lines [49]. Both complexes obtained a lower cytotoxic effect than cisplatin against normal cell lines. Thus, in vitro analysis suggests more selective effects of the compounds against cancer cell lines.

Ruarene complexes have been observed to be possible agents against cisplatin resistance with fewer side effects, demonstrating a different mechanism of action. Acharya et al. synthesized four ruarene complexes and characterized them using X-ray crystallography [50], and molecular docking was performed to demonstrate their ability to bind tubulin proteins (PDB: 1SA0). The cytotoxicity of the molecules was tested in vitro by MTT assays in three different cancer cell lines, including MDA-MB-231. The complexes that demonstrated a lower toxicity were selected to test their effect on the inhibition of the microtubule network in the MDA-MB-231 cell line. In silico and in vitro studies demonstrated good binding between the compounds and the tubulin, as well as antiproliferative action against advanced subtypes of cancer, such as triple-negative metastatic BC [50].

Glycyrrhiza glabra, an Indian therapeutic herb, contains a diglucopyranosiduronic acid of glycyrrhetinic acid (GA). GA plays a role in immune responses, cell cycle, apoptosis, and autophagy [51].

Shukla et al. [52] designed five glycyrrhetinic acid (GA) derivatives and analyzed their in vitro action in a metastatic breast cancer cell line (i.e., MDA-MB-231). Molecular docking studies have been carried out to investigate the action of compounds on BC targets such as glyoxalase-I (GLO-I). BC receives energy from glycolysis based on the Warburg effect, and GLO-1 is able to inhibit and inactivate methyl glyoxalases, a compound formed during glycolysis, making GLO-I inhibitors potential anticancer agents. It has been shown that GA-1 increases cytotoxic action, and a study using molecular docking confirmed the binding of GA derivatives with GLO-I [52].

As previous studies have described that epidermal growth factor receptor tyrosine kinase (EGFR-TK) is upregulated in BC, it has been suggested as a potential drug target for novel therapy agents. The EGFR-TK inhibitory activity of known 1,3,5-triazines derivatives against BC has been examined [53]. In addition, the effects of the compounds on several BC cell lines have been estimated, including MDA-MB-231, a metastatic BC cell line. The consequence of the compounds on β-catenin expression has also been evaluated: all designed compounds achieved a good binding energy against the target protein. Furthermore, in vitro experiments have confirmed the docking analysis since the synthesized derivatives demonstrated an inhibitory effect on EGFR-TK [53].

Rac family small GTPase 1 (RAC1) is involved in the migration and invasion of BC cells. Therefore, therapeutic strategies that silence RAC1 could be a new challenge [54]. A new series of carbazole derivatives have been designed for their antitumor properties. Vlaar et al. evaluated the role of the compounds via interactions with RAC1 through molecular docking: the molecular docking results indicated a favorable conformation of the receptor–ligand complex. The compounds also demonstrated moderate antiproliferative activity using in vitro techniques [54].

The main site of BC metastasis is bone, and bone metastases often lead to complications, including fractures, bone pain, and hypercalcemia. However, to date, no biomarkers have been identified that are able to predict bone metastases. A first circulating fragment of parathyroid hormone-related protein (PTHrP), PTHrP(12-48), has been proposed as a biomarker associated with the presence of bone metastases. Kamalakar et al. [55] investigated the biological processes and mechanisms of action of PTHrP(12-48). First of all, they predicted the tertiary structure of PTHrP(12-48) through bioinformatics analyses, and the molecular modeling found that PTHrP(12-48) interacts via a weak binding with the PTH1 receptor (PTHR1). In vitro analysis supported this model: PTHrP(12-48) treatment does not promote an increase in cAMP in PTHR1-expressing SaOS2 cells. In conclusion, these data indicate that PTHrP(12-48) acts in the regulation of the differentiation of hematopoietic cells and regulates the osteoclasts within the tumor–bone marrow microenvironment, possibly to induce bone metastasis [55].

Angiogenesis, a physiological process that produces new blood vessels from pre-existing blood cells, consists of a series of steps such as the production of protease, endothelial cell migration and proliferation, vascular tube formation, and the maturation of cells [56]. Accelerated angiogenesis is correlated with several diseases, including cancer. The inhibitors of angiogenesis that target several proteins, such as vascular endothelial growth factor (VEGF), epidermal growth factor (EGF), platelet-derived growth factor (PDGF), and basic fibroblast growth factor (bFGF), that are involved in the regulation of the process can be used as a therapeutic strategy. The main factor in the angiogenesis process is VEGF [56]. Aboul-Enein et al. [57] designed 7-chloro-4-(piperazin-1-yl)quinoline derivatives as VEGFR-II inhibitors, and the anticancer abilities of these compounds were tested in vitro using breast cancer and prostate cancer cell lines. These analyses demonstrated that compound **4q** is the most active on both cell lines, and molecular modeling revealed that **4q** has similar sites of binding as sorafenib and lenavatinib at the ATP binding site of VEGFR-II. Furthermore, the binding energy of **4q** is slightly better than that of lenavatinib but lower than that of sorafenib [57].

Apoptosis is one of the key hallmarks of cancer. The BCL-2 family is a group of proteins that have a fundamental role in the regulation of apoptosis. For this reason, BCL-2 has become one of the most studied drug targets over recent years. Ziedan et al. [58] designed a 3D pharmacophore model to act as an inhibitor of the antiapoptotic BCL-2 protein. The 15 oxadiazole derivative compounds were tested to demonstrate their inhibitory activity in two cancer cell lines (i.e., cervical HeLa and breast

MDA-MB-231). Molecular docking was carried out to evaluate the interaction scores between the compounds and the BCL-2 protein. Some of the proposed compounds obtained good antiproliferative activity and good binding energy with the BCL-2 protein (compounds **1** and **16j**) [58]. The simplicity of the synthesis of these compounds and their low molecular weight is promising; therefore, additional studies should be employed based on this novel class of BCL-2 inhibitors.

Oncogenic proteins such as tyrosine kinases, cell cycle regulators, and transcriptional factors are implicated in metastatic pathways in cancer. Many of them interact with heat shock protein 90 (Hsp90). Therefore, Hsp90 inhibitors have been proposed as novel cancer treatment methods, although they can lead to adverse effects in clinical trials. To overcome these disadvantages, Koca et al. [59] designed novel molecular Hsp90 inhibitors. Specifically, they designed novel pyrimidinyl acyl thiourea derivatives as Hsp90 inhibitors, and in vitro analyses revealed that these compounds can inhibit cell proliferation and demonstrate cytotoxic effects in BC and human bone osteosarcoma cell lines. Molecular docking confirmed the interaction of these compounds with the Hsp90 domain [59].

Previous studies have suggested KDM5A and KDM5B as oncogenic regulators [60]. The catalytic domain of KDM5 proteins has an unusual inclusion of an ARID and PHD1 domain that divides the catalytic domain into two subdomains—namely, JmjN and JmjC. Horton et al. demonstrated that a deletion of the ARID and PHD1 domains has a negative impact on the in vitro enzymatic kinetics of the KDM5 family. Thus, the challenge is finding inhibitors that act on the catalytic domain of the KDM5 family; to this end, the authors proposed GSK-J1 as an inhibitor of the KDM family through in silico studies [61].

EPH receptor A2 (EphA2) is a receptor tyrosine kinase that is involved in drug resistance and metastatic processes [62]. Gambini et al. designed agonistic peptides that target the ligand-binding domain of the EphA2 receptor, called 135H11 and 135H12 [63]. In vitro approaches and computational methods have suggested that both are effective agonistic EphA2 agents and are effective in inhibiting cell migration and invasion [63]. In addition, both dimeric agents are able to induce EphA2 receptor degradation.

The use of drugs to treat or manage the progression of BC is the best strategy. However, the efficacy of traditional drugs has been seriously compromised due to the phenomenon of resistance. Therefore, it is essential to discover new synthesized drugs that target novel sites and regulate biological processes involved in the progression of cancer.

Table 3. Characteristics of studies reported in the review categorized as "new synthesized molecules" used in metastatic breast cancer. The table reports the drug, its target with PDB ID, in silico/in vitro/in vivo methods used to test the drug, clinical trials, mechanism of action and reference.

Drug	Target	In Silico	In Vitro	In Vivo	Clinical Trials	Mechanism of Action	Ref.
new benzimidazole derivatives	PIN1 (PDB [1]: 4TYO)	Docking studies	MTT [2] and apoptosis assay	-	-	apoptosis	[46]
Dihydroquinoline derivate, M-CNP	ALDH1A1 (PDB ID: NA)	Docking studies	MTT assay	-	-	cell viability	[48]
Two new nickel (II) triphenylphosphine complexes	DNA (PDB ID:1Z3F), BSA (PDB ID: 4F5S)	Docking studies	CCK-8 [3] assay	-	-	antioxidant activity	[49]
Ruarene complexes	Tubulin (PDB ID: 1SA0)	Docking studies	MTT, Annexin-V/PE assays	-	-	proliferation	[50]
5 glycyrrhetinic acid (GA) derivates	GLO-I (PDB: 4PV5)	Docking studies	cytotoxicity assay	-	-	metabolism	[52]
1,3,5-triazine derivatives	epidermal growth factorreceptor-tyrosine kinase (EGFR-TK) (PDB ID: 1M17)	Docking studies	MTT and apoptosis assay	-	-	apoptosis	[53]
carbazole derivatives	RAC1 (PDB ID: NA)	Docking studies	Wound healing	-	-	migration	[54]
PTHrP(12-48)	PTH1 receptor (PDB ID: NA)	Docking studies	Immunofluorescence assays	-	NCT00051779	activity of osteoclasts	[55]
Certain 7-Chloro-4-(piperazin-1-yl)quinoline Derivatives	VEGFR-II (PDB ID: NA)	Docking studies	SRB [4] assay	-	-	proliferation	[57]
Oxadiazole derivates	BCL-2 protein (PDB ID: 1YSW)	Docking studies	MTT assay	-	-	apoptosis	[58]
novel pyrimidinyl acyl thiourea derivatives	Heat Shock Protein 90(Hsp90) (PDB ID: 1UYM)	Docking studies	XTT [5] assay	-	-	ATPase function	[59]
crystalstructure of the linked JmjN-JmjC domain	KDM5A and KDM5B (PDB ID: NA)	Docking studies	SRB assay	-	-	cell growth	[61]
135H11 and 135H12	EphA2 (PDB ID: 6B9L)	Docking studies	Wound healing	-	-	migration	[63]

[1] PDB: Protein Data Bank, [2] MTT: 3-[4, 5-dimethylthiazol-2-yl]-2, 5-diphenyltetrazolium bromide, [3] CCK-8 cell counting kit-8, [4] SRB: SulfoRhodamine-B stain, [5] XTT: 2,3-bis[2-methoxy-4-nitro-5-sulfophenyl]-2H-tetrazolium-5-carboxanilide.

3.4. Combination of Drugs

Current chemotherapeutic agents can lead to many adverse effects and can be toxic to healthy cells. For this reason, the identification of new agents that can effectively eradicate tumorigenic cells without damaging normal cells is necessary. A possible solution could be the use of combinations of drugs. Indeed, the efficacy of treatment could be improved, as individual drugs can target different biological pathways. Moreover, combinations of drugs could also potentially reduce drug resistance [64]. In silico and in vitro approaches have been used to test combinations of small molecules against cancer. Table 4 shows the studies in the last five years that have implemented in silico and in vitro models for drug discovery using combinations of drugs/molecules for metastatic BC.

For example, Nayak et al. [65] showed the ability of quinacrine and curcumin to regulate the apoptosis of cancer stem cells with an in vitro model. Curcumin is a diarylheptanoid, which is a natural phenol isolated from the *Curcuma longa* plant, and has multiple pleiotropic effects, such as the suppression of multiple signaling pathways, the inhibition of cell proliferation, and antimetastatic properties [66]. Quinacrine, a 9-aminoacridine (9-AA) derivative, shows anticancer properties against several cancers, such as breast, pancreatic, and lung cancers. Its antiapoptotic activity is shown by its ability to arrest the cell cycle in the S-phase via the inactivation of topoisomerase activity, the activation of p53 and p21, and the inhibition of NF-kβ [67]. The study of Nayak et al. [65] analyzed the anticancer effects of curcumin and quinacrine, as well as their combination, using in vitro and molecular modeling. Multiple BC cells were used to characterize a metastatic model that demonstrated the effects of the combination of molecules on decreasing the migration and invasion and inducing apoptosis. The cytotoxic and antiproliferative activity results showed the synergistic action of the drugs, and molecular docking showed a good affinity of the molecules with ABCG2, a biomarker of BC. Specifically, the binding site is in the transmembrane domain of ABCG2 [65].

Table 4. Characteristics of studies reported in the review categorized as "combination of drugs" used in metastatic breast cancer. The table reports the drug, its target with PDB ID, in silico/in vitro/in vivo methods used to test the drug, clinical trials, mechanism of action and reference.

Drug	Target	In Silico	In Vitro	In Vivo	Clinical Trials	Mechanism of Action	Ref.
Quinacrine and curcumin	ABCG2 (PDB [1] ID: NA)	Docking studies	MTT [2] assay	-	-	DNA damage and repair	[65]
ITH-47 and ESE-15-ol	bromodomain-containing protein 4 32(BRD4) (PDB ID: NA)	Docking and dynamics Studies	Annexin V-FITC [3] and caspase activation assays	-	-	apoptosis	[68]
Vitamin E and Paclitaxel	Bovine serum albumin: (PDB ID: 4OR0)	Docking studies	MTT assay	Tumor xenografts	-	proliferation	[69]

[1] PDB: Protein Data Bank, [2] MTT: 3-[4, 5-dimethylthiazol-2-yl]-2, 5-diphenyltetrazolium bromide [3] FITC Annexin-V fluorescein isothiocyanate.

In another study [68], the authors investigated the combination of two novel compounds, namely, ITH-47 (a BRD4 inhibitor) and ESE-15-ol (an antimitotic agent). The in vitro study revealed that the combination of these two compounds inhibits the growth of MDA-MB-231. To compare the binding energy of the two molecules, they performed molecular docking with a known drug—i.e., JQ1—which revealed that, compared to JQ1, the molecules can achieve similar binding energies and sites as bromodomain-containing protein 4 (BRD4). BRD4 plays a role in regulating c-Myc, a key regulator of cell growth and apoptosis [68].

Paclitaxel (PTX), also known as Taxol, is used as a drug in clinical treatment against different cancers. However, as it causes different side effects, Tang et al. [69] suggested vitamin E (VE)–albumin core–shell nanoparticles (NPs) to improve the efficacy of PTX in BC models. They also investigated the cytotoxicity with in vitro approaches on MCF-7 BC cell lines. Docking studies were performed to analyze the interaction between PTX or VE and BSA, and the results demonstrated a strong receptor–ligand interaction and PTX–VE NPs exhibited better cytotoxic effects than PTX NPs [69].

The antitumor effect was also studied using the xenograft model, showing that treatment with PTX–VE NPs is more effective and lowers the toxicity of the molecules.

3.5. Drug Latentiation

Drug latentiation is a procedure where a compound is chemically modified to improve its binding affinity with a target in order to increase its therapeutic activity.

Gefitinib is one of the more effective and specific epidermal growth factor receptor (EGFR) inhibitors, which interacts with the adenosine triphosphate (ATP)-binding site of the EGFR tyrosine kinase enzyme. Sharma et al. designed three gefitinib-based derivatives to improve the ligand–receptor interaction [70]. Molecular docking studies were also presented for the study of the interactions of the gefitinib derivatives with EGFR, DNA, and BSA. Synthesized compounds were further screened in different cancer cell lines, including MDA-MB-231, to evaluate the cytotoxicity of the new compounds. The results demonstrated a similar effect between experimental and molecular docking analyses, suggesting the important role of gefitinib-based derivatives [70]. Indeed, the in vitro cytotoxicity and antiproliferative activity demonstrated that the derivatives are more potent than gefitinib.

4. Conclusions

In this review, we reported recent studies that have used molecular docking and in vitro studies in metastatic BC for drug discovery. We divided the studies into five main categories: "Systematic repositioning of drugs/molecules", "natural drugs", "new synthesized molecules", "combinations of drugs", and "drug latentiation".

The studies in the systematic repositioning of drugs/molecules category generated new clinical indications for old known drugs or molecules, such as ropinirole, small molecular compound SYSU-21598, etoposide, thionine, and 2,4-diacetylphloroglucinol, reporting a new possible application in metastatic BC. In addition, natural products such as pseurotin A, atranorin, eugenol, astragalus membranaceus, 21 plant-derived inhibitors, noscapine, shikonin, plantamajoside, flavipin, and 13 new synthesized molecules were analyzed and proposed as effective drugs.

Another possible application of molecular docking is studying combinations of drugs and drug latentiation in metastatic BC. We proposed, as promising combinations of drugs, quinacrine and curcumin, ITH-47 and ESE-15-ol, and vitamin E and paclitaxel. The drug latentiation procedures demonstrated that the three gefitinib-based derivatives are more potent than gefitinib.

Overall, the computational and experimental results included herein reported a good consistency and demonstrated that molecular docking and in vitro studies should be used as complementary methods, which, together, can increase the knowledge for drug discovery and development. Indeed, molecular docking generates a binding score that shows the affinity between drugs and targets and in vitro studies investigate the biological responses. However, the binding affinity energy obtained by a single docking algorithm could be inaccurate due to incorrect ligand poses. A better strategy could be the use of combined scores obtained by two or more docking algorithms and/or the use of molecular dynamics. Molecular dynamics is a computational method that describes the dynamic behavior of a biological complex as a function of time. These methods could be more powerful approaches to investigate the novel biological aspects of disease mechanisms, providing a combined procedure to increase innovation in the pharmaceutical industry and to discovery novel therapies for unmet medical needs.

Author Contributions: Conceptualization, C.C.; methodology, C.C.; writing—original draft preparation, C.C.; supervision, I.C.; project administration, I.C.; funding acquisition, I.C. All authors have read and agreed to the published version of the manuscript.

Funding: This research received no external funding

Conflicts of Interest: The authors declare no conflict of interest.

References

1. Weigelt, B.; Peterse, J.L.; Veer, L.J.V. Breast cancer metastasis: Markers and models. *Nat. Rev. Cancer* **2005**, *5*, 591–602. [CrossRef]
2. Cava, C.; Pini, S.; Taramelli, D.; Castiglioni, I. Perturbations of pathway co-expression network identify a core network in metastatic breast cancer. *Comput. Biol. Chem.* **2020**, *87*, 107313. [CrossRef] [PubMed]
3. Bravatà, V.; Cava, C.; Minafra, L.; Cammarata, F.P.; Russo, G.; Gilardi, M.C.; Castiglioni, I.; Forte, G.I. Radiation-Induced Gene Expression Changes in High and Low Grade Breast Cancer Cell Types. *Int. J. Mol. Sci.* **2018**, *19*, 1084. [CrossRef] [PubMed]
4. Ponnusankar, S.; Mohan, A. Newer therapies for the treatment of metastatic breast cancer: A clinical update. *Indian J. Pharm. Sci.* **2013**, *75*, 251–261. [CrossRef] [PubMed]
5. Hughes, J.; Rees, S.; Kalindjian, S.; Philpott, K. Principles of early drug discovery. *Br. J. Pharmacol.* **2011**, *162*, 1239–1249. [CrossRef]
6. Cava, C.; Colaprico, A.; Bertoli, G.; Bontempi, G.; Mauri, G.; Castiglioni, I. How interacting pathways are regulated by miRNAs in breast cancer subtypes. *BMC Bioinform.* **2016**, *17*, 111–133. [CrossRef]
7. Cava, C.; Novello, C.; Martelli, C.; Lodico, A.; Ottobrini, L.; Piccotti, F.; Truffi, M.; Corsi, F.; Bertoli, G.; Castiglioni, I. Theranostic application of miR-429 in HER2+ breast cancer. *Theranostics* **2020**, *10*, 50–61. [CrossRef]
8. Yang, Y.; Adelstein, S.J.; Kassis, I.A. Target discovery from data mining approaches. *Drug Discov. Today* **2009**, *14*, 147–154. [CrossRef]
9. Meng, X.-Y.; Zhang, H.-X.; Mezei, M.; Cui, M. Molecular docking: A powerful approach for structure-based drug discovery. *Curr. Comput. Drug Des.* **2011**, *7*, 146–157. [CrossRef]
10. McConkey, B.J.; Sobolev, V.; Edelman, M. The performance of current methods in ligand-protein docking. *Curr. Sci.* **2002**, *83*, 845–855.
11. Csermely, P.; Korcsmáros, T.; Kiss, H.J.; London, G.; Nussinov, R. Structure and dynamics of molecular networks: A novel paradigm of drug discovery: A comprehensive review. *Pharmacol. Ther.* **2013**, *138*, 333–408. [CrossRef] [PubMed]
12. Ngo, H.X.; Garneau-Tsodikova, S. What are the drugs of the future? *MedChemComm* **2018**, *9*, 757–758. [CrossRef] [PubMed]
13. Lin, X.; Li, X.; Lin, X. A Review on Applications of Computational Methods in Drug Screening and Design. *Molecules* **2020**, *25*, 1375. [CrossRef] [PubMed]
14. Bafna, D.; Ban, F.; Rennie, P.S.; Singh, K.; Cherkasov, A. Computer-Aided Ligand Discovery for Estrogen Receptor Alpha. *Int. J. Mol. Sci.* **2020**, *21*, 4193. [CrossRef]
15. Xue, H.; Xie, H.; Xie, H.; Wang, Y. Review of Drug Repositioning Approaches and Resources. *Int. J. Biol. Sci.* **2018**, *14*, 1232–1244. [CrossRef]
16. Palacio-Rodríguez, K.; Lans, I.; Cavasotto, C.N.; Cossio, P. Exponential consensus ranking improves the outcome in docking and receptor ensemble docking. *Sci. Rep.* **2019**, *9*, 5142. [CrossRef]
17. Bajpai, M.; Esmay, J.D. In Vitro Studies in Drug Discovery and Development: An Analysis of Study Objectives and Application of Good Laboratory Practices (GLP). *Drug Metab. Rev.* **2002**, *34*, 679–689. [CrossRef]
18. Rymbai, E.; Sugumar, D.; Saravanan, J.; Divakar, S. Ropinirole, a potential drug for systematic repositioning based on side effect profile for management and treatment of Breast Cancer. *Med. Hypotheses* **2020**, *144*, 110156. [CrossRef]
19. Liu, Y.; Zheng, H.; Li, Q.; Li, S.; Lai, H.; Song, E.; Li, D.; Chen, J. Discovery of CCL18 antagonist blocking breast cancer metastasis. *Clin. Exp. Metastasis* **2019**, *36*, 243–255. [CrossRef]
20. Chen, J.; Yao, Y.; Gong, C.; Yu, F.; Su, S.; Chen, J.; Liu, B.; Deng, H.; Wang, F.; Lin, L.; et al. CCL18 from Tumor-Associated Macrophages Promotes Breast Cancer Metastasis via PITPNM3. *Cancer Cell* **2011**, *19*, 541–555. [CrossRef]
21. Das, S.; Tripathi, N.; Siddharth, S.; Nayak, A.; Nayak, D.; Sethy, C.; Bharatam, P.V.; Kundu, C.N. Etoposide and doxorubicin enhance the sensitivity of triple negative breast cancers through modulation of TRAIL-DR5 axis. *Apoptosis* **2017**, *22*, 1205–1224. [CrossRef] [PubMed]
22. Baldwin, E.L.; Osheroff, N. Etoposide, Topoisomerase II and Cancer. *Curr. Med. Chem. Anti Cancer Agents* **2005**, *5*, 363–372. [CrossRef] [PubMed]

23. Zhao, N.; Woodle, M.C.; Mixson, A.J. Advances in Delivery Systems for Doxorubicin. *J. Nanomed. Nanotechnol.* **2018**, *9*, 1–9. [CrossRef] [PubMed]

24. Manivel, P.; Paulpandi, M.; Murugan, K.; Benelli, G.; Ilanchelian, M. Probing the interaction of thionine with human serum albumin by multispectroscopic studies and its in vitro cytotoxic activity toward MCF-7 breast cancer cells. *J. Biomol. Struct. Dyn.* **2016**, *35*, 3012–3031. [CrossRef]

25. Dohno, C.; Stemp, E.D.A.; Barton, J.K. Fast Back Electron Transfer Prevents Guanine Damage by Photoexcited Thionine Bound to DNA. *J. Am. Chem. Soc.* **2003**, *125*, 9586–9587. [CrossRef]

26. Veena, V.K.; Popavath, R.N.; Kennedy, K.; Sakthivel, N. In vitro antiproliferative, pro-apoptotic, antimetastatic and anti-inflammatory potential of 2,4-diacteylphloroglucinol (DAPG) by Pseudomonas aeruginosa strain FP10. *Apoptosis* **2015**, *20*, 1281–1295. [CrossRef]

27. Veena, V.K.; Kennedy, K.; Lakshmi, P.; Krishna, R.; Sakthivel, N. Anti-leukemic, anti-lung, and anti-breast cancer potential of the microbial polyketide 2, 4-diacetylphloroglucinol (DAPG) and its interaction with the metastatic proteins than the antiapoptotic Bcl-2 proteins. *Mol. Cell. Biochem.* **2016**, *414*, 47–56. [CrossRef]

28. Harvey, A.L.; Edrada-Ebel, R.; Quinn, R.J. The re-emergence of natural products for drug discovery in the genomics era. *Nat. Rev. Drug Discov.* **2015**, *14*, 111–129. [CrossRef]

29. Nelson, E.R.; Chang, C.-Y.; McDonnell, D.P. Cholesterol and breast cancer pathophysiology. *Trends Endocrinol. Metab.* **2014**, *25*, 649–655. [CrossRef]

30. Bloch, P.; Tamm, C.; Bollinger, P.; Petcher, T.J.; Weber, H.P. Pseurotin, a new metabolite of Pseudeurotium ovalis Stolk having an unusual hetero-spirocyclic system. *Helv. Chim. Acta* **1976**, *7*, 133–137. [CrossRef]

31. Wenke, J.; Anke, H.; Sterner, O. Pseurotin A and 8-Odemethylpseurotin A from Aspergillus fumigatus and their inhibitory activities on chitin synthase. *Biosci. Biotech. Biochem.* **1993**, *57*, 961–964. [CrossRef]

32. Komagata, D.; Fujita, S.; Yamashita, N.; Saito, S.; Morino, T. Novel Neuritogenic Activities of Pseurotin A and Penicillic Acid. *J. Antibiot.* **1996**, *49*, 958–959. [CrossRef] [PubMed]

33. Wink, J.; Grabley, S.; Gareis, M.; Zeeck, A.; Phillips, S. Biologically active pseurotin A and D, new metabolites from Aspergillus fumigatus, process for their preparation and their use as apomorphine antagonists. *Eur. Pat. Appl.* **1993**, ep546475.

34. Abdelwahed, K.S.; Siddique, A.B.; Mohyeldin, M.M.; Qusa, M.H.; Goda, A.A.; Singh, S.S.; Ayoub, N.M.; King, J.A.; Jois, S.D.; El Sayed, K.A. Pseurotin A as a novel suppressor of hormone dependent breast cancer progression and recurrence by inhibiting PCSK9 secretion and interaction with LDL receptor. *Pharmacol. Res.* **2020**, *158*, 104847. [CrossRef] [PubMed]

35. Harikrishnan, A.; Veena, V.; Lakshmi, B.; Shanmugavalli, R.; Theres, S.; Prashantha, C.N.; Shah, T.; Oshin, K.; Togam, R.; Nandi, S.; et al. Atranorin, an antimicrobial metabolite from lichen Parmotrema rampoddense exhibited in vitro anti-breast cancer activity through interaction with Akt activity. *J. Biomol. Struct. Dyn.* **2020**, *9*, 1–11. [CrossRef]

36. Kaboli, P.J.; Salimian, F.; Aghapour, S.; Xiang, S.; Zhao, Q.; Li, M.; Wu, X.; Du, F.; Zhao, Y.; Shen, J.; et al. Akt-targeted therapy as a promising strategy to overcome drug resistance in breast cancer—A comprehensive review from chemotherapy to immunotherapy. *Pharmacol. Res.* **2020**, *156*, 104806. [CrossRef]

37. Choudhury, P.; Barua, A.; Roy, A.; Pattanayak, R.; Bhattacharyya, M.; Saha, P. Eugenol restricts Cancer Stem Cell population by degradation of β-catenin via N-terminal Ser37 phosphorylation-an in vivo and in vitro experimental evaluation. *Chem. Biol. Interact.* **2020**, *316*, 108938. [CrossRef]

38. Liu, C.; Wang, K.; Zhuang, J.; Gao, C.; Li, H.; Liu, L.; Feng, F.; Zhou, C.; Yao, K.; Deng, L.; et al. The Modulatory Properties of Astragalus membranaceus Treatment on Triple-Negative Breast Cancer: An Integrated Pharmacological Method. *Front. Pharmacol.* **2019**, *10*. [CrossRef]

39. Sinha, S.; Patel, S.; Athar, M.; Vora, J.; Chhabria, M.T.; Jha, P.C.; Shrivastava, N. Structure-based identification of novel sirtuin inhibitors against triple negative breast cancer: An in silico and in vitro study. *Int. J. Biol. Macromol.* **2019**, *140*, 454–468. [CrossRef]

40. Stanton, R.A.; Gernert, K.M.; Nettles, J.H.; Aneja, R. Drugs that target dynamic microtubules: A new molecular perspective. *Med. Res. Rev.* **2011**, *31*, 443–481. [CrossRef]

41. Maurya, N.; Maurya, J.K.; Singh, U.K.; Dohare, R.; Yab, Z.; Alam Rizvi, M.M.; Kumari, M.; Patel, R. In Vitro Cytotoxicity and Interaction of Noscapine with Human Serum Albumin: Effect on Structure and Esterase Activity of HSA. *Mol. Pharm.* **2019**, *16*, 952–966. [CrossRef] [PubMed]

42. Chen, H.M.; Wang, P.H.; Chen, S.S.; Wen, C.C.; Chen, Y.H.; Yang, W.C.; Yang, N.S. Shikonin induces immunogenic cell death in tumor cells and enhances dendritic cell-based cancer vaccine. *Cancer Immunol. Immunother.* **2012**, *61*, 1989–2002. [CrossRef] [PubMed]

43. Yin, S.-Y.; Efferth, T.; Jian, F.-Y.; Chen, Y.-H.; Liu, C.-I.; Wang, A.H.; Chen, Y.-R.; Hsiao, P.-W.; Yang, N.S. Immunogenicity of mammary tumor cells can be induced by shikonin via direct binding-interference with hnRNPA1. *Oncotarget* **2016**, *7*, 43629–43653. [CrossRef] [PubMed]

44. Pei, S.; Yang, X.; Wang, H.; Zhang, H.; Zhou, B.; Zhang, D.; Lin, D. Plantamajoside, a potential anti-tumor herbal medicine inhibits breast cancer growth and pulmonary metastasis by decreasing the activity of matrix metalloproteinase-9 and-2. *BMC Cancer* **2015**, *15*, 965. [CrossRef]

45. Hanieh, H.; Mohafez, O.; Hairul-Islam, V.I.; Alzahrani, A.; Ismail, M.B.; Thirugnanasambantham, K. Novel Aryl Hydrocarbon Receptor Agonist Suppresses Migration and Invasion of Breast Cancer Cells. *PLoS ONE* **2016**, *11*, e0167650. [CrossRef]

46. Nashaat, S.; Henen, M.A.; El-Messery, S.M.; Eisa, H. Synthesis, state-of-the-art NMR-binding and molecular modeling study of new benzimidazole core derivatives as Pin1 inhibitors: Targeting breast cancer. *Bioorganic Med. Chem.* **2020**, *28*, 115495. [CrossRef]

47. Bacharaju, K.; Jambula, S.R.; Sivan, S.; JyostnaTangeda, S.; Manga, V. Design, synthesis, molecular docking and biological evaluation of new dithiocarbamates substituted benzimidazole and chalcones as possible chemotherapeutic agents. *Bioorganic Med. Chem. Lett.* **2012**, *22*, 3274–3277. [CrossRef]

48. Vaz, W.F.; Custodio, J.M.F.; D'Oliveira, G.D.C.; Neves, B.J.; Junior, P.S.C.; Filho, J.T.M.; Andrade, C.H.; Perez, C.N.; Silveira-Lacerda, E.P.; Napolitano, H.B. Dihydroquinoline derivative as a potential anticancer agent: Synthesis, crystal structure, and molecular modeling studies. *Mol. Divers.* **2020**, 1–12. [CrossRef]

49. Li, Y.; Li, Y.; Wang, N.; Lin, D.; Liu, X.; Yang, Y.; Gao, Q. Synthesis, DNA/BSA binding studies and in vitro biological assay of nickel(II) complexes incorporating tridentate aroylhydrazone and triphenylphosphine ligands. *J. Biomol. Struct. Dyn.* **2019**, 1–20. [CrossRef]

50. Acharya, S.; Maji, M.; Ruturaj; Purkait, K.; Gupta, A.; Mukherjee, A. Synthesis, Structure, Stability, and Inhibition of Tubulin Polymerization by RuII–p-Cymene Complexes of Trimethoxyaniline-Based Schiff Bases. *Inorg. Chem.* **2019**, *58*, 9213–9224. [CrossRef]

51. Cai, Y.; Zhao, B.; Liang, Q.-Y.; Zhang, Y.; Cai, J.; Li, G. The selective effect of glycyrrhizin and glycyrrhetinic acid on topoisomerase IIα and apoptosis in combination with etoposide on triple negative breast cancer MDA-MB-231 cells. *Eur. J. Pharmacol.* **2017**, *809*, 87–97. [CrossRef] [PubMed]

52. Shukla, A.; Tyagi, R.; Meena, S.; Datta, D.; Srivastava, S.K.; Khan, F. 2D- and 3D-QSAR modelling, molecular docking and in vitro evaluation studies on 18β-glycyrrhetinic acid derivatives against triple-negative breast cancer cell line. *J. Biomol. Struct. Dyn.* **2019**, *38*, 168–185. [CrossRef] [PubMed]

53. Yan, W.; Zhao, Y.; He, J. Anti-breast cancer activity of selected 1,3,5-triazines via modulation of EGFR-TK. *Mol. Med. Rep.* **2018**, *18*, 4175–4184. [CrossRef] [PubMed]

54. Vlaar, C.P.; Castillo-Pichardo, L.; Medina, J.I.; Marrero-Serra, C.M.; Velez, E.; Ramos, Z.; Hernández, E. Design, synthesis and biological evaluation of new carbazole derivatives as anti-cancer and anti-migratory agents. *Bioorganic Med. Chem.* **2018**, *26*, 884–890. [CrossRef] [PubMed]

55. Kamalakar, A.; Washam, C.L.; Akel, N.S.; Allen, B.J.; Williams, D.K.; Swain, F.L.; Leitzel, K.; Lipton, A.; Gaddy, D.; Suva, L. PTHrP(12-48) Modulates the Bone Marrow Microenvironment and Suppresses Human Osteoclast Differentiation and Lifespan. *J. Bone Miner. Res.* **2017**, *32*, 1421–1431. [CrossRef]

56. Rajabi, M.; Mousa, S.A. The Role of Angiogenesis in Cancer Treatment. *Biomedicines* **2017**, *5*, 34. [CrossRef]

57. Aboul-Enein, M.N.; El Azzouny, A.; Ragab, F.A.-F.; Hamissa, M.F. Design, Synthesis, and Cytotoxic Evaluation of Certain 7-Chloro-4-(piperazin-1-yl)quinoline Derivatives as VEGFR-II Inhibitors. *Arch. Pharm.* **2017**, *350*, 1600377. [CrossRef]

58. Ziedan, N.I.; Hamdy, R.; Cavaliere, A.; Kourti, M.; Prencipe, F.; Brancale, A.; Jones, A.T.; Westwell, A.D. Virtual screening, SAR, and discovery of 5-(indole-3-yl)-2-[(2-nitrophenyl)amino] [1,3,4]-oxadiazole as a novel Bcl-2 inhibitor. *Chem. Biol. Drug Des.* **2017**, *90*, 147–155. [CrossRef]

59. Koca, I.; Özgür, A.; Er, M.; Gümüş, M.; Coşkun, K.A.; Tutar, Y. Design and synthesis of pyrimidinyl acyl thioureas as novel Hsp90 inhibitors in invasive ductal breast cancer and its bone metastasis. *Eur. J. Med. Chem.* **2016**, *122*, 280–290. [CrossRef]

60. Nie, Z.; Shi, L.; Lai, C.; O'Connell, S.M.; Xu, J.; Stansfield, R.K.; Hosfield, D.J.; Veal, J.M.; Stafford, J.A. Structure-based design and discovery of potent and selective KDM5 inhibitors. *Bioorganic Med. Chem. Lett.* **2018**, *28*, 1490–1494. [CrossRef]

61. Horton, J.R.; Engstrom, A.; Zoeller, E.L.; Liu, X.; Shanks, J.R.; Zhang, X.; Johns, M.A.; Vertino, P.M.; Fu, H.; Cheng, X. Characterization of a Linked Jumonji Domain of the KDM5/JARID1 Family of Histone H3 Lysine 4 Demethylases. *J. Biol. Chem.* **2015**, *291*, 2631–2646. [CrossRef] [PubMed]

62. Salem, A.F.; Wang, S.; Billet, S.; Chen, J.-F.; Udompholkul, P.; Gambini, L.; Baggio, C.; Tseng, H.-R.; Posadas, E.M.; Bhowmick, N.A.; et al. Reduction of Circulating Cancer Cells and Metastases in Breast-Cancer Models by a Potent EphA2-Agonistic Peptide–Drug Conjugate. *J. Med. Chem.* **2018**, *61*, 2052–2061. [CrossRef] [PubMed]

63. Gambini, L.; Salem, A.F.; Udompholkul, P.; Tan, X.-F.; Baggio, C.; Shah, N.; Aronson, A.; Song, J.; Pellecchia, M. Structure-Based Design of Novel EphA2 Agonistic Agents with Nanomolar Affinity in Vitro and in Cell. *ACS Chem. Biol.* **2018**, *13*, 2633–2644. [CrossRef]

64. Cava, C.; Bertoli, G.; Castiglioni, I. In silico identification of drug target pathways in breast cancer subtypes using pathway cross-talk inhibition. *J. Transl. Med.* **2018**, *16*, 154. [CrossRef]

65. Nayak, D.; Tripathi, N.; Kathuria, D.; Siddharth, S.; Nayak, A.; Bharatam, P.V.; Kundu, C.N. Quinacrine and curcumin synergistically increased the breast cancer stem cells death by inhibiting ABCG2 and modulating DNA damage repair pathway. *Int. J. Biochem. Cell Biol.* **2020**, *119*, 105682. [CrossRef]

66. Minafra, L.; Porcino, N.; Bravatà, V.; Gaglio, D.; Bonanomi, M.; Amore, E.; Cammarata, F.P.; Russo, G.; Militello, C.; Savoca, G.; et al. Radiosensitizing effect of curcumin-loaded lipid nanoparticles in breast cancer cells. *Sci. Rep.* **2019**, *9*, 11134. [CrossRef] [PubMed]

67. Gurova, K.V.; Hill, J.E.; Guo, C.; Prokvolit, A.; Burdelya, L.G.; Samoylova, E.; Khodyakova, A.V.; Ganapathi, R.; Tararova, N.D.; Bosykh, D.; et al. Small molecules that reactivate p53 in renal cell carcinoma reveal a NF-B-dependent mechanism of p53 suppression in tumors. *Proc. Natl. Acad. Sci. USA* **2005**, *102*, 17448–17453. [CrossRef] [PubMed]

68. Mqoco, T.; Stander, A.; Engelbrecht, A.-M.; Joubert, A.M. A Combination of an Antimitotic and a Bromodomain 4 Inhibitor Synergistically Inhibits the Metastatic MDA-MB-231 Breast Cancer Cell Line. *BioMed Res. Int.* **2019**, *2019*, 1850462. [CrossRef]

69. Tang, B.; Qian, Y.; Gou, Y.; Cheng, G.; Fang, G. VE-Albumin Core-Shell Nanoparticles for Paclitaxel Delivery to Treat MDR Breast Cancer. *Molecules* **2018**, *23*, 2760. [CrossRef]

70. Sharma, M.J.; Kumar, M.S.; Murahari, M.; Mayur, Y.C. Synthesis of novel gefitinib-based derivatives and their anticancer activity. *Arch. Pharm.* **2019**, *352*, e1800381. [CrossRef]

© 2020 by the authors. Licensee MDPI, Basel, Switzerland. This article is an open access article distributed under the terms and conditions of the Creative Commons Attribution (CC BY) license (http://creativecommons.org/licenses/by/4.0/).

Article

Improved Anticancer Activity of the Malloapelta B-Nanoliposomal Complex against Lung Carcinoma

Thi Thao Do [1,2,*], Thi Nga Nguyen [1], Thi Phuong Do [1], Thi Cuc Nguyen [1], Ha Phuong Trieu [1], Phuong Thi Thu Vu [1] and Tuan Anh Hoang Le [3]

[1]   Institute of Biotechnology, Vietnam Academy of Science and Technology, 18 Hoang Quoc Viet Road, Hanoi 10000, Vietnam; nguyenngaibt@gmail.com (T.N.N.); dothiphuong74@yahoo.com.vn (T.P.D.); nguyencuc125@gmail.com (T.C.N.); trieuhaphuong95@gmail.com (H.P.T.); vuthuphuong9x@gmail.com (P.T.T.V.)

[2]   Graduate University of Science and Technology, Vietnam Academy of Science and Technology, 18 Hoang Quoc Viet Road, Hanoi 10000, Vietnam

[3]   Center for Research and Technology Transfer, Vietnam Academy of Science and Technology, 18 Hoang Quoc Viet Road, Hanoi 10000, Vietnam; hoangletuananh.cretech@gmail.com

*   Correspondence: thaodo@ibt.ac.vn; Tel.: +84-24-38361774

Received: 17 October 2020; Accepted: 14 November 2020; Published: 17 November 2020

Featured Application: Anticancer Drug Activity and Underlying Mechanisms.

Abstract: Previous studies regarding malloapelta B (malB), a natural compound isolated from the Vietnamese medicinal plant, showed a strong NF-κB inhibitory effect, making it a promising source for the development of novel anticancer drugs. However, similar to many other natural compounds from plants, malB has several disadvantages for clinical applications, including high toxicity and low solubility. To improve its bioavailability, malB was conjugated into nanoliposomes, which are ideal drug carriers. The formulations with 1,2-dipalmitoyl-sn-glycero-3-phosphocholine, mPEG-cholesterol, malB, with or without cholesterol exhibited nanoliposomes with an average diameter of approximately 76.98 nm, PDI of 0.28, zeta potential of −5.53 mV, and the highest encapsulation efficiency of 78.73% ± 9.5%. These malB-nanoliposomes inhibited the survival of all lung cancer cell lines examined with IC_{50} values ranging from 11.86 to 13.12 μM. Moreover, malB-nanoliposomes showed stronger inhibition of A549 colony-forming activity compared to that of the free compound. The effects of malB and its nanoliposomal formulation may be mediated through activation of apoptosis by the significant induction of caspase 3 activity. The nanoliposomal formulations also showed potential to inhibit tumor growth (37.03%) and prolong survival (32.20 days) of tumor-bearing mice compared with the unloaded drug ($p < 0.05$). The improved antitumor activity of malB-nanoliposomes suggests their promising clinical applications.

Keywords: apoptosis; caspase 3; Lewis lung carcinoma; malloapelta B; nanoliposome

---

1. Introduction

Lung cancer is the leading cause of cancer-related death worldwide (18.4%), and is especially prevalent in male smokers [1]. The most common type of lung cancer is non-small cell lung cancer (NSCLC), accounting for 85% of cases of lung cancer [2]. Approximately 40% of lung cancers are adenocarcinomas, and most cases of lung cancer in smokers are of this type. Several therapeutic methods are available for treatment of lung adenocarcinomas, including surgery, radiofrequency ablation, radioactive therapy, chemotherapy, and immunotherapy, alone or in various combinations. However, these therapeutic options are not only expensive but also insufficiently effective. Therefore, there is an increasing need for new, safer, and more effective clinical treatments. Our previous study showed that 1-(5,7-dimetoxy-2,2dimetyl-2H-cromen-8-yl)-but-2-en-1-on (malloapelta B), isolated from *Mallotus apelta*, is a potential active anticancer compound [3]. This compound, malloapelta B (malB), inhibits the activation of nuclear factor kappa B (NF-κB) with an IC_{50} value of 5.0 μM. This IC_{50} value is much lower than that of parthenolide (PTN, 6.66 ± 0.07 μM) [4,5]. The compound was also shown to downregulate genes that contribute to inflammatory mechanisms, such as tumor necrosis factor (TNF)-α, interleukin (IL)-6, and Il-1β [6]. Additionally, as reported, the enone side chain may play an important role in the activity of this molecule [4]. However, the potential effects of malB against lung carcinomas in vitro and in vivo have not been studied in detail. In addition, its high toxicity and low solubility represent barriers to the development of malB as an anticancer drug for use in chemotherapy. Nanoliposomes have recently emerged as ideal drug carriers with a number of beneficial characteristics, including minimal immune response, biocompatibility, biodegradability, reduction of drug toxicity, etc. [7]. Nanoliposomes can be used to encapsulate hydrophobic drugs as a means of improving their solubility in water. Due to these advantages, a number of nanoliposomal formulations incorporating different anticancer drugs are available commercially, including daunorubicin (DaunoXome), doxorubicin in PEG-liposomes (Doxil), vincristine (Marqibo), topotecan (INX-0076), nystatin (Nyotran), and paclitaxel (LEP-ET) [8]. Therefore, in the present study, malB was entrapped in nanoliposomes at different concentrations, and anticancer activities of these malB-nanoliposomal formulations against lung carcinomas were examined as well as their anti-malignant potential in Lewis lung carcinoma (LLC) tumor-bearing mice.

2. Materials and Methods

2.1. Chemicals

1,2-Dipalmitoyl-sn-glycero-3-phosphocholine (DPPC) was purchased from Avanti Polar Lipids (Alabaster, AL, USA); cholesterol was produced at Acros Organics, a part of Thermo Fisher Scientific, (Merelbeke, Belgium), and mPEG-cholesterol was provided by Dr. Chun-Liang Lo, National Yang-Ming National University, Taipei, Taiwan. The compound malloapelta B (malB) was provided by the MienTrung Institute for Scientific Research, Vietnam Academy of Science and Technology (Hue, Vietnam). Other chemicals and cell culture reagents were from Sigma Chemical Co. (St. Louis, MO, USA) and Invitrogen (Carlsbad, CA, USA).

2.2. Animals

Male and female albino BALB/c mice (8–10 weeks old) were received from the Institute of Biotechnology, Vietnam Academy of Science and Technology (VAST, Hanoi, Vietnam). All mice were caged in a temperature-controlled room on a 12-h light/12-h dark cycle with food and water ad libitum. Experiments were performed in accordance with Vietnamese Ethical Laws, European Communities Council Directives of 24 November 1986 (86/609/EEC) guidelines and Approval from the Scientific Council of Institute of Biotechnology, Vietnam Academy of Science and Technology, for the care and use of laboratory animals.

2.3. MalB-Nanoliposome Preparation

2.3.1. Bangham Thin Film Method

Liposome production was carried out according to the Bangham thin film method [9] with some modifications. Briefly, a complex including lipids and malB was diluted in dichloromethane (DCM) solvent (Table 1). A magnetic stirrer was used to dissolve the complex at 200 rpm, room temperature (RT) for 20 min. DCM was then removed by a rotary evaporator for the formation of thin film. The solvent was thoroughly dispatched by nitrogen gas flushing. Subsequently, the thin film was hydrated with PBS (pH = 7.2) at 60 °C. To obtain nanoliposomes, the hydrated complex was then immediately subjected to a probe sonicator that directly inserted the probed head into the solution with an ultrasonic frequency of 2 atm, 20 s of ultrasound and 10 s of rest, repeated five times. The ultrasonic solution was shaken at 60 °C and 500 rpm for one hour before large size nanoliposomes were removed by filtering with a 0.22 μm polyvinylidene difluoride (PVDF) membrane. The filtered nanoliposomes were washed three times with PBS (pH = 7.2).

Table 1. Lipid components including 1,2-Dipalmitoyl-sn-glycero-3-phosphocholine (DPPC), mPEG-cholesterol, with/without cholesterol and malloapelta B (malB) compound ratio.

Types	DPPC (mg)	mPEG-Cholesterol (mg)	Cholesterol (mg)	malB (mg)	DCM (mL)
Formulation A	4	1	0.50	0.50	10
Formulation B	4	1	0.75	0.25	10
Formulation C	4	1	0	0.25	10
Formulation D	4	1	0	0.50	10
Blank	4	1	0.75	-	10

2.3.2. Nanoliposomal Characterization

The size (z-average), polydispersity indexes (PDIs), and zeta potential of the nanoparticles were measured using a dynamic light scattering instrument (DLS, Horiba Instruments Inc., Irvine, CA, USA). The malB-nanoliposomes were also stained with 4% uranyl acetate (UR 4%) dye on the surface of a thin carbon-coated copper plate to a thickness of 200–500 A° and dried at RT. Then, the plates were magnified 200× to capture the morphology of the prepared nanoparticles using high-solution transmission electronic microscopy (TEM) (Jeol 1200EX TEM, Jeol Company, Tokyo, Japan).

2.3.3. Encapsulated Efficiency

The encapsulated efficiency (EE) of the loaded malB was calculated by UV–spectrophotometry method. The pure malB compound was dissolved in DMSO (100%) as a standard curve with a two time diluted concentration range which was started from0.25 mg/mL. The malB-nanoliposomes were re-diluted in DMSO (100%) to release all packed malB. Then, 100 μL of both the standard curve and the nanoliposomal samples were placed into a 96-well plate, repeated three times to ensure accuracy. The optical density values (ODs) were measured at 280 nm on a ThermoScientific™ Varioskan™ Flash Multimode Reader. The standard curve drawn using Microsoft Office Excel 2016 software was used to calculate the amount of active ingredient that was packed into the nanoliposomes. The encapsulated efficiency (EE) of the process was determined using the following equation:

$$EE(\%) = 100 \times (\text{weight of conjugated malB})/(\text{weight of initial malB})$$

2.4. Cell Lines and Cell Culture

A549, SK-LU-1, CL-141, and LLC lung cancer cell lines were provided by Prof. Chi-Ying Huang, Institute of Biopharmaceutical Sciences, National Yang Ming University, Taipei, Taiwan and Prof. J Meier, Milan University, Milan, Italy. Cells were cultured in DMEM medium supplemented with 10% fetal bovine serum (FBS), 100 μg/mL streptomycin, 100 IU/mL penicillin, and 2mM L-glutamine. All cells were maintained in a humidified incubator with 5% CO_2 at 37 °C, and harvested with trypsin-EDTA.

2.5. Cytotoxic Activities of malB-Conjugated Nanoliposomes

Cells were seeded in 96-well plates at a density of 2000 cells per well in triplicate for 24 h before sample exposure. Cells were then treated with different unloaded malB concentrations and corresponding malB-nanoliposomes of indicated treatments for 48 h. Cytotoxicity was assessed by using the sulforhodamine B (SRB) assay [10]. Briefly, the medium was discarded, and adherent cells were fixed by 100 μL/well of cold 10% trichloroacetic acid (*w/v*) for 1 h at 4 °C. After fixation, cells were stained with 0.4% SRB solution (*w/v* in 1% acetic acid) for 30 min at RT, and then washed twice with 1% acetic acid. After air-drying, 100 μL of 20 mM Tris-base were added to each well and the absorbance was measured at 540 nm. Cytotoxicity is expressed as the percent of cells relative to the number of cells in the solvent only as control (set to 100%). Each experiment was performed independently at least 3 times.

2.6. Clonogenic Inhibition of malB-Conjugated Nanoliposomes

Clonogenic assay was performed to determine the ability of a single cell to grow into a colony under the compound treatment. In detail, 800 cells were seeded in each well of 6-well plates at 37 °C, 5% CO_2. The addition of drugs was manipulated after 24 h of seeding and triplication of experiments was carried out. The medium including drugs was changed each 3 days. The cells were harvested after 8 days of treatment. After the treatment, the cells were washed with 1 mL phosphate-buffered saline (PBS), and fixed with 1mL mixture of methanol and acetic acid (3:1) in 15 min at RT. Subsequently, the colonies were stained with 0.5 mL of 0.5% crystal violet in methanol for 15 min. Then, crystal violet was discarded and the plate were rinsed under tap water. Only colonies consisting of approximately 50 cells were counted. The percentage of cell survival after drug treatment was expressed as a percentage of the control-colony efficiency.

2.7. Caspase 3 Inducible Activities of malB-Conjugated Nanoliposome

Caspase 3 activity was performed using a Caspase-3 Colorimetric Assay Kit (BioVision Inc., USA). According to the assaying protocol of the kit manufacturer, 1×10^6 cells treated with malB or malB-nanoliposome at different concentrations for 24h were lysed with chilled lysis buffer. The lysates were then centrifuged to collect the supernatant. After measurement of protein content using Bradford reagent, caspase 3 activity was specified by mixing 50 μL of cell lysis supernatant with 50 μL of 10 mM dithiothreitol (DTT) and 5 μL of the 4 mM DEVD-p-nitroaniline (DEVD-pNA) substrate (from the kit). The caspase 3 activity, which corresponds with the presence of the chromophore pNA, a product formed from the cleavage of the DEVD-pNA substrate in enzyme reaction, can be detected by using a spectrophotometer at 405 nm.

2.8. In Vivo Antitumor Activity

The experiment was carried out using 36 healthy BALB/c mice at 20–25 g weight. Mice were subcutaneously injected with 1×10^6 LLC cells to induce tumors. After 5 days, tumorized mice were randomly distributed to 6 groups (n = 6). Group 1 served as the negative control that received normal saline. Group 2 was treated with blank nanoliposome (with lipid components only). Groups 3 and 4 received malB-nanoliposome at doses of 5 mg/kg and 2.5 mg/kg body weight (b.w.) by intraperitoneal injection (i.p.) every 2 days for 14 days continuously. Group 5 was i.p. injected with unconjugated malB compound at a dose of 5 mg/kg b.w. every 2 days for the same duration. Group 6 was treated with the reference control (doxorubicin 5 mg/kg b.w. i.p. injection). The tumor size of each mouse was measured every 7 days for 28 days and tumor volume was calculated by the following equation:

$$V = (W^2 \times L)/2,$$

where V is the volume of the tumor, W is the width of the tumor, and L is the length of the tumor.

The survival time of tumorized mice in all experimental groups was also determined. It was calculated from the day of LLC cell inoculation to the day of death and percentage increase in average life span (ILS) was calculated by the following equation:

$$\% \text{ ILS} = (A/B - 1) \times 100,$$

where A is the survival time of the treated group, B is the mean survival time of the control group, and ILS is the increase in the average life span group.

2.9. Statistical Analyses

All data were expressed as means ± standard error of the mean (SEM). Statistical differences were analyzed by two-tailed paired Student's *t*-test or one-way analysis of variance (one-way ANOVA). A value of $p < 0.05$ was considered a statistically significant difference.

3. Results

3.1. Preparation and Morphology of malB-Encapsulated Nanoliposomes

The main components used to produce drug-encapsulating nanoliposomes included malB compound at different concentrations, DPPC, mPEG-cholesterol, with or without cholesterol. The physicochemical characteristics of the nanoliposomes thus formed were assessed by determining the mean diameter (nm), polydispersity index (PDI), zeta potential, and encapsulation efficiency (EE).

Based on the results shown in Table 2, all of the nanoliposomal formulations had not only a mean diameter <200 nm, but also PDI <0.3, and carried a low negative charge. However, the formulations without cholesterol as a component showed significantly higher packing efficiency ($p < 0.01$). The EE of formulations A and B containing cholesterol were only about 10%, whereas the EEs of the cholesterol-free structures were >50%. The EE was also dependent on the malB concentration; a lower malB concentration of 0.25 mg was associated with higher EE (78.73%), whereas a higher concentration of 0.5 mg showed a lower EE (51.33%). Due to the highest efficient effect, structure C was selected for further studies.

Table 2. Characteristics of obtained malB-loaded nanoliposomes.

Types	Size (nm)	PDI	Zeta Potential (mV)	EE (%)
Formulation A	130.53 ± 4.38	0.26 ± 0.04	−2.02 ± 0.95	12.80 ± 1.55
Formulation B	101.53 ± 3.89	0.28 ± 0.03	−0.52 ± 0.04	9.85 ± 0.21
Formulation C	76.98 ± 13.18	0.28 ± 0.04	−5.53 ± 0.21	78.73 ** ± 9.52
Formulation D	124.41 ± 35.41	0.30 ± 0.03	−2.70 ± 0.13	51.33 ** ± 6.53
Blank	119.70 ± 22.17	0.22 ± 0.02	−2.80 ± 0.24	-

Note: ** $p < 0.01$ compared to the formulation without cholesterol; blank liposome including DPPC, cholesterol, and mPEG-cholesterol.

The morphology of the malB-nanoliposomal formulation C was examined by transmission electron microscopy (TEM) (Figure 1). The malB-nanoliposomes were small unilamellar vesicles, likely spherical in shape, and fairly homogeneous in size with an average diameter of 76.98 ± 9.57 nm (Table 2). Based on morphological assessment, the mal B-nanoliposomal formulation C was suitable to manipulate in the further experiments for anticancer activities.

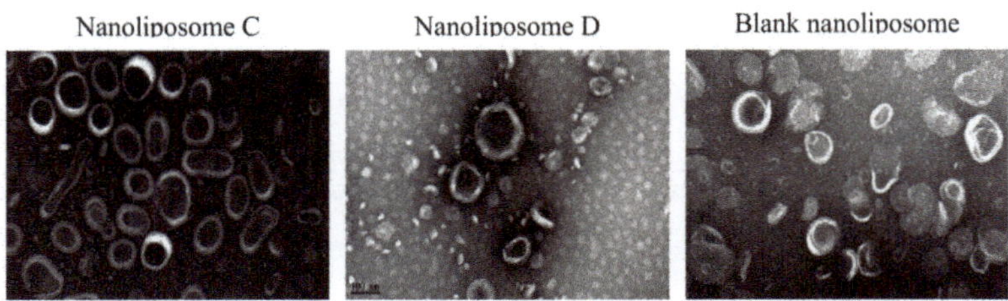

Figure 1. TEM morphological images of malB-loaded nanoliposomes: nanoliposome C (malB entrapped in nanoliposomes at 0.25 mg, EE = 78.73% ± 9.52%); nanoliposome D (malB entrapped in nanoliposomes at 0.50 mg, EE = 51.33% ± 16.53%). Blank nanoliposomal particles.

3.2. Compound Encapsulation by Nanoliposomes and Their Cytotoxicity

To determine the cytotoxicity of free malB and malB-nanoliposomes on non-small cell lung cancer cell lines, the viability of different cell lines was examined by sulforhodamine B (SRB) assay with malB treatment at several concentrations. As shown in Figure 2 and Table 3, both free compound and malB-nanoliposomes inhibited cell growth in a dose-dependent manner and significantly compared with the control ($p < 0.005$). High doses of malB and malB-nanoliposomes (7.5 μM and 15 μM) remarkably inhibited viability of all tested lung cancer cell lines. However, the cytotoxic effects of malB and malB-nanoliposomes on all cell lines were decreased at a dose of 3.75 μM, and minimal effects were observed at a dose of 1.875 μM. Based on the IC_{50} values, malB-nanoliposomes exhibited the strongest cytotoxicity against the A549 cell line. Free malB showed stronger cytotoxic activity than that of the malB-nanoliposomal formulation on all examined cell lines. However, the difference was insignificant ($p > 0.05$). Blank nanoliposomes showed no significant effects on cell growth (cell survival >90%). Therefore, the IC_{50} value of blank liposomes could not be determined.

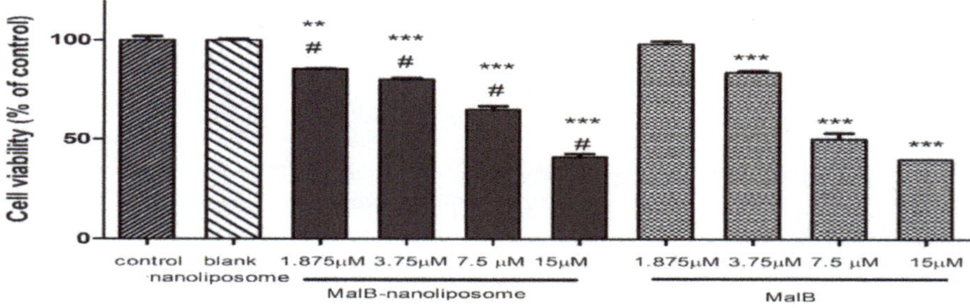

Figure 2. *Cont.*

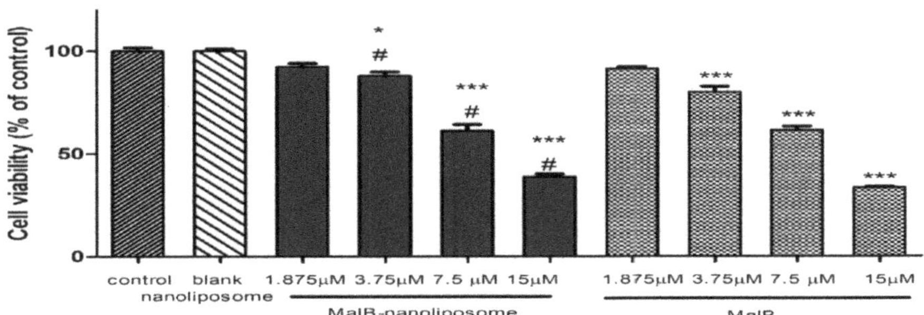

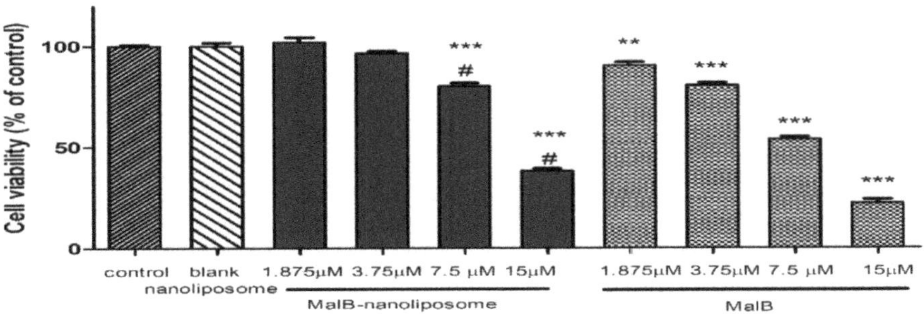

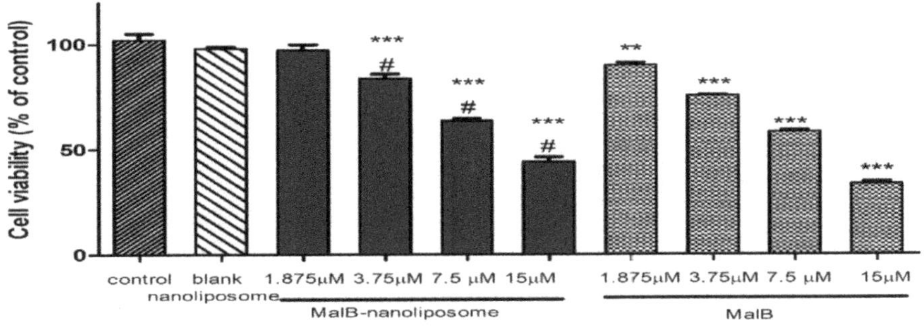

Figure 2. Cell viability after 48 h under the treatment of malB-nanoliposome C or free compound on different lung cancer cell lines. Cultured cells (2×10^3 cells/well) were treated with different concentrations of either malB-nanoliposomal or free malB. Normal saline served as the negative control. Each value represents the mean ± SEM. One-way ANOVA was used for analyzing statistical differences between groups. * $p < 0.05$, ** $p < 0.01$, *** $p < 0.001$ when compared with negative control. # $p < 0.01$ when compared with blank liposome.

Table 3. Cytotoxic activities of malB-nanoliposomes and untrapped compound malB on different lung cancer cell lines.

Samples	IC$_{50}$ Values (µM)			
	SK-LU-1	LLC	CL141	A549
MalB compound	8.67 ± 0.89	9.59 ± 0.52	9.81 ± 0.81	10.35 ± 0.32
MalB-nanoliposomes	13.12 ± 1.47	12.35 ± 1.45	12.24 ± 1.38	11.86 ± 0.99
Blank nanoliposomes	NA	NA	NA	NA

Note: NA means not available.

3.3. Inhibition of Colony Formation malB–Nanoliposome Complex

One of the most important characteristics of cancer is self-renewal ability, which allows tumor cells to proliferate and grow from a single cancer stem cell (CSC). To evaluate the effects of malB and malB-nanoliposomes on cellular self-renewal, a colony-forming assay was performed on the A549 cell line, cells of which were able to form colonies in vitro. The outcome after 8 days of treatment showed that both malB-liposomes and free malB decreased the numbers of colonies derived from A549 cells (Figure 3). The number of colonies decreased markedly at 5 µM and 2.5 µM malB in comparison to the controls. However, a lower concentration of malB had a minimal effect on the number of colonies. In addition, malB-nanoliposome C also inhibited colony formation by >50% at concentrations of 5 µM and 2.5 µM. The IC$_{50}$ values of malB and malB-nanoliposomes were calculated as 2.28 µM and 1.75 µM, respectively. These inhibitory activities of the compound in both forms were significant compared to the control ($p < 0.001$). The results suggested that malB encapsulated in nanoliposomes had a stronger inhibitory effect on the formation of A549 lung adenocarcinomas than the free form. In addition, the effects of free malB and malB-nanoliposomes were different as the concentration increased. In detail, at the same dose of 0.75 µM, malB-nanoliposomes inhibited the formation of A549 colonies by up to 40.66%, whereas the free malB showed only 17.21% inhibition of A549 colony formation. The same was also true at a dose of 1.25 µM.

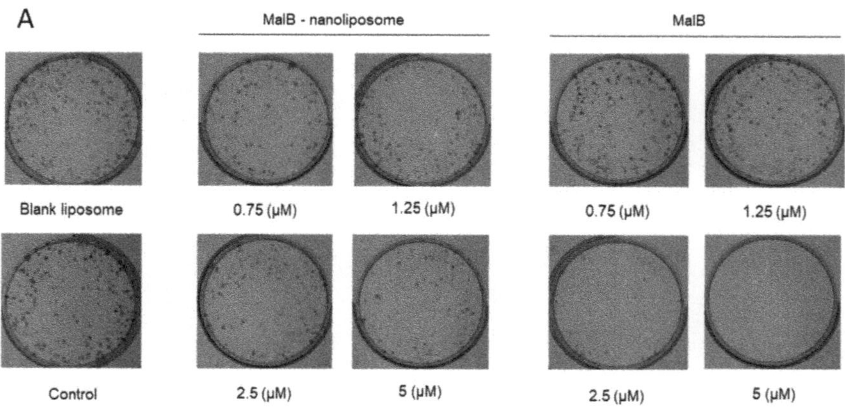

Figure 3. *Cont.*

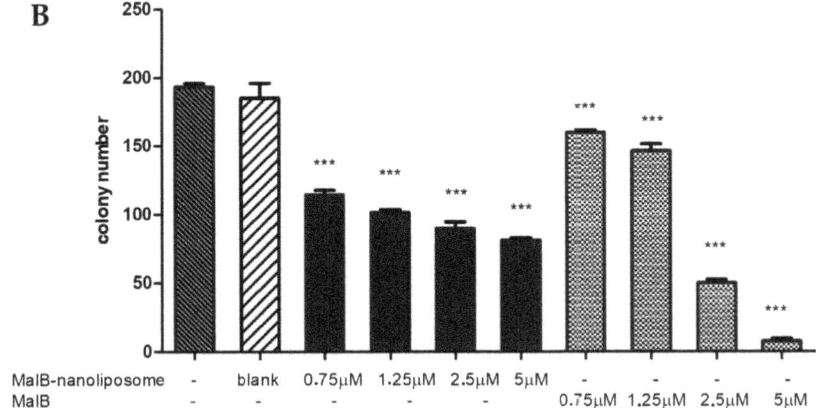

Figure 3. Antitumorigenic activity of malB-nanoliposome and free compound on A549 cells. (**A**) Colony images after treatment of malB-nanoliposome and free-compound at different concentrations compared to the negative control (normal saline) and blank liposome on day 8. (**B**) Colony number after treated with blank nanoliposome, malB-nanoliposome, and malB compared to the negative control. Each value represents the mean ± SEM. *** $p < 0.001$.

3.4. Caspase 3 Inductive Activities of malB and malB-Nanoliposome

Caspase 3 is a typical enzyme related to the apoptosis of mammalian cells. To determine the apoptosis-inducing activity of malB and malB-liposomes, caspase 3 activity was measured under conditions of treatment with the free compound or its nanoliposomes using a caspase 3 colorimetric kit. As shown in Figure 4, malB and structure C malB-nanoliposomes significantly increased the changes in caspase 3 activity at doses of 10 µM and 5 µM in comparison with the control ($p < 0.01$ and $p < 0.05$, respectively). However, no significant differences in caspase 3 induction were observed between malB and malB-nanoliposomes.

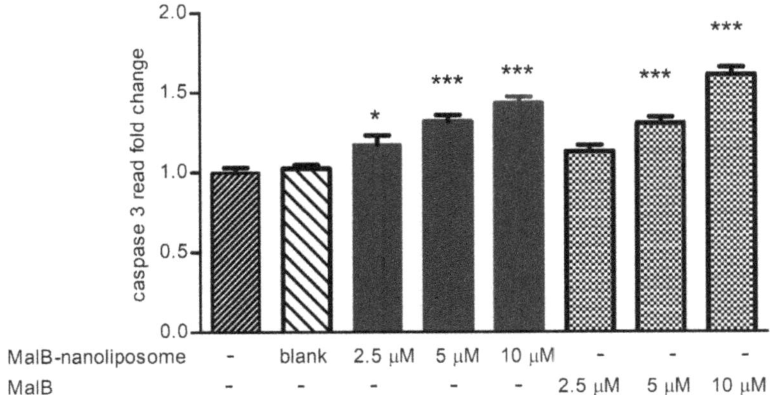

Figure 4. Caspase 3 inducible effects of unconjugated malB and malB-nanoliposomes at different treated concentrations ranging from 2.5 to 10 µM on the A549 cells after 24 h of incubation. Normal saline served as the negative control. Each value represents the mean ± SEM. *** $p < 0.001$ and * $p < 0.05$ compared to negative control.

3.5. Tumor Inhibition Activity

The antitumor activity of malB-liposomes was examined in the LLC tumor model in BALB/c mice. MalB and malB-liposome had no significant change in body weight of mice at both initial time and the end time of treatment (Figure 5). As shown in Figure 6, animals treated with free malB at a concentration of 5 mg/kg body weight (b.w.) did not show a significant reduction in tumor size compared to negative controls. However, malB-nanoliposomes significantly inhibited tumor growth at the same concentration (5 mg/kg b.w.) ($p < 0.05$). After 28 days of treatment with malB-nanoliposomes, the tumor size was decreased by 37.03% compared to the saline-treated negative control, whereas the free form reduced the tumor size by only 13.09%. When the concentration was decreased, the tumor suppression effect of malB-nanoliposomes was markedly reduced and was not significant compared with the control. Moreover, blank liposome did not affect the tumors' growth at any tested time points.

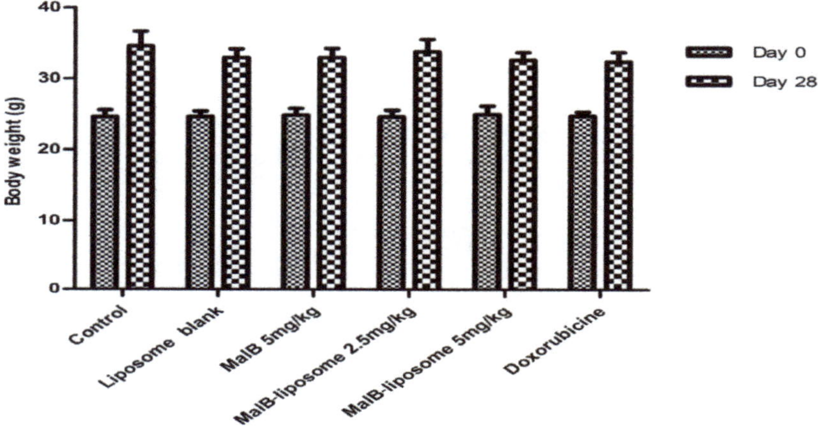

Figure 5. Effects of malB or malB-nanoliposomes on body weight of BALB/c mice harboring a malignant tumor induced by LLC cells ($n = 6$). Error bars represent standard error of the mean (SEM).

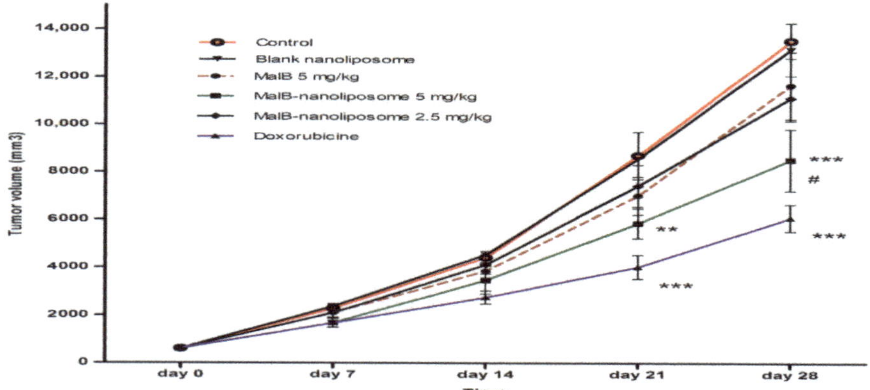

Figure 6. Anti-tumor effects of malB or malB-nanoliposomes on BALB/c mice harboring a malignant tumor induced by LLC cells ($n = 6$). Liposome-conjugated malB at the dose 5.0 mg/kg b.w. significantly inhibited tumor growth after 21 and 28 days compared with the negative control (normal saline treated) (** $p < 0.01$ and *** $p < 0.005$, respectively). # $p < 0.05$ compared to the free malB at the same dose. Error bars represent standard error of the mean (SEM).

In addition, the administration of malB-nanoliposomes prolonged the survival of the mice but still insignificantly ($p > 0.05$) (Table 4). The median survival of mice treated with the malB-nanoliposomal formulation was 32.20 ± 0.97 days, which was 6.62% longer than that of control mice. Survival was also increased slightly (31.00 ± 0.84 days) in the free malB-treated group at the treated dose.

Table 4. Effect of malB-nanoliposome and free malB on survival time of LLC-induced tumor-bearing mice from different experimented groups (mean ± SEM) ($n = 6$).

Groups	Mean Survival Time (Days)	% ILS
Control group (normal saline)	30.20 ± 0.86	-
Blank nanoliposome	30.67 ± 0.80	1.56
malB free (5 mg/kg b.w.)	31.00 ± 0.84	2.65
malB-nanoliposomes (2.5 mg/kg b.w.)	31.20 ± 0.86	3.31
malB-nanoliposomes (5 mg/kg b.w.)	32.20 ± 0.97	6.62
Doxorubicin (5 mg/kg b.w.)	36.20 * ± 0.86	19.86

* $p < 0.05$ in comparison with the control group (normal saline treated).

4. Discussion

In addition to novel treatments for cancer, such as antibodies and gene therapy, a number of drugs based on natural bioactive compounds have been shown to be efficacious in chemotherapy and the prevention of cancer. Therefore, there is much interest in identifying potential new anticancer compounds, such as the large-scale programs of the National Cancer Institute for the discovery and screening of natural products for development as anticancer drugs [11]. These studies have identified a number of categories of novel compounds with anticancer activities. However, the use of these compounds has been hampered by their toxicity and poor solubility, problems that have also prevented the development of malB for clinical use. Previous reports regarding its promising anticancer activities suggested that malB was a promising candidate for drug development. Microarray analysis demonstrated that this compound has anticancer capacity in vitro by regulating gene expression similar to the commercial anticancer drug, withanolide A [12]. This compound was also shown to inhibit the activation of NF-κB with an IC_{50} value in the range of 3.5–5.0 μM [3,5]. The compound affected NF-κB activation by inhibiting the activation of IκB kinase (IKK) [6]. However, malB shows high toxicity and is insoluble, which has limited its pharmaceutical applications. The use of nanoliposomal carriers for drug design to reduce toxicity and enhance the bioavailability of natural compounds is appropriate for anticancer drug development. Drugs packed into liposomes show improved blood circulation activity, promotion of deposition in tumors, protection from metabolism, direct distribution of the drug into tumors, as well as enhanced uptake in adenocarcinoma, mononuclear macrophages, and the intracellular lattice system (liver, spleen, bone marrow). Furthermore, liposomal-encapsulated drugs show reduced uptake in the kidneys, myocardium, and brain tissues [13,14].

In the present study, malB was experimentally incorporated into liposomal nanocarriers. The nanoliposomes containing malB were small unilamellar vesicles <200 nm in diameter with a slight negative charge and low PDI. Magin reported that nanoliposomes in the size range of 50–200 nm would be the most suitable for liposomal stability in the body [15–17]. However, liposomes containing cholesterol had much lower EE in comparison with the non-cholesterol formulations. Similar results were reported previously for some lipophilic drugs, such as ciprofloxacin [18] and dexamethasone [19], as well as natural compounds, such as ascorbic acid [20]. Cholesterol molecules are normally located in the space between lipid bilayer membranes. Therefore, it was supposed that cholesterol competed out malB molecules and some positions in the bilayers were occupied, preventing successful incorporation of the test compound. Furthermore, cholesterol makes the bilayer more rigid, which would make the incorporation of malB molecules difficult. In this study, the addition of cholesterol caused a noticeable decrease in EE, which was taken to indicate that EE depends on the normal structure of the liposomal bilayer [21].

The investigation of liposomal stability in this study showed that malB-nanoliposomes were stable at 4 °C in PBS for approximately 30 days, as the average size of the liposomes was still <200 nm and they had a slight negative charge. However, the PDI of liposomes was significantly increased at around 30 days to almost twice that on day 1 (data not shown). The PDI parameter is an indicator of whether the formed liposomes are monodispersed and have an average size distribution. The PDI should be as low as possible, as higher PDI indicates a broader size distribution of nanoparticles. Thus, the results of the present study suggested that malB-liposomes exhibited an appropriate PDI and this characteristic seems to change after 30 days of storage at 4 °C in PBS.

With regard to assessment of bioactivity, malB encapsulated in nanoliposomes still showed strong anticancer activities in vitro that were comparable to those of the free form. Indeed, the nanoliposomal formulation inhibited cancer cell survival after 48 h with IC_{50} values ranging from 11.86 to 13.12 µM, which was slightly higher than that of free malB (8.67 − 10.35 µM). However, the anti-clonogenic activity of malB-nanoliposomes was more obvious than that of the free compound in the A549 colony-forming assay. In addition, malB and malB-nanoliposomes affected the activity of caspase 3, a crucial protease that mediates apoptosis. This caspase catalyzes the specific cleavage of many key cellular proteins, and plays an important role in apoptotic chromatin condensation and DNA fragmentation [22,23]. The significant caspase 3 induction and activation by malB and malB-nanoliposomes, which contributed to both intrinsic and extrinsic apoptotic pathways, were reported here for the first time.

Furthermore, the improved cytotoxic and antitumorigenic effects of malB-nanoliposomes were also verified in tumor-bearing mice. The malB-nanoliposomes showed the ability to effectively inhibit tumor growth in comparison with negative controls as well as with free malB at the same concentration of 5 mg/kg. There have been many studies on loading of potential anticancer agents into nanoliposomes to improve their activities. Those studies showed that the encapsulation of drugs into liposomes reduces toxicity, improves bioactivity, and increases the circulation time of drugs in the body, leading to increased effectiveness of these drugs in clinical treatment [24]. The intraperitoneal (i.p.) administration of liposomes also has advantages, such as reduction of local toxicity [25]. In addition, when administered by the i.p. route, the drug is absorbed into the organs through the peritoneum or the lymphatic system [26]. Drugs packaged in liposomes could be stable for longer period in the abdominal cavity or in the lymph vessels, thus improving the effectiveness of treatment for peritoneal carcinomas, such as ovarian cancer and liver cancer [27]. However, the uptake of the drug into the blood via the abdominal cavity also depends on the size of the liposomes. Feng reported that liposomes approximately 100 nm in diameter show a high concentration in blood equivalent to intravenous (i.v.) administration, and could therefore be used to treat cancer far from the peritoneum [28]. On the other hand, PEGylation of liposomes plays a central role in prolonging their circulation in the blood. PEGylation helps protect drug-nanoliposomes from mononuclear phagocytes by impeding the absorption of opsonin protein in the circulation on the liposome surface, which would minimize clearance of the drug by these cells [29,30]. The characteristics of PEGylated liposomes allow them to circulate longer in the bloodstream than free drugs, leading to increased effectiveness in cancer treatment. In the present study, the PEG-cholesterol components in the malB-nanoliposome complex may have enhanced the antitumor activity in comparison with the free drug.

5. Conclusions

The results of this study confirmed that malB has the potential for development as an anticancer drug after nanoliposomal encapsulation. The malB-encapsulated formulation had typical characteristics of nanoliposomes, such as size <200 nm, negative charge, and PDI <0.3. The free form of malB and malB-nanoliposomes showed activities against all lung cancer cell lines examined. Both malB forms significantly inhibited colony formation and increased caspase 3 activity of A549 cancer cells. The malB-nanoliposomes showed significantly stronger tumor growth inhibition compared with the free form. The malB-nanoliposomes could slightly prolong the survival period of tumor-bearing mice in comparison with the untreated group.

Author Contributions: Conceptualization, T.T.D.; Methodology, T.N.N., T.P.D., T.C.N., H.P.T., P.T.T.V., and T.A.H.L.; Software, validation, formal analysis, investigation, resources, and data curation, T.N.N. and P.T.T.V.; Writing—original draft preparation: P.T.T.V. and T.N.N.; Writing—review and editing, T.T.D. All authors have read and agreed to the published version of the manuscript.

Funding: This research was funded by the Vietnam National Foundation for Science and Technology Development (NAFOSTED) under grant number 106.02-2017.20.

Acknowledgments: The authors wish to thank Chi-Ying Huang, Lu-Yi Yu, Chun-Liang Lo, Department of Biomedical Engineering, National Yang-Ming University, Taipei, Taiwan, for their kind guidance and their support with all equipment, especially the Jeol 1200EX Transmission Electronic Microscope (TEM system) (Jeol Ltd., Japan).

Conflicts of Interest: The authors declare no conflict of interest.

References

1. Bray, F.; Ferlay, J.; Soerjomataram, I.; Siegel, R.L.; Torre, L.A.; Jemal, A. Global cancer statistics 2018: GLOBOCAN estimates of incidence and mortality worldwide for 36 cancers in 185 countries. *CA Cancer J. Clin.* **2018**, *68*, 394–424. [CrossRef]

2. Kim, J.S.; Cho, M.S.; Nam, J.H.; Kim, H.-J.; Choi, K.-W.; Ryu, J.-S. Prognostic impact of EGFR mutation in non-small-cell lung cancer patients with family history of lung cancer. *PLoS ONE* **2017**, *12*, e0177015. [CrossRef]

3. Van Kiem, P.; Dang, N.H.; Bao, H.V.; Huong, H.T.; Van Minh, C.; Lee, J.J.; Kim, Y.H. New cytotoxic benzopyrans from the leaves of Mallotus apelta. *Arch. Pharm. Res.* **2005**, *28*, 1131–1134. [CrossRef]

4. Van Luu, C.; Van Chau, M.; Lee, J.-J.; Jung, S.-H. Exploration of essential structure of malloapelta B for the inhibitory activity against TNF induced NF-κB activation. *Arch. Pharmacal Res.* **2006**, *29*, 840–844. [CrossRef]

5. Nam, N.H.; Dang, N.H.; Van Kiem, P.; Van Chinh, L.; Binh, P.T.; La Dinh, M.; Van Minh, C. Study on benzopyrans and other isolated compounds from *Mallotus apelta*. *J. Chem.* **2007**, *45*, 111–121.

6. Ma, J.; Shi, H.; Mi, C.; Li, H.L.; Lee, J.J.; Jin, X. Malloapelta B suppresses LPS-induced NF-κB activation and NF-κB-regulated target gene products. *Int. J. Immunopharmacol.* **2015**, *24*, 147–152. [CrossRef] [PubMed]

7. Gurunathan, S.; Kang, M.-H.; Qasim, M.; Kim, J.-H. Nanoparticle-mediated combination therapy: Two-in-one approach for cancer. *Int. J. Mol. Sci.* **2018**, *19*, 3264. [CrossRef] [PubMed]

8. ElBayoumi, T.A.; Torchilin, V.P. Current trends in liposome research. *Methods Mol. Biol.* **2010**, *605*, 1–27.

9. Bangham, A.D.; Horne, R. Negative staining of phospholipids and their structural modification by surface-active agents as observed in the electron microscope. *Adv. Drug Deliv. Rev.* **1964**, *8*, 660-IN10. [CrossRef]

10. Skehan, P.; Storeng, R.; Scudiero, D.; Monks, A.; McMahon, J.; Vistica, D.; Warren, J.T.; Bokesch, H.; Kenney, S.; Boyd, M.R. New colorimetric cytotoxicity assay for anticancer-drug screening. *J. Natl. Cancer Inst.* **1990**, *82*, 1107–1112. [CrossRef]

11. Nobili, S.; Lippi, D.; Witort, E.; Donnini, M.; Bausi, L.; Mini, E.; Capaccioli, S. Natural compounds for cancer treatment and prevention. *Pharmacol. Res.* **2009**, *59*, 365–378. [CrossRef] [PubMed]

12. Thao, D.T.; Phuong, D.T.; Trang, N.T.; Nga, N.T.; Chi, H.Y. Study the anticancer mechanism of the promising compound 2B2D by using microarray technique. *Vietnam J. Sci. Technol.* **2012**, *50*, 267.

13. Lomis, N.; Westfall, S.; Farahdel, L.; Malhotra, M.; Shum-Tim, D.; Prakash, S. Human serum albumin nanoparticles for use in cancer drug delivery: Process optimization and in vitro characterization. *J. Nanomater.* **2016**, *6*, 116. [CrossRef] [PubMed]

14. Prathyusha, K.; Muthukumaran, M.; Krishnamoorthy, B. Liposomes as targetted drug delivery systems present and future prospectives: A review. *J. Drug Deliv. Ther.* **2013**, *3*. 195–201. [CrossRef]

15. Lin, C.-M.; Li, C.-S.; Sheng, Y.-J.; Wu, D.T.; Tsao, H.-K. Size-dependent properties of small unilamellar vesicles formed by model lipids. *Langmuir* **2012**, *28*, 689–700. [CrossRef] [PubMed]

16. Olusanya, T.O.; Haj Ahmad, R.R.; Ibegbu, D.M.; Smith, J.R.; Elkordy, A.A.J.M. Liposomal drug delivery systems and anticancer drugs. *Molecules* **2018**, *23*, 907. [CrossRef]

17. Magin, R.L.; Hunter, J.M.; Niesman, M.R.; Bark, G.A. Effect of vesicle size on the clearance, distribution, and tumor uptake of temperature-sensitive liposomes. *Cancer Drug Deliv.* **1986**, *3*, 223–237. [CrossRef]

18. Hosny, K.M. Ciprofloxacin as ocular liposomal hydrogel. *AASP PharmSciTech* **2010**, *11*, 241–246. [CrossRef]

19. Tsotas, V.-A.; Mourtas, S.; Antimisiaris, S.G. Dexamethasone incorporating liposomes: Effect of lipid composition on drug trapping efficiency and vesicle stability. *Drug Deliv.* **2007**, *14*, 441–445. [CrossRef]

20. Tabandeh, H.; Mortazavi, S.A. An investigation into some effective factors on encapsulation efficiency of alpha-tocopherol in MLVs and the release profile from the corresponding liposomal gel. *Iran J. Pharm. Res.* **2013**, *12*, 21.

21. Epand, R.M.; Epand, R.F.; Maekawa, S. The arrangement of cholesterol in membranes and binding of NAP-22. *Chem. Phys. Lipids* **2003**, *122*, 33–39. [CrossRef]

22. Porter, A.G.; Jänicke, R.U. Emerging roles of caspase-3 in apoptosis. *Cell Death Differ.* **1999**, *6*, 99–104. [CrossRef] [PubMed]

23. Chang, H.Y.; Yang, X. Proteases for Cell Suicide: Functions and Regulation of Caspases. *Microbiol. Mol. Biol. Rev.* **2000**, *64*, 821–846. [CrossRef]

24. Allen, T.M.; Cullis, P.R. Liposomal drug delivery systems: From concept to clinical applications. *Adv. Drug Deliv. Rev.* **2013**, *65*, 36–48. [CrossRef] [PubMed]

25. De Smet, L.; Ceelen, W.; Remon, J.P.; Vervaet, C. Optimization of drug delivery systems for intraperitoneal therapy to extend the residence time of the chemotherapeutic agent. *Sci. World J.* **2013**, *2013*, 720858. [CrossRef] [PubMed]

26. Al Shoyaib, A.; Archie, S.R.; Karamyan, V.T. Intraperitoneal Route of Drug Administration: Should it Be Used in Experimental Animal Studies? *J. Pharm. Res.* **2020**, *37*, 12. [CrossRef] [PubMed]

27. Mirahmadi, N.; Babaei, M.; Vali, A.; Dadashzadeh, S. Effect of liposome size on peritoneal retention and organ distribution after intraperitoneal injection in mice. *Int. J. Pharm.* **2010**, *383*, 7–13. [CrossRef] [PubMed]

28. Feng, J.; Iyer, A.; Seo, Y.; Broaddus, C.; Liu, B.; VanBrocklin, H.; He, J. Effects of size and targeting ligand on biodistribution of liposome nanoparticles in tumor mice. *J. Nucl. Med.* **2013**, *54*, 1339.

29. Singhania, A.; Wu, S.Y.; McMillan, N.A. Effective delivery of PEGylated siRNA-containing lipoplexes to extraperitoneal tumours following intraperitoneal administration. *J. Drug Deliv.* **2011**, *2011*, 192562. [CrossRef]

30. Mohamed, M.; Abu Lila, A.S.; Shimizu, T.; Alaaeldin, E.; Hussein, A.; Sarhan, H.A.; Szebeni, J.; Ishida, T. PEGylated liposomes: Immunological responses. *Sci. Technol. Adv. Mater.* **2019**, *20*, 710–724. [CrossRef]

Publisher's Note: MDPI stays neutral with regard to jurisdictional claims in published maps and institutional affiliations.

 © 2020 by the authors. Licensee MDPI, Basel, Switzerland. This article is an open access article distributed under the terms and conditions of the Creative Commons Attribution (CC BY) license (http://creativecommons.org/licenses/by/4.0/).

Hypothesis

Ibrutinib Could Suppress CA-125 in Ovarian Cancer: A Hypothesis

Julian Matthias Metzler [1,*], Daniel Fink [2] and Patrick Imesch [1]

[1] Department of Gynecology, University Hospital Zurich, Frauenklinikstrasse 10, 8091 Zurich, Switzerland; Patrick.Imesch@usz.ch

[2] Faculty of Medicine, University of Zurich, Pestalozzistrasse 3/5, 8091 Zurich, Switzerland; D.Fink@hin.ch

* Correspondence: Julian.Metzler@usz.ch; Tel.: +41-442551111

Abstract: Ibrutinib is a small-molecule inhibitor of Bruton's tyrosine kinase, an enzyme central in B cell development. It is indicated as a therapy for certain hematological diseases such as chronic lymphocytic leukemia (CLL), but also exerts off-target effects on several receptors and kinases. In this paper, we hypothesize that ibrutinib may suppress the tumor marker CA-125 in ovarian cancer. The hypothesis is based on an observation of CA-125 normalization in a patient with low-grade serous ovarian cancer who received ibrutinib for concurrent CLL. We propose a mechanistic model explaining this possible drug effect as a foundation for further research.

Keywords: ibrutinib; Bruton's tyrosine kinase; Btk; kinase inhibitor; ovarian cancer; gynecological oncology; solid tumors; CA-125

Citation: Metzler, J.M.; Fink, D.; Imesch, P. Ibrutinib Could Suppress CA-125 in Ovarian Cancer: A Hypothesis. *Appl. Sci.* **2021**, *11*, 222. https://doi.org/10.3390/app11010222

Received: 23 November 2020
Accepted: 21 December 2020
Published: 28 December 2020

Publisher's Note: MDPI stays neutral with regard to jurisdictional claims in published maps and institutional affiliations.

Copyright: © 2020 by the authors. Licensee MDPI, Basel, Switzerland. This article is an open access article distributed under the terms and conditions of the Creative Commons Attribution (CC BY) license (https://creativecommons.org/licenses/by/4.0/).

1. Introduction

Ibrutinib is a small-molecule inhibitor of Bruton's tyrosine kinase (Btk), an essential enzyme in B cell differentiation. Indications for oral ibrutinib include certain hematological diseases such as chronic lymphocytic leukemia (CLL), mantle cell lymphoma (MCL), or marginal zone lymphoma (MZL). Apart from its affinity to Btk, ibrutinib exerts off-target effects on several receptors and kinases such as the epidermal growth factor receptors (EGFRs 1-4), interleukin-2-inducible T-cell kinase (ITK), or Janus kinase 3 (JAK3). Influencing these additional molecular targets, ibrutinib has been shown to be effective not only in hematological malignancies but also in solid tumors. Nevertheless, evidence of ibrutinib affecting ovarian cancer is scarce and limited to preclinical trials [1–5].

CA-125 is a heavily glycosylated protein and a tumor-associated antigen. It is the epitope of MUC16, a high-molecular-weight transmembrane mucine, occurring in the pleura and the peritoneum as well as in the female reproductive tract epithelia, the ocular surface, and the respiratory tract [6]. Physiologically functioning as a protecting and lubricating agent, the altered expression and glycosylation of MUC16 have been identified in ovarian carcinoma [6] and MUC16 downregulation has been linked with increased cisplatin sensitivity [7]. Even though CA-125 is the most widely used tumor marker in ovarian cancer, elevated serum levels have a limited specificity, allowing for a broad range of benign differential diagnosis. This serum increase in non-malignant or inflammatory conditions is known to be mediated by a series of cytokines such as IL-1β, IL-6, IL-8, IL-17, TNFα, and IFNγ [6].

Nonetheless, there are insufficient data about the regulation and expression of CA-125. Regarding malignancy, recent research has demonstrated an important role of mesothelial cells in MUC16 production and has suggested malignant ascites as a strong modulator of its expression [6]. Upregulation of MUC16 has been shown to be regulated via the KRAS/ERK axis [8]. A thorough literature search revealed no information about ibrutinib's influence on its expression or regulation.

As a clinical introduction, we first present the case of a patient with ovarian cancer, whose CA-125 levels normalized after the initiation of ibrutinib therapy for concurrent CLL. To the best of our knowledge, this synchronism has not been described in the literature until now, providing evidence for our hypothesis of a possible, formerly unknown drug-effect. Secondly, the hypothesis is presented and evaluated considering the available literature.

2. Case Report

A 61-year-old woman was referred to our department in 2010 for further treatment of an enlarged ovary, which had been an incidental finding in a follow-up CT scan for CLL. The CLL had previously been treated with chlorambucil and prednisone. Secondary diagnosis included arterial hypertension and dyslipidemia. Upon referral, transvaginal sonography revealed an ovarian tumor of 57×30 mm containing solid and cystic components. Blood serum analysis was negative for CA-125. Surgical treatment was performed as follows: hysterectomy, adnexectomy, pelvic and para-aortal lymphadenectomy, incidental appendectomy, and radical omentectomy. Histological workup revealed a papillary serous borderline tumor with transition to a low-grade ovarian cancer, stage FIGO (Fédération internationale de gynécologie obstétrique) IIIb. Postoperatively, an adjuvant platinum-based therapy (6 cycles of carboplatin/paclitaxel) was administered as standard of care.

In 2011, the patient was diagnosed with sarcoidosis. Systemic steroids were administered from December 2011 until April 2012, resulting in total remission. In 2013, the patient suffered her first recurrence of CLL. The treatment regimen included 2 cycles of ribomustin and rituximab.

In 2014, rising levels of CA-125 marked recurrent ovarian cancer, and a re-laparotomy for tumor-debulking and partial colon resection with end-to-end anastomosis was performed in October 2014. Second-line chemotherapy contained 6 cycles of carboplatin and gemcitabine until March 2015. In July 2016, sixteen months after second-line chemotherapy, elevated CA-125 levels were recorded, without further clinical or radiological findings, allowing for expectant management of the asymptomatic patient (Figure 1). Gynecological follow-up visits were continued every 3 months.

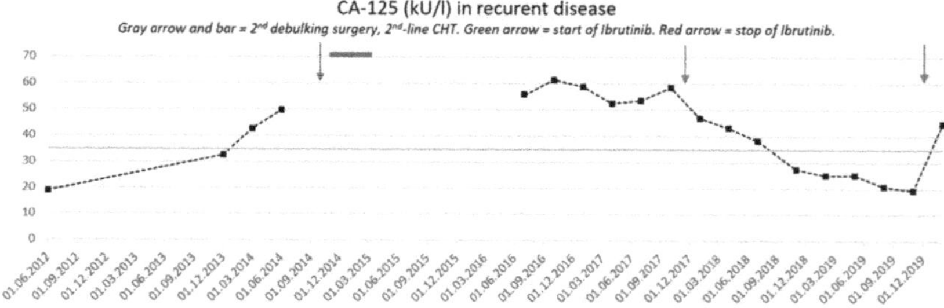

Figure 1. Development of CA-125 levels in recurrent disease. An increase was seen during the course of the disease, and debulking surgery with second-line chemotherapy was performed (grey arrow and grey bar). CA-125 monitoring was not continued postoperatively until July 2016, when elevated levels suggested recurrent (subclinical) disease. During expectant management, CA-125 dropped sharply after the initiation of ibrutinib therapy for a different indication (green arrow). After termination of ibrutinib intake (red arrow), a marked increase was seen.

A second recurrence of CLL was diagnosed in May 2017. Repeated therapy with ribomustin and rituximab showed minimal response. Therefore, an oral therapy with ibrutinib (140 mg, alternating $2\times/3\times$ daily) was established in November 2017, which allowed for hematological disease control.

Following the initiation of ibrutinib, the CA-125 levels started to decrease continuously, with normalization after 12 months of therapy. During the continued therapy, the patient had no laboratory, radiological, or clinical signs of recurring ovarian cancer.

We initiated genetic testing of the cancer sampled in the second surgery via Foundation One CDx (Foundation Medicine Inc., Cambridge, MA, USA; Hoffmann La Roche AG, Basel, Switzerland). The molecular profile showed the genomic signature of a microsatellite-stable carcinoma with a loss-of-heterozygosity score of 0.0% and a low tumor mutational burden (3 Muts/Mb). Gene alterations were found in ARID1A and MUTYH, as well as an activating oncogenic KRAS G12V mutation. None of the mutations offered any therapeutic consequences, as confirmed by our molecular tumor board.

In January 2020, ibrutinib was stopped by the treating hematologist after 26 months of therapy due to laboratory signs of hepatitis. The patient was hospitalized due to a deterioration of general condition shortly thereafter. The signs and symptoms were interpreted in the context of relapsing sarcoidosis and treated accordingly. In February 2020, six weeks after discontinuation of ibrutinib, an elevated CA-125 level was registered, without any clinically or radiologically apparent tumor.

3. Hypothesis

In our case report, we describe a patient with low-grade ovarian cancer experiencing a prolonged normalization in CA-125 under concurrent ibrutinib treatment. Although not necessarily causal, the timely fashion of the protein decline and subsequently incline after termination of treatment is staggering.

From these observations, the next step was to formulate a bio-molecular hypothesis that could mechanistically explain a decrease in CA-125 after ibrutinib intake. We proposed the following model, as depicted in Figure 2. Ibrutinib inhibits the phosphorylation of EGFR (1–4) as well as Src [9] and the FGFR in mesenchymal cells [10]. This blocks downstream effectors and stops EGFR/Src from activating RAS [11]. Subsequently, the oncogenic RAS pathway, which has been shown to encourage MUC16 upregulation via RAF, MEK, and ERK [8], is not activated. Finally, MUC16 production and CA-125 shedding diminish. MEK inhibitors affect proteins further downstream the pathway and have been described to normalize CA-125 levels [12].

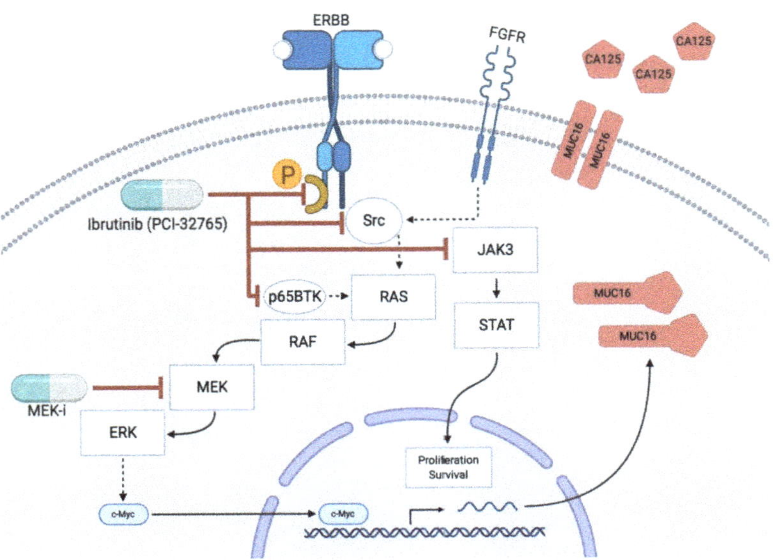

Figure 2. Proposed mechanism of ibrutinib-mediated suppression of CA-125 production along the RAS–RAF–MEK–ERK pathway including additional inhibitory sites (created with Biorender.com).

4. Discussion

There are several limitations to our hypothesis. It is important to stress that the observed correlation does not imply causation. Nevertheless, the absence of other concomitant oncological medication during the time of the CA-125 decrease in our case report suggests at least a partial effect of ibrutinib.

Randomly elevated CA-125 levels due to sarcoidosis have sporadically been described in case reports [13]. These reports are rare, implying a correlation due to the absence of differential diagnosis or due to involvement of serosal surfaces. In our patient, sarcoidosis was primarily cutaneous and pulmonal, without pleural or peritoneal effusions, and CA-125 was not elevated during the initial diagnosis of sarcoidosis in 2011. The tumor marker has not since been established for monitoring sarcoidosis activity. Concerning the observed decline of CA-125, spontaneous regression is extremely rare in epithelial ovarian cancer. A PubMed search (1 May, 2020) revealed only one case report of spontaneous regression in recurrent disease after radiation of a single nodal metastasis [14]. As spontaneous regression and sarcoidosis fail to explain the serum marker fluctuation, the kinase inhibitor's possible role needs to be investigated.

Few in vitro experiments have studied ibrutinib's effect on ovarian cancer. Papillary serous cells, such as in the present case, displayed drug response in ex vivo drug sensitivity testing [15]. Ovarian cancer stem cells have been shown to express Btk, with ibrutinib diminishing their self-renewal capacities [5].

Regarding other kinase inhibitors, recent case reports describe disease response and declining CA-125 after therapy with trametinib, a MEK inhibitor affecting the MAPK/ERK pathway [12]. Furthermore, we found a case report about a patient harboring the same KRAS mutation who experienced an impressive clinical response after treatment with binimetinib, another MEK inhibitor [16]. Animal studies found overexpression of p65BTK, an isoform of Btk, in KRAS-mutant cell lines. Btk inhibitors proving effective against cell viability in these experiments offers another molecular explanation for our findings [17].

CA-125 expression shows a high correlation with serous subtype borderline and malignant ovarian tumors [18], and we conclude that the varying levels of CA-125 in our patient were a direct result of ibrutinib's effect on subclinical ovarian cancer. The absence of a radiologically or clinically detectable tumor during this period is consistent with this conclusion, as low tumor volumes typically elevate serum levels before they become clinically evident. The observed decrease in CA-125 during ibrutinib intake can be explained by a cytotoxic effect, a metabolic effect (suppression of CA-125 production in sub-apoptotic serum levels), or a combination of these.

Further research is needed to test this hypothesis in low- and high-grade ovarian cancer, and we encourage the conduction of in vitro experiments exploiting the underlying pathways for tumor suppression. In this context, our work can be seen as an additional puzzle piece in this widely unexplored area, bridging the gap toward more clinically relevant research.

5. Conclusions

To the best of our knowledge, the described case is the first published report of ibrutinib possibly suppressing CA-125 levels, implying a surrogate ovarian cancer suppression. The temporal context of starting ibrutinib intake and the subsequent decrease in tumor marker levels further support the preclinical evidence that this kinase inhibitor's multiple anti-neoplastic effects can be used against ovarian cancer. We proposed a possible mechanism of action leading to CA-125 suppression. Future research should focus on ibrutinib's capabilities in CA-125 suppression and the underlying processes involved. This may ultimately lead to an extended clinical use of this drug.

Author Contributions: Conceptualization, J.M.M. and D.F.; writing—original draft preparation, J.M.M.; writing—review and editing, P.I.; visualizations, J.M.M.; validation, P.I.; supervision, P.I. and D.F.; project administration, all authors. All authors have read and agreed to the published version of the manuscript.

Funding: This research received no external funding.

Institutional Review Board Statement: According to Swiss law, this is not a research project under the Swiss Human Research Act (Humanforschungsgesetz, HFG), and therefore, no authorization is required. Written informed consent was obtained from the patient for publication of the case report and the accompanying images.

Informed Consent Statement: A copy of the written consent is available for review by the Editor-in-Chief of this journal upon request.

Data Availability Statement: The datasets used and/or analyzed during the current study are available from the corresponding author on reasonable request, to the limit where individual privacy could be compromised.

Acknowledgments: Thanks to Carla Trachsel for proofreading the article.

Conflicts of Interest: The authors declare no conflict of interest.

Abbreviations

Btk	Bruton's tyrosine kinase
CLL	Chronic lymphocytic leukemia
EGFR	Epidermal growth factor receptor
FIGO	Fédération internationale de gynécologie obstétrique
FGFR	Fibroblast growth factor receptor
HPMC	Human peritoneal mesothelial cell
IFN	Interferon
IL	Interleukin
ITK	interleukin-2-inducible T-cell kinase
MEK(-i)	mitogen-activated protein kinase (-inhibitor)
MZL	Marginal zone lymphoma

References

1. Chen, J.; Kinoshita, T.; Sukbuntherng, J.; Chang, B.Y.; Elias, L. Ibrutinib Inhibits ERBB Receptor Tyrosine Kinases and HER2-Amplified Breast Cancer Cell Growth. *Mol. Cancer Ther.* **2016**, *15*, 2835–2844. [CrossRef] [PubMed]
2. Grabinski, N.; Ewald, F. Ibrutinib (ImbruvicaTM) potently inhibits ErbB receptor phosphorylation and cell viability of ErbB2-positive breast cancer cells. *Investig. New Drugs* **2014**, *32*, 1096–1104. [CrossRef] [PubMed]
3. Wang, A.; Yan, X.E.; Wu, H.; Wang, W.; Hu, C.; Chen, C.; Zhao, Z.; Zhao, P.; Li, X.; Wang, L.; et al. Ibrutinib targets mutant-EGFR kinase with a distinct binding conformation. *Oncotarget* **2016**, *7*, 69760–69769. [CrossRef] [PubMed]
4. Gao, W.; Wang, M.; Wang, L.; Lu, H.; Wu, S.; Dai, B.; Ou, Z.; Zhang, L.; Heymach, J.V.; Gold, K.A.; et al. Selective antitumor activity of ibrutinib in EGFR-mutant non-small cell lung cancer cells. *J. Natl. Cancer Inst.* **2014**, *106*. [CrossRef] [PubMed]
5. Zucha, M.A.; Wu, A.T.; Lee, W.H.; Wang, L.S.; Lin, W.W.; Yuan, C.C.; Yeh, C.T. Bruton's tyrosine kinase (Btk) inhibitor ibrutinib suppresses stem-like traits in ovarian cancer. *Oncotarget* **2015**, *6*, 13255–13268. [CrossRef] [PubMed]
6. Matte, I.; Garde-Granger, P.; Bessette, P.; Piche, A. Ascites from ovarian cancer patients stimulates MUC16 mucin expression and secretion in human peritoneal mesothelial cells through an Akt-dependent pathway. *BMC Cancer* **2019**, *19*, 406. [CrossRef] [PubMed]
7. Boivin, M.; Lane, D.; Piche, A.; Rancourt, C. CA125 (MUC16) tumor antigen selectively modulates the sensitivity of ovarian cancer cells to genotoxic drug-induced apoptosis. *Gynecol. Oncol.* **2009**, *115*, 407–413. [CrossRef] [PubMed]
8. Liang, C.; Qin, Y.; Zhang, B.; Ji, S.; Shi, S.; Xu, W.; Liu, J.; Xiang, J.; Liang, D.; Hu, Q.; et al. Oncogenic KRAS Targets MUC16/CA125 in Pancreatic Ductal Adenocarcinoma. *Mol. Cancer Res.* **2017**, *15*, 201–212. [CrossRef] [PubMed]
9. Katopodis, P.; Chudasama, D.; Wander, G.; Sales, L.; Kumar, J.; Pandhal, M.; Anikin, V.; Chatterjee, J.; Hall, M.; Karteris, E. Kinase Inhibitors and Ovarian Cancer. *Cancers (Basel)* **2019**, *11*, 1357. [CrossRef] [PubMed]
10. Uitdehaag, J.C.M.; Kooijman, J.J.; de Roos, J.; Prinsen, M.B.W.; Dylus, J.; Willemsen-Seegers, N.; Kawase, Y.; Sawa, M.; de Man, J.; van Gerwen, S.J. Combined Cellular and Biochemical Profiling to Identify Predictive Drug Response Biomarkers for Kinase Inhibitors Approved for Clinical Use between 2013 and 2017. *Mol. Cancer Ther.* **2019**, *18*, 470–481. [CrossRef] [PubMed]
11. Pinilla-Macua, I.; Grassart, A.; Duvvuri, U.; Watkins, S.C.; Sorkin, A. EGF receptor signaling, phosphorylation, ubiquitylation and endocytosis in tumors in vivo. *Elife* **2017**, *6*. [CrossRef] [PubMed]
12. Champer, M.; Miller, D.; Kuo, D.Y. Response to trametinib in recurrent low-grade serous ovarian cancer with NRAS mutation: A case report. *Gynecol. Oncol. Rep.* **2019**, *28*, 26–28. [CrossRef] [PubMed]

13. Kalluri, M.; Judson, M.A. Sarcoidosis associated with an elevated serum CA 125 level: Description of a case and a review of the literature. *Am. J. Med. Sci.* **2007**, *334*, 441–443. [CrossRef] [PubMed]

14. Fujiwaki, R.; Sawada, K. Spontaneous regression in recurrent epithelial ovarian cancer. *Arch. Gynecol. Obstet.* **2007**, *275*, 389–391. [CrossRef] [PubMed]

15. Lohse, I.; Azzam, D.J.; Al-Ali, H.; Volmar, C.H.; Brothers, S.P.; Ince, T.A.; Wahlestedt, C. Ovarian Cancer Treatment Stratification Using Ex Vivo Drug Sensitivity Testing. *Anticancer Res.* **2019**, *39*, 4023–4030. [CrossRef] [PubMed]

16. Han, C.; Bellone, S.; Zammataro, L.; Schwartz, P.E.; Santin, A.D. Binimetinib (MEK162) in recurrent low-grade serous ovarian cancer resistant to chemotherapy and hormonal treatment. *Gynecol. Oncol. Rep.* **2018**, *25*, 41–44. [CrossRef] [PubMed]

17. Giordano, F.; Vaira, V.; Cortinovis, D.; Bonomo, S.; Goedmakers, J.; Brena, F.; Cialdella, A.; Ianzano, L.; Forno, I.; Cerrito, M.G.; et al. p65BTK is a novel potential actionable target in KRAS-mutated/EGFR-wild type lung adenocarcinoma. *J. Exp. Clin. Cancer Res.* **2019**, *38*, 260. [CrossRef] [PubMed]

18. Hogdall, E.V.; Christensen, L.; Kjaer, S.K.; Blaakaer, J.; Kjaerbye-Thygesen, A.; Gayther, S.; Jacobs, I.J.; Høgdall, C.K. CA125 expression pattern, prognosis and correlation with serum CA125 in ovarian tumor patients. From The Danish "MALOVA" Ovarian Cancer Study. *Gynecol. Oncol.* **2007**, *104*, 508–515. [CrossRef] [PubMed]

Review

An Updated Review of Smac Mimetics, LCL161, Birinapant, and GDC-0152 in Cancer Treatment

Yung-Chieh Chang [1] and Chun Hei Antonio Cheung [1,2,*

[1] Institute of Basic Medical Sciences, College of Medicine, National Cheng Kung University, No. 1 University Road, Tainan 701, Taiwan; stu11706123@gmail.com

[2] Department of Pharmacology, College of Medicine, National Cheng Kung University, No. 1 University Road, Tainan 701, Taiwan

* Correspondence: acheung@mail.ncku.edu.tw

Abstract: Inhibitor of apoptosis proteins (IAPs) are suggested as therapeutic targets for cancer treatment. Smac/DIABLO is a natural IAP antagonist in cells; therefore, Smac mimetics have been developed for cancer treatment in the past decade. In this article, we review the anti-cancer potency and novel molecular targets of LCL161, birinapant, and GDC-0152. Preclinical studies demonstrated that Smac mimetics not only induce apoptosis but also arrest cell cycle, induce necroptosis, and induce immune storm in vitro and in vivo. The safety and tolerance of Smac mimetics are evaluated in phase 1 and phase 2 clinical trials. In addition, the combination of Smac mimetics and chemotherapeutic compounds was reported to improve anti-cancer effects. Interestingly, the novel anti-cancer molecular mechanism of action of Smac mimetics was reported in recent studies, suggesting that many unknown functions of Smac mimetics still need to be revealed. Exploring these currently unknown signaling pathways is important to provide hints for the modification and combination therapy of further compounds.

Keywords: inhibitor of apoptosis proteins (IAPs); Smac/DIABLO; LCL161; birinapant; GDC-0152

Citation: Chang, Y.-C.; Cheung, C.H.A. An Updated Review of Smac Mimetics, LCL161, Birinapant, and GDC-0152 in Cancer Treatment. *Appl. Sci.* **2021**, *11*, 335. https://doi.org/10.3390/app11010335

Received: 5 December 2020
Accepted: 27 December 2020
Published: 31 December 2020

Publisher's Note: MDPI stays neutral with regard to jurisdictional claims in published maps and institutional affiliations.

Copyright: © 2020 by the authors. Licensee MDPI, Basel, Switzerland. This article is an open access article distributed under the terms and conditions of the Creative Commons Attribution (CC BY) license (https://creativecommons.org/licenses/by/4.0/).

1. Introduction

Inhibitor of apoptosis proteins (IAPs), including cellular inhibitor of apoptosis protein 1 (cIAP1), cellular inhibitor of apoptosis protein 2 (cIAP2), melanoma inhibitor of apoptosis (ML-IAP/Livin), testis-specific inhibitor of apoptosis (Ts-IAP/ILP-2), neuronal apoptosis inhibitory protein (NAIP), X-linked inhibitor of apoptosis protein (XIAP), survivin, and BIR repeat containing ubiquitin-conjugating enzyme (BRUCE), are known for their anti-apoptotic effects [1,2]. Members of the IAP family are characterized by the presence of the baculoviral IAP repeat (BIR) domain, which physically interacts with caspase proteins and inhibits the activity of caspases. BIR domains are grouped into two types, based on the presence or absence of the IAP binding motif (IBM) on the BIR domain [3]. Only type II BIR domain, which is with IBM, can interact with caspases. The BIR1-BIR2 linker of XIAP interacts with caspase-3 and -7 [3]. The BIR3 of XIAP inhibits the activity of caspase-9 by interacting with the N-terminal tetrapeptide of caspase-9 [4]. IAPs play important roles in mediating a variety of cellular processes, including apoptosis, mitosis, autophagy, and DNA damage in cancer cells [5–14]. Therefore, the dysregulation of IAPs promotes tumorigenesis, metastasis, angiogenesis, and therapeutic resistance, including chemotherapy and radiotherapy [15–18]. Currently, many IAP-targeting treatments, such as small-molecule inhibitors (i.e., ASTX660, Embelin, and YM155) [19–21], anti-sense oligonucleotides (i.e., LY2181308) [22], and Smac mimetic compounds (i.e., birinapant, LCL161, and GDC-0152) [23–25], have been developed. Despite the anti-cancer potency of birinapant, LCL161, and GDC-0152, they are still under investigation in preclinical and clinical studies, and these compounds have already received much attention in recent years. In this review, we mainly focus on describing the current development of LCL161,

61

birinapant, and GDC-0152 as anti-cancer agents, and we discuss the potential of using these agents for the treatment of cancer in the future.

2. Smac/DIABLO Inhibits IAPs in Cancer

Second mitochondria-derived activator of apoptosis/direct inhibitor of apoptosis-binding protein with low pI (Smac/DIABLO) physically interacts with IAPs and antagonizes the anti-apoptotic activity of IAPs in cells, resulting in apoptosis (Figure 1) [26–28]. In the presence of apoptosis stimuli, mature Smac/DIABLO is released from the mitochondria to cytosol [26,27]. Smac/DIABLO interacts with the BIR domain of IAPs by its particular NH$_2$-terminal motif consisting of four amino acids (Ala-Val-Pro-Ile), and it releases caspases from IAPs, thereby inducing caspase-dependent apoptosis [26]. Previous studies demonstrated that Smac/DIABLO interacts with the BIR2 and BIR3 domains of XIAP, and caspase-3 and -9 are released from XIAP, respectively (Figure 1) [4,29]. Smac/DIABLO not only mediates the cellular function of XIAP but also regulates cIAP1 and cIAP2. Smac/DIABLO induces the ubiquitination and degradation of cIAP1 and cIAP2 (Figure 1) [30]. However, Smac/DIABLO does not degrade XIAP [31]. Interestingly, a recent study found that a Smac/DIABLO isoform, Smac3, induces the autoubiquitination and degradation of XIAP [32]. On the other hand, cIAP1 and cIAP2 can ubiquitinate receptor-interacting kinase 1 (RIPK1), resulting in the inhibition of caspase-dependent apoptosis and necroptosis [1,33,34]. Moreover, cIAP1 and cIAP2 promote cell proliferation, migration, and invasion by activating the canonical nuclear factor kappa-light-chain-enhancer of the activated B cell (NF-κB) signaling pathway [1]. Therefore, upregulation of Smac/DIABLO can induce caspase-dependent apoptosis through the de-ubiquitination of receptor-interacting serine/threonine-protein kinase 1 (RIPK1) and necroptosis by activating the RIPK1/receptor-interacting serine/threonine-protein kinase 3 (RIPK3)/mixed lineage kinase domain-like protein (MLKL) signaling pathway [35]. Pathologically, the protein expression level of Smac is frequently downregulated in renal carcinoma, colorectal cancer, bladder cancer, lung cancer, hepatocellular carcinoma, testicular germ cell tumors, and pancreatic cancer compared with normal tissues, but not in cervical cancer [36–45]. For these reasons, the use of Smac mimetics was suggested as a potential approach for cancer treatment.

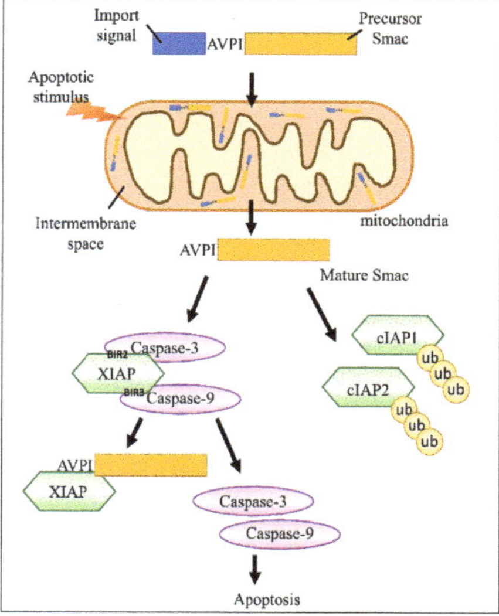

Figure 1. Smac/DIABLO inhibits IAPs in cancer cells. The precursor of Smac (yellow) is transported to the intermembrane space of mitochondria by the import signal (blue box). Then, mature Smac is released from the mitochondria into the cytosol and subsequently inhibits the cellular functions of IAPs. AVPI stands for "Ala-Val-Pro-Ile". BIR stands for "baculoviral IAP repeat". Ub stands for "ubiquitin".

3. Smac Mimetics for Cancer Treatment

The first Smac mimetic compound with eight amino acids was studied in 2000 [46]. Currently, eight Smac mimetics have been developed, and their anti-cancer potency has been evaluated in different preclinical and clinical studies. Smac mimetics are classified into two groups based on the number of Smac-mimicking moieties; for example, monovalent compounds contain one Smac-mimicking moiety (i.e., LCL161, AT-406 (Debio1143), GDC-0152, and GDC-0917 (CUDC-427)), and bivalent compounds are Smac-mimicking elements connected via a linker (i.e., birinapant (TL32711), BV6, and SM-164) (Figure 2) [23,47–52]. The anti-cancer potency of six of them has been elucidated in clinical trials. Bivalent compounds exhibit higher anti-cancer potency than monovalent compounds because the former possess better binding affinity with IAPs and higher potency to induce caspase-dependent apoptosis than the latter [53]. The anti-cancer potency of Smac mimetics is also dependent on their specificity to IAPs. For example, birinapant and AT406 preferentially target cIAP1 and cIAP2 rather than XIAP. LCL161 and GDC-0152 are pan-IAP inhibitors that have similar affinities to XIAP, cIAP1, and cIAP2 [23,48,49,51]. Among these Smac mimetic compounds, LCL161, birinapant, and GDC-0152 are currently the most popular, in which their therapeutic effectiveness and the molecular mechanism of actions have been studied extensively in pre-clinical and clinical studies. Therefore, we summarize current LCL161, birinapant, and GDC-0152-related findings in the following sections.

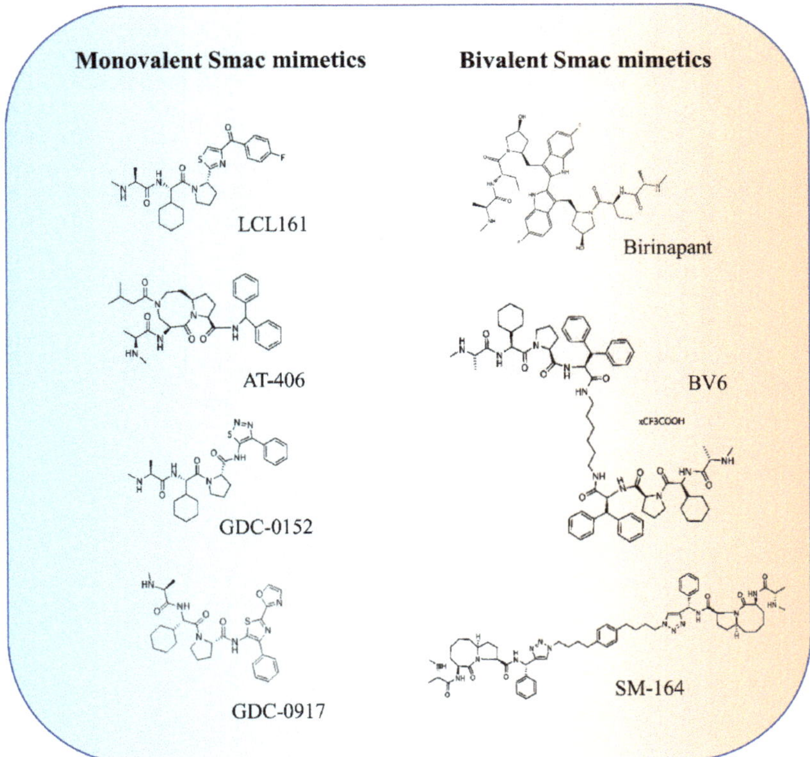

Figure 2. Chemical structure of Smac mimetics for cancer treatment. BV6 was modified with trifluoroacetic acid (xCF3COOH) to improve the solubility of BV6 to DMSO and water.

4. Anti-Cancer Molecular Mechanism of LCL161

LCL161 is an orally bioavailable monovalent Smac mimetic developed by Novartis Pharmaceuticals. LCL161 shows both pro-apoptotic and anti-proliferation effects in cancer cells [47]. In many preclinical studies, the anti-cancer potency of LCL161 has been established in multiple myeloma, glioblastoma, hepatocellular carcinoma, oral squamous carcinoma, neuroblastoma, osteosarcoma, sarcoma, triple-negative breast cancer (TNBC), leukemia, cervical cancer, non-small cell lung cancer, and head and neck cancer cells [24,54–56]. At the molecular level, LCL161 binds to the BIR3 domain of cIAP1 and cIAP2 with high affinity, and induces cIAP1 and cIAP2 autoubiquitination and proteasome degradation, resulting in the activation of non-canonical NF-κB signaling pathways and production of tumor necrosis factor α (TNF-α) (Figure 3) [57]. LCL161 also induces apoptosis and necroptosis through the RIP1/RIP3/MLKL signaling pathway (Figure 3) [58,59]. Surprisingly, a few studies showed that LCL161 does not induce apoptosis, but instead it induces G2/M phase arrest by downregulating cIAP1 and activating the p21 signaling pathway in medulloblastoma and neuroblastoma cells (Figure 3) [60,61]. Moreover, it has been shown that LCL161 does not kill cancer cells directly, but it promotes the activity of immune cells and enhances cytokine secretion instead. Mechanistically, LCL161 promotes the activity of T cells by inducing cytokine secretion and dendritic cell maturation (Figure 3) [62–64]. In multiple myeloma, cIAP1 and cIAP2 are usually deleted, but LCL161 still inhibits the growth of multiple myeloma through the upregulation of tumor cell autonomous type I interferon signaling, and strong acute inflammatory signaling (promoting macrophagy, M-spike, and dendritic cell maturation) in transgenic Vk*MYC mice [65]. Recent studies demonstrated the anti-cancer potency of LCL161 is correlated with the protein expression levels of B-cell lymphoma 2 (BCL-2) and TNF-α and NF-κB activity in cancer cells. In neuroblastoma, Langemann et al. suggested that the anti-cancer potency of LCL161 is based on the cells' sensitivity to TNF-α and modulation of NF-κB [61]. Interestingly, in our recent study, we found that LCL161 interacts with p-glycoprotein (also called multidrug resistance protein 1 (MDR1)) and inhibits the MDR1 multi-drug efflux activity, resulting in increased sensitivity to the MDR1 substrates, such as paclitaxel and YM155 in cervical and bladder cancer cells [66]. Moreover, we found that LCL161 downregulates the protein expression level of survivin in MDR1-expressing cervical and bladder cancer cells [66].

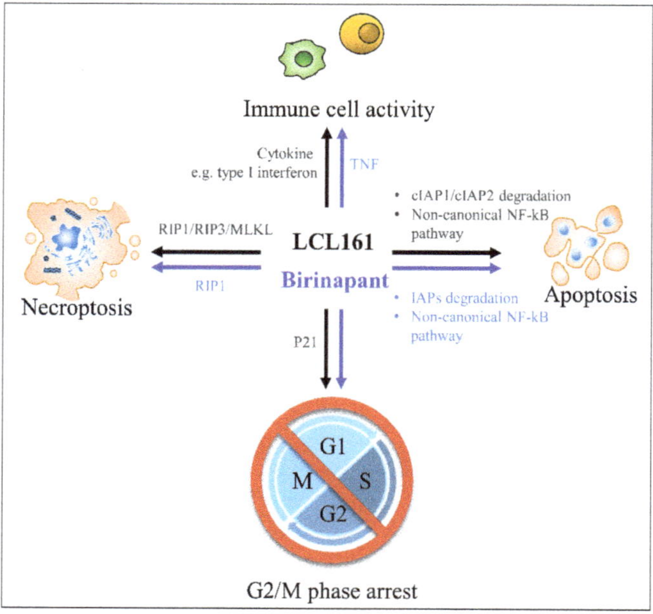

Figure 3. Anti-cancer effects and the molecular mechanisms of action of LCL161 (black) and birinapant (blue) in cancer cells.

5. LCL161 Combination Treatment

Many preclinical studies have explored the anti-cancer effects of LCL161 in combination with different anti-cancer agents. It has been demonstrated that LCL161 sensitizes cancer cells to paclitaxel, Fas ligand, vincristine, and obatoclax (BCL-2 inhibitor) in different types of cancer [54,55,60,61,67–70]. Moreover, LCL161 improves the anti-cancer effects of radiotherapy in head and neck squamous and esophageal carcinoma cells [71,72]. Immunotherapy is a popular treatment for cancer. Chesi et al. noted improvements in the anti-cancer effects of immunotherapy when combined with LCL161 in multiple myeloma in vivo [65]. Maintaining the balance of reactive oxygen species (ROS) plays an important role in tumorigenesis. Targeting redox homeostasis with erastin and auranofin, which are ROS inducers, can improve LCL161-induced cell death in acute lymphoblastic leukemia cells [73].

6. Current Status of LCL161 in Clinical Trials

LCL161 is currently involved in many phase 1 and phase 2 clinical trials (Table 1), and results showed that it has favorable pharmacological properties, such as good tolerability and minor toxicity. Infante et al. showed that LCL161 is well tolerated at doses of up to 1800 mg [74]. Some LCL161-treated patients may present with vomiting, nausea, asthenia, and anorexia side effects, but these effects are not severe after treatment once a week for 21 days [74]. Only 3 out of 53 patients studied demonstrated cytokine release syndrome [74]. LCL161 was rapidly absorbed at the time to reach maximum plasma concentration (i.e., 0.5–2 h), and the plasma concentration declined within the range of 4–16 h [74].

Table 1. Clinical status of LCL161 for cancer treatment.

ClinicalTrials.Gov Identifier	Phase	Condition or Disease (in Patients)	Combination Therapy
NCT01968915	Phase 1 (completed)	Neoplasms	-
NCT01955434	Phase 2 (completed)	Recurrent plasma cell myeloma Refractory plasma cell myeloma	Cyclophophamide
NCT01934634	Phase 1	Metastatic pancreatic cancer	Gemcitabine Nab-Paclitaxel
NCT02098161	Phase 2	Polycythemia vera, post-polycythemic myelofibrosis phase Primary myelofibrosis Secondary myelofibrosis	-
NCT02649673	Phase 1/2	Small cell lung cancer Ovarian cancer	Topotecan Pegylated granulocyte colony stimulating factor
NCT01617668	Phase 2 (completed)	Breast cancer	Paclitaxel
NCT01240655	Phase 1 (completed)	Solid tumors	Paclitaxel
NCT01098838	Phase 1 (completed)	Advanced solid tumors (lung, skin, colon, pancreas, and others)	-
NCT03111992	Phase 1 (completed)	Multiple myeloma	PDR001 CJM112
NCT02890069	Phase 1	Colorectal cancer Non-small cell lung carcinoma (Adenocarcinoma) Triple-negative breast cancer Renal cell carcinoma	PDR001 Evrolimus Panobinostat QBM076 HDM201

Combination therapy of LCL161 with chemotherapeutic drugs such as paclitaxel and gemcitabine has been evaluated in patients with cancer (Table 1). Bardia et al. found that (1) LCL161 combined with paclitaxel can improve the pathologic complete response of TNF-α-positive TNBC, (2) the expression level of TNF-α might be a biomarker for predicting the anti-cancer effects of LCL161 and paclitaxel combination treatment, and (3) the anti-cancer potency of this combination is not correlated with the alternation of breast cancer type 1 susceptibility protein (BRCA1) and breast cancer type 2 susceptibility protein (BRCA2) in TNBC [75]. Some adverse effects were observed during combination treatment, including pyrexia (5 of 209 patients studied), pneumonia (4 of 209 patients studied), and pneumonitis (4 of 209 patients studied).

7. Anti-Cancer Molecular Mechanisms of Birinapant (TL32711)

Birinapant is a second-generation bivalent Smac mimetic, first synthesized in 2014 [76,77]. Birinapant has better tolerability for treating solid tumors compared with the first-generation bivalent Smac mimetic, and it preferentially targets cIAP1, relative to cIAP2 and XIAP [76]. The anti-cancer potency of birinapant has been investigated in acute myeloid leukemia, melanoma, colorectal cancer, ovarian cancer, breast cancer, head and neck squamous cell carcinoma, hepatocellular carcinoma, glioblastoma, and breast cancer [23,77–84]. At the molecular level, birinapant binds the BIR domains of IAPs and promotes the degradation of TNF receptor-associated factor (TRAF)-bound IAPs, resulting in the induction of caspase-8-dependent apoptosis (Figure 3) [85]. In addition, birinapant degrades IAPs, activates the non-canonical NF-κB signaling pathway once the IAPs are degraded, and stabilizes mitogen-activated protein kinase 14 (MAP3K14) (NF-κB-inducing kinase), resulting in the activation of the non-canonical NF-κB signaling pathway [84,86]. Mak et al. demonstrated that apoptosis repressor with caspase recruitment domain, an apoptosis suppressor, decreases the anti-cancer effect of birinapant by inhibiting the cIPA1/MAP3K14 signaling pathway in acute myeloid leukemia [87]. Birinapant induces cell cycle G2/M arrest in head and neck squamous cell carcinoma; however, the underlying molecular mechanism of action remains unclear (Figure 3) [79,88]. In acute lymphoblastic leukemia, birinapant induces RIP-1-dependent necroptosis (Figure 3) [83]. When birinapant is combined with emricasan, a caspase-8 inhibitor, necroptosis is induced in acute myeloid leukemia [89]. Kearney et al. demonstrated that birinapant-induced TNF activates CD8+ T cells and natural killer cells, resulting in cancer cell death (Figure 3) [90]. Targeting programmed death-ligand 1 (PD-1) upregulates TNF production; therefore, the combination of birinapant and PD-1 blockade increases TNF production and promotes the anti-cancer potency of immune therapy [90].

8. Combination Therapy with Birinapant for Cancer Treatment

Overexpression of IAPs induces chemotherapeutic drug resistance in many cancer types. Targeting IAPs sensitizes drug-resistant cancer cells to chemotherapeutic drugs. Birinapant as an IAP antagonist has been combined with many chemotherapeutic drugs for cancer treatment. Birinapant has been combined with norcanthanidin, docetaxel, or TNF-related apoptosis-inducing ligand (TRAIL) for breast cancer treatment [23,91,92]. In addition, birinapant has been combined with paclitaxel, demethylating agent (5-AZ), TNF-α, Fas ligand (FasL), gemcitabine, 5-flurouracie (5-FU), and oxaliplatin for different types of cancer treatment, such as pancreatic cancer, head and neck squamous cell carcinoma, melanoma, acute myeloid leukemia, and colon cancer [79,80,84,86,88,93–97]. Birinapant improves the anti-cancer effects of radiotherapy for treating Fas-associated protein with death domain (FADD) and cIAP1-overexpressed head and neck cancers [88]. Kearney et al. demonstrated that birinapant activates CD8 T-cell and natural killer cells, resulting in cancer cell death [90]. The anti-cancer effects of chimeric antigen receptor (CAR) T-cell therapy correlate directly with the level of TNF. To improve the anti-cancer potency of CAR T cells, Michie et al. combined CAR T-cell therapy with birinapant for cancer treatment. Results showed that this combination approach significantly reduced cancer growth [98].

9. Current Status of Birinapant in Clinical Trials

Clinical trials of birinapant are listed in Table 2. A phase 1 trial study reported the maximum tolerated dose, safety, and pharmacokinetic properties of birinapant in solid tumors [99]. The maximum tolerated dose of birinapant is 47 mg/m^2 [99]. The half-life of birinapant is 30–35 h [99]. A phase 2 clinical trial demonstrated that birinapant has a plasma half-life of 31 h and a tumor tissue half-life of 52 h [100]. Birinapant accumulates in tumor tissues, resulting in the downregulated protein expression level of cIAP1 and apoptosis induced in peripheral blood mononuclear cells and cancer cells [99]. However, birinapant does not increase the protein level of TNF, monocyte chemoattractant protein-1 (MCP-1), or interleukin 1, 6, and 8 [99]. For treating patients with cancer, birinapant is used as a single agent or in combination therapy with chemotherapeutic drugs (i.e., pembrolizumab, carboplatin, docetaxel, and gemcitabine) and radiotherapy. However, some birinapant clinical trials were terminated because birinapant lacked anti-cancer efficacy (NCT02147873, NCT02587962, NCT01681368), or the sponsors did not fund further work (NCT01573780).

Table 2. Clinical status of birinapant (TL32711) for cancer treatment.

ClinicalTrials.Gov Identifier	Phase	Condition or Disease (in Patients)	Combination Therapy
NCT03803774	Phase 1	Head and neck squamous cell carcinoma	Radiation therapy
NCT02587962	Phase 1/2 (terminated)	Solid tumor	Pembrolizumab
NCT02147873	Phase 2 (terminated)	Myelodysplastic syndrome Chronic myelomonocytic leukemia	Azacitidine Placebo
NCT01940172	Phase 1 (completed)	Relapsed epithelial ovarian cancer Relapsed primary peritoneal cancer Relapsed fallopian tube cancer	Conatumumab
NCT01188499	Phase 1/2 (completed)	Cancer	Carboplatin Paclitaxel Irinotecan Docetaxel Gemcitabine Liposomal Doxorubicin
NCT01681368	Phase 2 (terminated)	Epithelial ovarian cancer Peritoneal neoplasms Fallopian tube neoplasms	-
NCT00993239	Phase 1 (completed)	Cancer	-
NCT01573780	Phase 1 (terminated)	Adult solid tumor	Gemcitabine
NCT01486784	Phase 1/2 (terminated)	Acute myelogenous leukemia	-

10. GDC-0152 for Cancer Treatment

The development of GDC-0152, a pan-IAP antagonist, as an anti-cancer agent was first reported in 2012 [49]. The anti-cancer potency of GDC-0152 was evaluated in glioblastoma, leukemia, osteosarcoma, and glioblastoma [49,101–104]. At the molecular level, GDC-0152 induces intrinsic caspase-dependent apoptosis and inhibits the phosphoinositide 3-kinase (PI3K)/protein kinase B (AKT) signaling pathway, resulting in cancer cell death [101,102]. Tchoghandjian et al. demonstrated that GDC-0152 not only downregulates the protein expression level of cIAP1, cIAP2, and XIAP, but it also downregulates ML-IAP in glioblastoma cells in vitro and in vivo [104]. Mantik et al. found that the pharmacological properties of GDC-0152 are correlated with pH [105]. They demonstrated that the presence of succinic acid and hydroxypropyl-β-cyclodextrin can increase the solubility of GDC-0152, thereby improving blood compatibility, reducing hemolysis, and increasing the maximum tolerated dose [105]. Additionally, they remain stable in human plasma for up to 25 h [106]. Only one phase 1 clinical trial has studied the safety and pharmacokinetics of GDC-0152 (NCT00977067); however, it was terminated for reasons unrelated to patient safety or anti-cancer activity in 2017.

11. Conclusions and Future Directions

As dysregulation of IAPs has been found in a variety of tumors, and overexpression of IAPs promotes tumorigenesis and tumor metastasis, the use of Smac mimetics is suggested as a potential therapeutic approach for the treatment of cancer. The anti-cancer potency of Smac mimetics has been demonstrated in different types of malignant tumors in preclinical studies. Smac mimetics induce apoptosis and the non-canonical NF-kB signaling pathway by downregulating the protein expression level of IAPs. However, many unknown anti-cancer molecular mechanisms of action of Smac mimetics are still yet to be discovered. Exploring these signaling pathways is important to provide hints for the modification and combination therapy of further compounds.

The safety and tolerance of Smac mimetics are currently investigated in many clinical studies. Most studies report that Smac mimetics, such as LCL161, are safe and have good tolerance in patients with cancer. However, other Smac mimetics have shown low anti-cancer potency in the treatment of patients with cancer. The underlying molecular mechanism of action is still unclear. The anti-cancer potency of most Smac mimetics is on the micromolar level. Thus, many studies have combined Smac mimetics with chemotherapeutic drugs, radiotherapy, and immune therapy for cancer treatment. Results showed that the synergistic effect of these combinations can decrease the dosage of Smac mimetics. With the advancement of nanotechnology, delivering chemotherapeutic drugs with various types of nanoparticles can improve anti-cancer potency and cancer-targeting specificity through active or passive cancer targeting. For instance, Nikkhoo et al. synthesized a chitosan-based nanoparticle for co-delivering STAT3 siRNA and BV6, a bivalent Smac mimetic, to treat breast cancer cells, colorectal carcinoma cells, and melanoma cells, and their nanoparticles were shown to suppress cancer cell progression through caspase-dependent apoptosis in vitro and in vivo [107]. Therefore, nanotechnology can potentially be applied to improve the anti-cancer potency of Smac mimetics.

Author Contributions: Literature research: Y.-C.C. and C.H.A.C.; Figure and table preparation: Y.-C.C.; Manuscript writing and proofreading: Y.-C.C. and C.H.A.C. All authors have read and agreed to the published version of the manuscript.

Funding: The APC was funded by Ministry of Science and Technology, Taiwan (MOST 109-2320-B-006-031).

Institutional Review Board Statement: Not applicable.

Informed Consent Statement: Not applicable.

Data Availability Statement: Not applicable.

Conflicts of Interest: The authors declare no conflict of interest.

Abbreviations

IAPs: Inhibitor of apoptosis proteins; BRCA1: breast cancer type 1 susceptibility protein; BRCA2: breast cancer type 2 susceptibility protein; BRUCE: BIR repeat containing ubiquitin-conjugating enzyme; cIAP1: cellular inhibitor of apoptosis protein 1; cIAP2: cellular inhibitor of apoptosis 2; DIABLO: direct inhibitor of apoptosis-binding protein with low pI; ML-IAP: melanoma inhibitor of apoptosis; Smac: second mitochondria-derived activator of apoptosis; NAIP: neuronal apoptosis inhibitory protein; TNBC: triple-negative breast cancer; TNF-α: tumor necrosis factor α; Ts-IAP: testis-specific inhibitor of apoptosis.

References

1. Fulda, S.; Vucic, D. Targeting IAP proteins for therapeutic intervention in cancer. *Nat. Rev. Drug Discov.* **2012**, *11*, 109–124. [CrossRef] [PubMed]
2. Wei, Y.; Fan, T.; Yu, M. Inhibitor of apoptosis proteins and apoptosis. *Acta Biochim. Biophys. Sin.* **2008**, *40*, 278–288. [CrossRef] [PubMed]
3. Eckelman, B.P.; Drag, M.; Snipas, S.J.; Salvesen, G.S. The mechanism of peptide-binding specificity of IAP BIR domains. *Cell Death Differ.* **2008**, *15*, 920–928. [CrossRef] [PubMed]
4. Shiozaki, E.N.; Chai, J.; Rigotti, D.J.; Riedl, S.J.; Li, P.; Srinivasula, S.M.; Alnemri, E.S.; Fairman, R.; Shi, Y. Mechanism of XIAP-mediated inhibition of caspase-9. *Mol. Cell* **2003**, *11*, 519–527. [CrossRef]
5. Lin, T.Y.; Chan, H.H.; Chen, S.H.; Sarvagalla, S.; Chen, P.S.; Coumar, M.S.; Cheng, S.M.; Chang, Y.C.; Lin, C.H.; Leung, E.; et al. BIRC5/Survivin is a novel ATG12-ATG5 conjugate interactor and an autophagy-induced DNA damage suppressor in human cancer and mouse embryonic fibroblast cells. *Autophagy* **2020**, *16*, 1296–1313. [CrossRef] [PubMed]
6. Fulda, S.; Debatin, K.M. Extrinsic versus intrinsic apoptosis pathways in anticancer chemotherapy. *Oncogene* **2006**, *25*, 4798–4811. [CrossRef] [PubMed]
7. Gyrd-Hansen, M.; Meier, P. IAPs: From caspase inhibitors to modulators of NF-kappaB, inflammation and cancer. *Nat. Rev. Cancer* **2010**, *10*, 561–574. [CrossRef] [PubMed]
8. Wertz, I.E.; Dixit, V.M. Regulation of death receptor signaling by the ubiquitin system. *Cell Death Differ.* **2010**, *17*, 14–24. [CrossRef] [PubMed]
9. Dogan, T.; Harms, G.S.; Hekman, M.; Karreman, C.; Oberoi, T.K.; Alnemri, E.S.; Rapp, U.R.; Rajalingam, K. X-linked and cellular IAPs modulate the stability of C-RAF kinase and cell motility. *Nat. Cell Biol.* **2003**, *10*, 1447–1455. [CrossRef] [PubMed]
10. Uren, A.G.; Beilharz, T.; O'Connell, M.J.; Bugg, S.J.; van Driel, R.; Vaux, D.L.; Lithgow, T. Role for yeast inhibitor of apoptosis (IAP)-like proteins in cell division. *Proc. Natl. Acad. Sci. USA* **1999**, *96*, 10170–10175. [CrossRef]
11. Ebner, P.; Poetsch, I.; Deszcz, L.; Hoffmann, T.; Zuber, J.; Ikeda, F. The IAP family member BRUCE regulates autophagosome-lysosome fusion. *Nat. Commun.* **2018**, *9*, 599. [CrossRef] [PubMed]
12. Sauer, M.; Reiners, K.S.; Hansen, H.P.; Engert, A.; Gasser, S.; von Strandmann, E.P. Induction of the DNA damage response by IAP inhibition triggers natural immunity via upregulation of NKG2D ligands in Hodgkin lymphoma in vitro. *Biol. Chem.* **2013**, *394*, 1325–1331. [CrossRef] [PubMed]
13. Cheng, S.M.; Chang, Y.C.; Liu, C.Y.; Lee, J.Y.; Chan, H.H.; Kuo, C.W.; Lin, K.Y.; Tsai, S.L.; Chen, S.H.; Li, C.F.; et al. YM155 down-regulates survivin and XIAP, modulates autophagy and induces autophagy-dependent DNA damage in breast cancer cells. *Br. J. Pharmacol.* **2015**, *172*, 214–234. [CrossRef] [PubMed]
14. Ghobrial, I.M.; Witzig, T.E.; Adjei, A.A. Targeting apoptosis pathways in cancer therapy. *CA Cancer J. Clin.* **2005**, *55*, 178–194. [CrossRef] [PubMed]
15. Mehrotra, S.; Languino, L.R.; Raskett, C.M.; Mercurio, A.M.; Dohi, T.; Altieri, D.C. IAP regulation of metastasis. *Cancer Cell* **2010**, *17*, 53–64. [CrossRef] [PubMed]
16. Liu, J.; Zhang, D.; Luo, W.; Yu, Y.; Yu, J.; Li, J.; Zhang, X.; Zhang, B.; Chen, J.; Wu, X.R.; et al. X-linked inhibitor of apoptosis protein (XIAP) mediates cancer cell motility via Rho GDP dissociation inhibitor (RhoGDI)-dependent regulation of the cytoskeleton. *J. Biol. Chem.* **2011**, *286*, 15630–15640. [CrossRef]
17. Tran, J.; Rak, J.; Sheehan, C.; Saibil, S.D.; LaCasse, E.; Korneluk, R.G.; Kerbel, R.S. Marked induction of the IAP family antiapoptotic proteins survivin and XIAP by VEGF in vascular endothelial cells. *Biochem. Biophys. Res. Commun.* **1999**, *264*, 781–788. [CrossRef]
18. Rathore, R.; McCallum, J.E.; Varghese, E.; Florea, A.M.; Busselberg, D. Overcoming chemotherapy drug resistance by targeting inhibitors of apoptosis proteins (IAPs). *Apoptosis* **2017**, *22*, 898–919. [CrossRef]
19. Ward, G.A.; Lewis, E.J.; Ahn, J.S.; Johnson, C.N.; Lyons, J.F.; Martins, V.; Munck, J.M.; Rich, S.J.; Smyth, T.; Thompson, N.T.; et al. ASTX660, a Novel Non-peptidomimetic Antagonist of cIAP1/2 and XIAP, Potently Induces TNFalpha-Dependent Apoptosis in Cancer Cell Lines and Inhibits Tumor Growth. *Mol. Cancer Ther.* **2018**, *17*, 1381–1391. [CrossRef]
20. Ahn, K.S.; Sethi, G.; Aggarwal, B.B. Embelin, an inhibitor of X chromosome-linked inhibitor-of-apoptosis protein, blocks nuclear factor-kappaB (NF-kappaB) signaling pathway leading to suppression of NF-kappaB-regulated antiapoptotic and metastatic gene products. *Mol. Pharmacol.* **2007**, *71*, 209–219. [CrossRef]
21. Nakahara, T.; Kita, A.; Yamanaka, K.; Mori, M.; Amino, N.; Takeuchi, M.; Tominaga, F.; Hatakeyama, S.; Kinoyama, I.; Matsuhisa, A.; et al. YM155, a novel small-molecule survivin suppressant, induces regression of established human hormone-refractory prostate tumor xenografts. *Cancer Res.* **2007**, *67*, 8014–8021. [CrossRef] [PubMed]
22. Talbot, D.C.; Ranson, M.; Davies, J.; Lahn, M.; Callies, S.; Andre, V.; Kadam, S.; Burgess, M.; Slapak, C.; Olsen, A.L.; et al. Tumor survivin is downregulated by the antisense oligonucleotide LY2181308 A proof-of-concept, first-in-human dose study. *Clin. Cancer Res.* **2010**, *16*, 6150–6158. [CrossRef] [PubMed]
23. Allensworth, J.L.; Sauer, S.J.; Lyerly, H.K.; Morse, M.A.; Devi, G.R. Smac mimetic Birinapant induces apoptosis and enhances TRAIL potency in inflammatory breast cancer cells in an IAP-dependent and TNF-alpha-independent mechanism. *Breast Cancer Res. Treat.* **2013**, *137*, 359–371. [CrossRef] [PubMed]
24. Houghton, P.J.; Kang, M.H.; Reynolds, C.P.; Morton, C.L.; Kolb, E.A.; Gorlick, R.; Keir, S.T.; Carol, H.; Lock, R.; Maris, J.M.; et al. Initial testing (stage 1) of LCL161, a SMAC mimetic, by the Pediatric Preclinical Testing Program. *Pediatr. Blood Cancer* **2012**, *58*, 636–639. [CrossRef] [PubMed]

25. Zhen, M.C.; Wang, F.Q.; Wu, S.F.; Zhao, Y.L.; Liu, P.G.; Yin, Z.Y. Identification of mTOR as a primary resistance factor of the IAP antagonist AT406 in hepatocellular carcinoma cells. *Oncotarget* **2017**, *8*, 9466–9475. [CrossRef] [PubMed]

26. Du, C.; Fang, M.; Li, Y.; Li, L.; Wang, X. Smac, a mitochondrial protein that promotes cytochrome c-dependent caspase activation by eliminating IAP inhibition. *Cell* **2000**, *102*, 33–42. [CrossRef]

27. Verhagen, A.M.; Ekert, P.G.; Pakusch, M.; Silke, J.; Connolly, L.M.; Reid, G.E.; Moritz, R.L.; Simpson, R.J.; Vaux, D.L. Identification of DIABLO, a mammalian protein that promotes apoptosis by binding to and antagonizing IAP proteins. *Cell* **2000**, *102*, 43–53. [CrossRef]

28. Kominsky, D.J.; Bickel, R.J.; Tyler, K.L. Reovirus-induced apoptosis requires mitochondrial release of Smac/DIABLO and involves reduction of cellular inhibitor of apoptosis protein levels. *J. Virol.* **2002**, *76*, 11414–11424. [CrossRef]

29. Scott, F.L.; Denault, J.B.; Riedl, S.J.; Shin, H.; Renatus, M.; Salvesen, G.S. XIAP inhibits caspase-3 and -7 using two binding sites: Evolutionarily conserved mechanism of IAPs. *EMBO J.* **2005**, *24*, 645–655. [CrossRef]

30. Darding, M.; Feltham, R.; Tenev, T.; Bianchi, K.; Benetatos, C.; Silke, J.; Meier, P. Molecular determinants of Smac mimetic induced degradation of cIAP1 and cIAP2. *Cell Death Differ.* **2011**, *18*, 1376–1386. [CrossRef]

31. Yang, Q.H.; Du, C. Smac/DIABLO selectively reduces the levels of c-IAP1 and c-IAP2 but not that of XIAP and livin in HeLa cells. *J. Biol. Chem.* **2004**, *279*, 16963–16970. [CrossRef] [PubMed]

32. Fu, J.; Jin, Y.; Arend, L.J. Smac3, a novel Smac/DIABLO splicing variant, attenuates the stability and apoptosis-inhibiting activity of X-linked inhibitor of apoptosis protein. *J. Biol. Chem.* **2003**, *278*, 52660–52672. [CrossRef] [PubMed]

33. Tenev, T.; Bianchi, K.; Darding, M.; Broemer, M.; Langlais, C.; Wallberg, F.; Zachariou, A.; Lopez, J.; MacFarlane, M.; Cain, K.; et al. The Ripoptosome, a signaling platform that assembles in response to genotoxic stress and loss of IAPs. *Mol. Cell* **2011**, *43*, 432–448. [CrossRef] [PubMed]

34. Feoktistova, M.; Geserick, P.; Kellert, B.; Dimitrova, D.P.; Langlais, C.; Hupe, M.; Cain, K.; MacFarlane, M.; Hacker, G.; Leverkus, M. cIAPs block Ripoptosome formation, a RIP1/caspase-8 containing intracellular cell death complex differentially regulated by cFLIP isoforms. *Mol. Cell* **2011**, *43*, 449–463. [CrossRef]

35. Akara-Amornthum, P.; Lomphithak, T.; Choksi, S.; Tohtong, R.; Jitkaew, S. Key necroptotic proteins are required for Smac mimetic-mediated sensitization of cholangiocarcinoma cells to TNF-alpha and chemotherapeutic gemcitabine-induced necroptosis. *PLoS ONE* **2020**, *15*, e0227454. [CrossRef] [PubMed]

36. Mizutani, Y.; Nakanishi, H.; Yamamoto, K.; Li, Y.N.; Matsubara, H.; Mikami, K.; Okihara, K.; Kawauchi, A.; Bonavida, B.; Miki, T. Downregulation of Smac/DIABLO expression in renal cell carcinoma and its prognostic significance. *J. Clin. Oncol.* **2005**, *23*, 448–454. [CrossRef]

37. Kempkensteffen, C.; Hinz, S.; Christoph, F.; Krause, H.; Magheli, A.; Schrader, M.; Schostak, M.; Miller, K.; Weikert, S. Expression levels of the mitochondrial IAP antagonists Smac/DIABLO and Omi/HtrA2 in clear-cell renal cell carcinomas and their prognostic value. *J. Cancer Res. Clin. Oncol.* **2008**, *134*, 543–550. [CrossRef]

38. Endo, K.; Kohnoe, S.; Watanabe, A.; Tashiro, H.; Sakata, H.; Morita, M.; Kakeji, Y.; Maehara, Y. Clinical significance of Smac/DIABLO expression in colorectal cancer. *Oncol. Rep.* **2009**, *21*, 351–355. [CrossRef]

39. Mizutani, Y.; Katsuoka, Y.; Bonavida, B. Prognostic significance of second mitochondria-derived activator of caspase (Smac/DIABLO) expression in bladder cancer and target for therapy. *Int. J. Oncol.* **2010**, *37*, 503–508. [CrossRef]

40. Mizutani, Y.; Katsuoka, Y.; Bonavida, B. Low circulating serum levels of second mitochondria-derived activator of caspase (Smac/DIABLO) in patients with bladder cancer. *Int. J. Oncol.* **2012**, *40*, 1246–1250. [CrossRef]

41. Sekimura, A.; Konishi, A.; Mizuno, K.; Kobayashi, Y.; Sasaki, H.; Yano, M.; Fukai, I.; Fujii, Y. Expression of Smac/DIABLO is a novel prognostic marker in lung cancer. *Oncol. Rep.* **2004**, *11*, 797–802. [CrossRef] [PubMed]

42. Bao, S.T.; Gui, S.Q.; Lin, M.S. Relationship between expression of Smac and Survivin and apoptosis of primary hepatocellular carcinoma. *Hepatobiliary Pancreat Dis. Int.* **2006**, *5*, 580–583. [PubMed]

43. Kempkensteffen, C.; Jager, T.; Bub, J.; Weikert, S.; Hinz, S.; Christoph, F.; Krause, H.; Schostak, M.; Miller, K.; Schrader, M. The equilibrium of XIAP and Smac/DIABLO expression is gradually deranged during the development and progression of testicular germ cell tumours. *Int. J. Androl.* **2007**, *30*, 476–483. [CrossRef] [PubMed]

44. Liang, W.; Liao, Y.; Zhang, J.; Huang, Q.; Luo, W.; Yu, J.; Gong, J.; Zhou, Y.; Li, X.; Tang, B.; et al. Heat shock factor 1 inhibits the mitochondrial apoptosis pathway by regulating second mitochondria-derived activator of caspase to promote pancreatic tumorigenesis. *J. Exp. Clin. Cancer Res.* **2017**, *36*, 64. [CrossRef]

45. Arellano-Llamas, A.; Garcia, F.J.; Perez, D.; Cantu, D.; Espinosa, M.; De la Garza, J.G.; Maldonado, V.; Melendez-Zajgla, J. High Smac/DIABLO expression is associated with early local recurrence of cervical cancer. *BMC Cancer* **2006**, *6*, 256. [CrossRef] [PubMed]

46. Chai, J.; Du, C.; Wu, J.W.; Kyin, S.; Wang, X.; Shi, Y. Structural and biochemical basis of apoptotic activation by Smac/DIABLO. *Nature* **2000**, *406*, 855–862. [CrossRef]

47. Weisberg, E.; Ray, A.; Barrett, R.; Nelson, E.; Christie, A.L.; Porter, D.; Straub, C.; Zawel, L.; Daley, J.F.; Lazo-Kallanian, S.; et al. Smac mimetics: Implications for enhancement of targeted therapies in leukemia. *Leukemia* **2010**, *24*, 2100–2109. [CrossRef]

48. Cai, Q.; Sun, H.; Peng, Y.; Lu, J.; Nikolovska-Coleska, Z.; McEachern, D.; Liu, L.; Qiu, S.; Yang, C.Y.; Miller, R.; et al. A potent and orally active antagonist (SM-406/AT-406) of multiple inhibitor of apoptosis proteins (IAPs) in clinical development for cancer treatment. *J. Med. Chem.* **2011**, *54*, 2714–2726. [CrossRef]

49. Flygare, J.A.; Beresini, M.; Budha, N.; Chan, H.; Chan, I.T.; Cheeti, S.; Cohen, F.; Deshayes, K.; Doerner, K.; Eckhardt, S.G.; et al. Discovery of a potent small-molecule antagonist of inhibitor of apoptosis (IAP) proteins and clinical candidate for the treatment of cancer (GDC-0152). *J. Med. Chem.* **2012**, *55*, 4101–4113. [CrossRef]

50. Wong, H.; Gould, S.E.; Budha, N.; Darbonne, W.C.; Kadel, E.E., III; La, H.; Alicke, B.; Halladay, J.S.; Erickson, R.; Portera, C.; et al. Learning and confirming with preclinical studies: Modeling and simulation in the discovery of GDC-0917, an inhibitor of apoptosis proteins antagonist. *Drug Metab. Dispos.* **2013**, *41*, 2104–2113. [CrossRef]

51. Varfolomeev, E.; Blankenship, J.W.; Wayson, S.M.; Fedorova, A.V.; Kayagaki, N.; Garg, P.; Zobel, K.; Dynek, J.N.; Elliott, L.O.; Wallweber, H.J.; et al. IAP antagonists induce autoubiquitination of c-IAPs, NF-kappaB activation, and TNFalpha-dependent apoptosis. *Cell* **2007**, *131*, 669–681. [CrossRef] [PubMed]

52. Lu, J.; Bai, L.; Sun, H.; Nikolovska-Coleska, Z.; McEachern, D.; Qiu, S.; Miller, R.S.; Yi, H.; Shangary, S.; Sun, Y.; et al. SM-164: A novel, bivalent Smac mimetic that induces apoptosis and tumor regression by concurrent removal of the blockade of cIAP-1/2 and XIAP. *Cancer Res.* **2008**, *68*, 9384–9393. [CrossRef] [PubMed]

53. Fulda, S. Promises and Challenges of Smac Mimetics as Cancer Therapeutics. *Clin. Cancer Res.* **2015**, *21*, 5030–5036. [CrossRef] [PubMed]

54. Ramakrishnan, V.; Painuly, U.; Kimlinger, T.; Haug, J.; Rajkumar, S.V.; Kumar, S. Inhibitor of apoptosis proteins as therapeutic targets in multiple myeloma. *Leukemia* **2014**, *28*, 1519–1528. [CrossRef]

55. Brands, R.C.; Herbst, F.; Hartmann, S.; Seher, A.; Linz, C.; Kubler, A.C.; Muller-Richter, U.D.A. Cytotoxic effects of SMAC-mimetic compound LCL161 in head and neck cancer cell lines. *Clin. Oral. Investig.* **2016**, *20*, 2325–2332. [CrossRef]

56. Ren, K.; Ma, L.; Chong, D.; Zhang, Z.; Zhou, C.; Liu, H.; Zhao, S. Effects of LCL161, a Smac mimetic on the proliferation and apoptosis in hepatocellular carcinoma cells. *Zhong Nan Da Xue Xue Bao Yi Xue Ban* **2016**, *41*, 898–904.

57. Shekhar, T.M.; Miles, M.A.; Gupte, A.; Taylor, S.; Tascone, B.; Walkley, C.R.; Hawkins, C.J. IAP antagonists sensitize murine osteosarcoma cells to killing by TNFalpha. *Oncotarget* **2016**, *7*, 33866–33886. [CrossRef]

58. Jin, G.; Lan, Y.; Han, F.; Sun, Y.; Liu, Z.; Zhang, M.; Liu, X.; Zhang, X.; Hu, J.; Liu, H.; et al. Smac mimeticinduced caspaseindependent necroptosis requires RIP1 in breast cancer. *Mol. Med. Rep.* **2016**, *13*, 359–366. [CrossRef]

59. Jin, G.; Liu, Y.; Xu, P.; Jin, G. Induction of Necroptosis in Human Breast Cancer Drug-Resistant Cells by SMAC Analog LCL161 After Caspase Inhibition Requires RIP3. *Pharmazie* **2019**, *74*, 363–368.

60. Chen, S.M.; Lin, T.K.; Tseng, Y.Y.; Tu, C.H.; Lui, T.N.; Huang, S.F.; Hsieh, L.L.; Li, Y.Y. Targeting inhibitors of apoptosis proteins suppresses medulloblastoma cell proliferation via G2/M phase arrest and attenuated neddylation of p21. *Cancer Med.* **2018**, *7*, 3988–4003. [CrossRef]

61. Langemann, D.; Trochimiuk, M.; Appl, B.; Hundsdoerfer, P.; Reinshagen, K.; Eschenburg, G. Sensitization of neuroblastoma for vincristine-induced apoptosis by Smac mimetic LCL161 is attended by G2 cell cycle arrest but is independent of NFkappaB, RIP1 and TNF-alpha. *Oncotarget* **2017**, *8*, 87763–87772. [CrossRef] [PubMed]

62. Dougan, M.; Dougan, S.; Slisz, J.; Firestone, B.; Vanneman, M.; Draganov, D.; Goyal, G.; Li, W.; Neuberg, D.; Blumberg, R.; et al. IAP inhibitors enhance co-stimulation to promote tumor immunity. *J. Exp. Med.* **2010**, *207*, 2195–2206. [CrossRef] [PubMed]

63. Knights, A.J.; Fucikova, J.; Pasam, A.; Koernig, S.; Cebon, J. Inhibitor of apoptosis protein (IAP) antagonists demonstrate divergent immunomodulatory properties in human immune subsets with implications for combination therapy. *Cancer Immunol. Immunother.* **2013**, *62*, 321–335. [CrossRef] [PubMed]

64. Muller-Sienerth, N.; Dietz, L.; Holtz, P.; Kapp, M.; Grigoleit, G.U.; Schmuck, C.; Wajant, H.; Siegmund, D. SMAC mimetic BV6 induces cell death in monocytes and maturation of monocyte-derived dendritic cells. *PLoS ONE* **2011**, *6*, e21556. [CrossRef] [PubMed]

65. Chesi, M.; Mirza, N.N.; Garbitt, V.M.; Sharik, M.E.; Dueck, A.C.; Asmann, Y.W.; Akhmetzyanova, I.; Kosiorek, H.E.; Calcinotto, A.; Riggs, D.L.; et al. IAP antagonists induce anti-tumor immunity in multiple myeloma. *Nat. Med.* **2016**, *22*, 1411–1420. [CrossRef] [PubMed]

66. Chang, Y.C.; Kondapuram, S.K.; Yang, T.H.; Syed, S.B.; Cheng, S.M.; Lin, T.Y.; Lin, Y.C.; Coumar, M.S.; Chang, J.Y.; Leung, E.; et al. The SMAC mimetic LCL161 is a direct ABCB1/MDR1-ATPase activity modulator and BIRC5/Survivin expression down-regulator in cancer cells. *Toxicol. Appl. Pharmacol.* **2020**, *401*, 115080. [CrossRef] [PubMed]

67. Tian, A.; Wilson, G.S.; Lie, S.; Wu, G.; Hu, Z.; Hebbard, L.; Duan, W.; George, J.; Qiao, L. Synergistic effects of IAP inhibitor LCL161 and paclitaxel on hepatocellular carcinoma cells. *Cancer Lett.* **2014**, *351*, 232–241. [CrossRef]

68. Scheurer, M.J.J.; Seher, A.; Steinacker, V.; Linz, C.; Hartmann, S.; Kubler, A.C.; Muller-Richter, U.D.A.; Brands, R.C. Targeting inhibitors of apoptosis in oral squamous cell carcinoma in vitro. *J. Craniomaxillofac. Surg.* **2019**, *47*, 1589–1599. [CrossRef]

69. Najem, S.; Langemann, D.; Appl, B.; Trochimiuk, M.; Hundsdoerfer, P.; Reinshagen, K.; Eschenburg, G. Smac mimetic LCL161 supports neuroblastoma chemotherapy in a drug class-dependent manner and synergistically interacts with ALK inhibitor TAE684 in cells with ALK mutation F1174L. *Oncotarget* **2016**, *7*, 72634–72653. [CrossRef]

70. Ramakrishnan, V.; Gomez, M.; Prasad, V.; Kimlinger, T.; Painuly, U.; Mukhopadhyay, B.; Haug, J.; Bi, L.; Rajkumar, S.V.; Kumar, S. Smac mimetic LCL161 overcomes protective ER stress induced by obatoclax, synergistically causing cell death in multiple myeloma. *Oncotarget* **2016**, *7*, 56253–56265. [CrossRef]

71. Yang, L.; Kumar, B.; Shen, C.; Zhao, S.; Blakaj, D.; Li, T.; Romito, M.; Teknos, T.N.; Williams, T.M. LCL161, a SMAC-mimetic, Preferentially Radiosensitizes Human Papillomavirus-negative Head and Neck Squamous Cell Carcinoma. *Mol. Cancer Ther.* **2019**, *18*, 1025–1035. [CrossRef] [PubMed]

72. Qin, Q.; Zuo, Y.; Yang, X.; Lu, J.; Zhan, L.; Xu, L.; Zhang, C.; Zhu, H.; Liu, J.; Liu, Z.; et al. Smac mimetic compound LCL161 sensitizes esophageal carcinoma cells to radiotherapy by inhibiting the expression of inhibitor of apoptosis protein. *Tumour. Biol.* **2014**, *35*, 2565–2574. [CrossRef] [PubMed]

73. Hass, C.; Belz, K.; Schoeneberger, H.; Fulda, S. Sensitization of acute lymphoblastic leukemia cells for LCL161-induced cell death by targeting redox homeostasis. *Biochem. Pharmacol.* **2016**, *105*, 14–22. [CrossRef] [PubMed]

74. Infante, J.R.; Dees, E.C.; Olszanski, A.J.; Dhuria, S.V.; Sen, S.; Cameron, S.; Cohen, R.B. Phase I dose-escalation study of LCL161, an oral inhibitor of apoptosis proteins inhibitor, in patients with advanced solid tumors. *J. Clin. Oncol.* **2014**, *32*, 3103–3110. [CrossRef]

75. Bardia, A.; Parton, M.; Kummel, S.; Estevez, L.G.; Huang, C.S.; Cortes, J.; Ruiz-Borrego, M.; Telli, M.L.; Martin-Martorell, P.; Lopez, R.; et al. Paclitaxel With Inhibitor of Apoptosis Antagonist, LCL161, for Localized Triple-Negative Breast Cancer, Prospectively Stratified by Gene Signature in a Biomarker-Driven Neoadjuvant Trial. *J. Clin. Oncol.* **2018**, JCO2017748392. [CrossRef] [PubMed]

76. Condon, S.M.; Mitsuuchi, Y.; Deng, Y.; LaPorte, M.G.; Rippin, S.R.; Haimowitz, T.; Alexander, M.D.; Kumar, P.T.; Hendi, M.S.; Lee, Y.H.; et al. Birinapant, a smac-mimetic with improved tolerability for the treatment of solid tumors and hematological malignancies. *J. Med. Chem.* **2014**, *57*, 3666–3677. [CrossRef] [PubMed]

77. Benetatos, C.A.; Mitsuuchi, Y.; Burns, J.M.; Neiman, E.M.; Condon, S.M.; Yu, G.; Seipel, M.E.; Kapoor, G.S.; Laporte, M.G.; Rippin, S.R.; et al. Birinapant (TL32711), a bivalent SMAC mimetic, targets TRAF2-associated cIAPs, abrogates TNF-induced NF-kappaB activation, and is active in patient-derived xenograft models. *Mol. Cancer Ther.* **2014**, *13*, 867–879. [CrossRef]

78. Po Yee Mak, D.H.M.; Ruvolo, V.; Jacamo, R.; Steven, M.K.; Andreeff, M.; Bing, Z. Carter, Apoptosis suppressor ARC modulates SMAC mimetic-induced cell death through BIRC2/MAP3K14 signalling in acute myeloid leukaemia. *Br. Kournal Haematol.* **2014**, *163*, 9.

79. Eytan, D.F.; Snow, G.E.; Carlson, S.G.; Schiltz, S.; Chen, Z.; Van Waes, C. Combination effects of SMAC mimetic birinapant with TNFalpha, TRAIL, and docetaxel in preclinical models of HNSCC. *Laryngoscope* **2015**, *125*, E118–E124. [CrossRef]

80. Krepler, C.; Chunduru, S.K.; Halloran, M.B.; He, X.; Xiao, M.; Vultur, A.; Villanueva, J.; Mitsuuchi, Y.; Neiman, E.M.; Benetatos, C.; et al. The novel SMAC mimetic birinapant exhibits potent activity against human melanoma cells. *Clin. Cancer Res.* **2013**, *19*, 1784–1794. [CrossRef]

81. Ding, J.; Qin, D.; Zhang, Y.; Li, Q.; Li, Y.; Li, J. SMAC mimetic birinapant inhibits hepatocellular carcinoma growth by activating the cIAP1/TRAF3 signaling pathway. *Mol. Med. Rep.* **2020**, *21*, 1251–1257. [CrossRef] [PubMed]

82. Zakaria, Z.; Tivnan, A.; Flanagan, L.; Murray, D.W.; Salvucci, M.; Stringer, B.W.; Day, B.W.; Boyd, A.W.; Kogel, D.; Rehm, M.; et al. Patient-derived glioblastoma cells show significant heterogeneity in treatment responses to the inhibitor-of-apoptosis-protein antagonist birinapant. *Br. J. Cancer* **2016**, *114*, 188–198. [CrossRef] [PubMed]

83. McComb, S.; Aguade-Gorgorio, J.; Harder, L.; Marovca, B.; Cario, G.; Eckert, C.; Schrappe, M.; Stanulla, M.; von Stackelberg, A.; Bourquin, J.P.; et al. Activation of concurrent apoptosis and necroptosis by SMAC mimetics for the treatment of refractory and relapsed ALL. *Sci. Transl. Med.* **2016**, *8*, 339ra70. [CrossRef] [PubMed]

84. Crawford, N.; Salvucci, M.; Hellwig, C.T.; Lincoln, F.A.; Mooney, R.E.; O'Connor, C.L.; Prehn, J.H.; Longley, D.B.; Rehm, M. Simulating and predicting cellular and in vivo responses of colon cancer to combined treatment with chemotherapy and IAP antagonist Birinapant/TL32711. *Cell Death Differ.* **2018**, *25*, 1952–1966. [CrossRef] [PubMed]

85. Hamacher, R.; Schmid, R.M.; Saur, D.; Schneider, G. Apoptotic pathways in pancreatic ductal adenocarcinoma. *Mol. Cancer* **2008**, *7*, 64. [CrossRef] [PubMed]

86. Carter, B.Z.; Mak, P.Y.; Mak, D.H.; Shi, Y.; Qiu, Y.; Bogenberger, J.M.; Mu, H.; Tibes, R.; Yao, H.; Coombes, K.R.; et al. Synergistic targeting of AML stem/progenitor cells with IAP antagonist birinapant and demethylating agents. *J. Natl. Cancer Inst.* **2014**, *106*, djt440. [CrossRef]

87. Mak, D.Y.; Fraser, I.; Ferris, R.; James, K.; Liu, M.; Thomas, S.D.; McKenzie, M.; Lefresne, S. Comparison of Rapid to Standard Volumetric Modulated Arc Therapy for Palliative Radiotherapy in Lung Cancer Patients. *Cureus* **2020**, *12*, e10055. [CrossRef]

88. Eytan, D.F.; Snow, G.E.; Carlson, S.; Derakhshan, A.; Saleh, A.; Schiltz, S.; Cheng, H.; Mohan, S.; Cornelius, S.; Coupar, J.; et al. SMAC Mimetic Birinapant plus Radiation Eradicates Human Head and Neck Cancers with Genomic Amplifications of Cell Death Genes FADD and BIRC2. *Cancer Res.* **2016**, *76*, 5442–5454. [CrossRef]

89. Brumatti, G.; Ma, C.; Lalaoui, N.; Nguyen, N.Y.; Navarro, M.; Tanzer, M.C.; Richmond, J.; Ghisi, M.; Salmon, J.M.; Silke, N.; et al. The caspase-8 inhibitor emricasan combines with the SMAC mimetic birinapant to induce necroptosis and treat acute myeloid leukemia. *Sci. Transl. Med.* **2016**, *8*, 339ra69. [CrossRef]

90. Kearney, C.J.; Lalaoui, N.; Freeman, A.J.; Ramsbottom, K.M.; Silke, J.; Oliaro, J. PD-L1 and IAPs co-operate to protect tumors from cytotoxic lymphocyte-derived TNF. *Cell Death Differ.* **2017**, *24*, 1705–1716. [CrossRef]

91. Zhao, L.; Yang, G.; Bai, H.; Zhang, M.; Mou, D. NCTD promotes Birinapant-mediated anticancer activity in breast cancer cells by downregulation of c-FLIP. *Oncotarget* **2017**, *8*, 26886–26895. [CrossRef] [PubMed]

92. Lalaoui, N.; Merino, D.; Giner, G.; Vaillant, F.; Chau, D.; Liu, L.; Kratina, T.; Pal, B.; Whittle, J.R.; Etemadi, N.; et al. Targeting triple-negative breast cancers with the Smac-mimetic birinapant. *Cell Death Differ.* **2020**, *27*, 2768–2780. [CrossRef] [PubMed]

93. Orozco, C.A.; Martinez-Bosch, N.; Guerrero, P.E.; Vinaixa, J.; Dalotto-Moreno, T.; Iglesias, M.; Moreno, M.; Djurec, M.; Poirier, F.; Gabius, H.J.; et al. Targeting galectin-1 inhibits pancreatic cancer progression by modulating tumor-stroma crosstalk. *Proc. Natl. Acad. Sci. USA* **2018**, *115*, E3769–E3778. [CrossRef] [PubMed]

94. Brands, R.C.; Scheurer, M.J.J.; Hartmann, S.; Seher, A.; Kubler, A.C.; Muller-Richter, U.D.A. Apoptosis-sensitizing activity of birinapant in head and neck squamous cell carcinoma cell lines. *Oncol. Lett.* **2018**, *15*, 4010–4016. [CrossRef] [PubMed]
95. Zhu, X.; Shen, X.; Qu, J.; Straubinger, R.M.; Jusko, W.J. Proteomic Analysis of Combined Gemcitabine and Birinapant in Pancreatic Cancer Cells. *Front. Pharmacol.* **2018**, *9*, 84. [CrossRef] [PubMed]
96. Zhu, X.; Straubinger, R.M.; Jusko, W.J. Mechanism-based mathematical modeling of combined gemcitabine and birinapant in pancreatic cancer cells. *J. Pharmacokinet. Pharmacodyn.* **2015**, *42*, 477–496. [CrossRef]
97. Cekay, M.J.; Roesler, S.; Frank, T.; Knuth, A.K.; Eckhardt, I.; Fulda, S. Smac mimetics and type II interferon synergistically induce necroptosis in various cancer cell lines. *Cancer Lett.* **2017**, *410*, 228–237. [CrossRef]
98. Michie, J.; Beavis, P.A.; Freeman, A.J.; Vervoort, S.J.; Ramsbottom, K.M.; Narasimhan, V.; Lelliott, E.J.; Lalaoui, N.; Ramsay, R.G.; Johnstone, R.W.; et al. Antagonism of IAPs Enhances CAR T-cell Efficacy. *Cancer Immunol. Res.* **2019**, *7*, 183–192. [CrossRef]
99. Amaravadi, R.K.; Schilder, R.J.; Martin, L.P.; Levin, M.; Graham, M.A.; Weng, D.E.; Adjei, A.A. A Phase I Study of the SMAC-Mimetic Birinapant in Adults with Refractory Solid Tumors or Lymphoma. *Mol. Cancer Ther.* **2015**, *14*, 2569–2575. [CrossRef]
100. Senzer, N.N.; LoRusso, P.; Martin, L.P.; Schilder, R.J.; Amaravadi, R.K.; Papadopoulos, K.P.; Segota, Z.E.; Weng, D.E.; Graham, M.; Adjei, A.A. Phase II clinical activity and tolerability of the SMAC-mimetic birinapant (TL32711) plus irinotecan in irinotecan-relapsed/refractory metastatic colorectal cancer. *J. Clin. Oncol.* **2013**, *31* (Suppl. 15), 3621. [CrossRef]
101. Hu, R.; Li, J.; Liu, Z.; Miao, M.; Yao, K. GDC-0152 induces apoptosis through down-regulation of IAPs in human leukemia cells and inhibition of PI3K/Akt signaling pathway. *Tumour Biol.* **2015**, *36*, 577–584. [CrossRef] [PubMed]
102. Yang, L.; Shu, T.; Liang, Y.; Gu, W.; Wang, C.; Song, X.; Fan, C.; Wang, W. GDC-0152 attenuates the malignant progression of osteosarcoma promoted by ANGPTL2 via PI3K/AKT but not p38MAPK signaling pathway. *Int. J. Oncol.* **2015**, *46*, 1651–1658. [CrossRef] [PubMed]
103. Shekhar, T.M.; Burvenich, I.J.G.; Harris, M.A.; Rigopoulos, A.; Zanker, D.; Spurling, A.; Parker, B.S.; Walkley, C.R.; Scott, A.M.; Hawkins, C.J. Smac mimetics LCL161 and GDC-0152 inhibit osteosarcoma growth and metastasis in mice. *BMC Cancer* **2019**, *19*, 924. [CrossRef] [PubMed]
104. Tchoghandjian, A.; Souberan, A.; Tabouret, E.; Colin, C.; Denicolai, E.; Jiguet-Jiglaire, C.; El-Battari, A.; Villard, C.; Baeza-Kallee, N.; Figarella-Branger, D. Inhibitor of apoptosis protein expression in glioblastomas and their in vitro and in vivo targeting by SMAC mimetic GDC-0152. *Cell Death Dis.* **2016**, *7*, e2325. [CrossRef]
105. Mantik, P.; Xie, M.; Wong, H.; La, H.; Steigerwalt, R.W.; Devanaboyina, U.; Ganem, G.; Shih, D.; Flygare, J.A.; Fairbrother, W.J.; et al. Cyclodextrin Reduces Intravenous Toxicity of a Model Compound. *J. Pharm. Sci.* **2019**, *108*, 1934–1943. [CrossRef] [PubMed]
106. Shin, Y.G.; Jones, S.A.; Murakami, S.C.; Budha, N.; Ware, J.; Wong, H.; Buonarati, M.H.; Dean, B.; Hop, C.E. Validation and application of a liquid chromatography-tandem mass spectrometric method for the determination of GDC-0152 in human plasma using solid-phase extraction. *Biomed. Chromatogr.* **2013**, *27*, 102–110. [CrossRef]
107. Nikkhoo, A.; Rostami, N.; Farhadi, S.; Esmaily, M.; Moghadaszadeh Ardebili, S.; Atyabi, F.; Baghaei, M.; Haghnavaz, N.; Yousefi, M.; Aliparasti, M.R.; et al. Codelivery of STAT3 siRNA and BV6 by carboxymethyl dextran trimethyl chitosan nanoparticles suppresses cancer cell progression. *Int. J. Pharm.* **2020**, *581*, 119236. [CrossRef]

applied sciences

MDPI

Review

Diarylureas as Antitumor Agents

Alessia Catalano [1,*], **Domenico Iacopetta** [2], **Maria Stefania Sinicropi** [2] **and Carlo Franchini** [1]

[1] Department of Pharmacy-Drug Sciences, University of Bari Aldo Moro, 70126 Bari, Italy;
carlo.franchini@uniba.it

[2] Department of Pharmacy, Health and Nutritional Sciences, University of Calabria,
87036 Arcavacata di Rende, Italy; domenico.iacopetta@unical.it (D.I.); s.sinicropi@unical.it (M.S.S.)

* Correspondence: alessia.catalano@uniba.it; Tel.: +39-080-5442731; Fax: ++39-080-5442724

Abstract: The diarylurea is a scaffold of great importance in medicinal chemistry as it is present in numerous heterocyclic compounds with antithrombotic, antimalarial, antibacterial, and anti-inflammatory properties. Some diarylureas, serine-threonine kinase or tyrosine kinase inhibitors, were recently reported in literature. The first to come into the market as an anticancer agent was sorafenib, followed by some others. In this review, we survey progress over the past 10 years in the development of new diarylureas as anticancer agents.

Keywords: diarylureas; antitumor agents; bis-aryl ureas; hepatocellular carcinoma (HCC); renal cell carcinoma (RCC); gastrointestinal stromal tumors (GISTs); metastatic colorectal cancer (mCRC); B-cell lymphomas

check for
updates

Citation: Catalano, A.; Iacopetta, D.; Sinicropi, M.S.; Franchini, C. Diarylureas as Antitumor Agents. *Appl. Sci.* **2021**, *11*, 374. https://doi.org/10.3390/app11010374

Received: 24 November 2020
Accepted: 28 December 2020
Published: 2 January 2021

Publisher's Note: MDPI stays neutral with regard to jurisdictional claims in published maps and institutional affiliations.

Copyright: © 2021 by the authors. Licensee MDPI, Basel, Switzerland. This article is an open access article distributed under the terms and conditions of the Creative Commons Attribution (CC BY) license (https://creativecommons.org/licenses/by/4.0/).

1. Introduction

Ureas (R-NHCONH-R') are known organic compounds that possess biological activities and serve as templates for numerous medicinal chemistry researches [1]. Barbital is a diethylmalonyl urea discovered at the beginning of 1900, used as sleep aid and hypnotic [2]. In the following century, the urea scaffold has represented the pharmacophore the backbone motif for entire classes of therapeutic agents [3]. This review focuses on diarylureas, i.e., ureas substituted with two aromatic moieties also known as bis-aryl ureas. Diarylureas are found in numerous heterocyclic compounds with various biological activities [4], such as antithrombotic, antimalarial, antibacterial, antinflammatory, and anticancer [5,6]. In particular, diarylurea is a prominent pharmacophore in anticancer drugs. This activity is due to its near-perfect binding with certain acceptors. The NH moiety behaves as hydrogen bond donor and the urea oxygen atom acts as acceptor (Figure 1) [7].

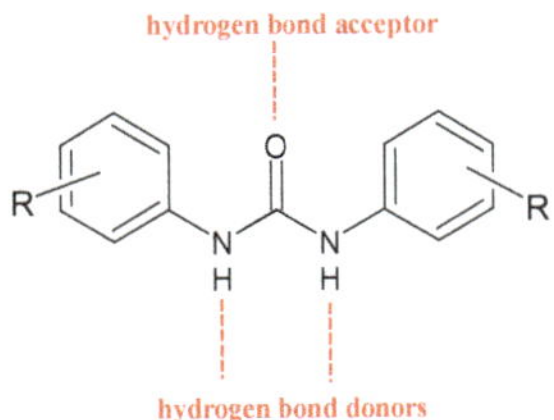

Figure 1. H-bonds in diarylureas.

This structure provides urea derivatives endowed with capability of binding several enzymes and receptors [8–10]. Moreover, it may link different pharmacophore fragments of new biological active compounds. A urea linker has been used to overcome the poor solubility of some phenyl N-mustards [11]. In this way, the authors obtained water soluble N-mustards, some of which showing high anticancer activity against various human tumor xenograft models and were able to introduce cross-linking within the DNA double strand.

Diarylurea-based compounds present strong inhibitory activity against kinases, including RAF kinases [12], platelet derived growth factor receptor (PDGF) [13], vascular endothelial growth factor receptor 2 (VEGFR-2) [14], receptor tyrosine kinase (RTKs) [15], and Aurora kinases [16]. The diarylurea moiety is, in fact, widespread in type II kinase inhibitors. These compounds circumvent kinases in an inactive state, the so-called DFG-out, and occupy a hydrophobic pocket next to the ATP-binding site. The diarylurea fragment is able to link the hinge-binding moiety with the portion that occupies the hydrophobic pocket that is in the inactive conformation of kinases [17] (Figure 2).

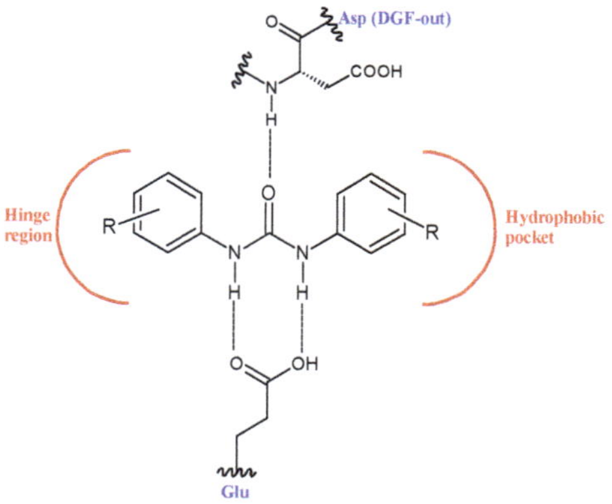

Figure 2. Diarylureas in the type II kinase inhibitor.

Diarylureas represent the skeleton of the main systemic therapies for several cancers, as advanced, metastatic hepatocellular carcinoma (HCC) [18], advanced renal cell carcinoma (RCC) [19], gastrointestinal stromal tumors (GISTs) [20], metastatic colorectal cancer (mCRC) [21]. Sorafenib is a multi-targeted small molecule tyrosine protein kinase that improves median survival over placebo for unresectable HCC patients [22]. In 2008 it obtained Food and Drug Administration (FDA) and European Medicinal Agency (EMEA) approval for the treatment of RCC and HCC [23]. Basing on sorafenib as the lead compound, several other diarylurea derivatives, such as regorafenib, linifanib, tivozanib, and ripretinib have been synthesized and evaluated as kinase inhibitors. Regorafenib was approved by the FDA in the United States in February 2013 in patients with advanced GISTs for those who had failed on imatinib and sunitinib [24,25]. Linifanib is currently being studied in HCC clinical trials [26]. Unlike the other inhibitors of VEGFR and PDGFR, linifanib seems to be also involved in adipocyte browning, thus being considered for the treatment of obesity [27]. Tivozanib is a diarylurea which is to be considered as third and fourth line therapy in patients with metastatic RCC in phase 3 study [28]. By end of 2019, the Chinese and European regulatory authorities have marketed five small-molecule protein kinase inhibitors (PKIs), including tivozanib [29]. Ripretinib has been suggested as a promising treatment for advanced GISTs [30]. In this paper, these already known drugs bearing a diarylurea skeleton are reviewed, along with new synthetic diarylureas described in the literature as promising agents for the treatment of diverse type of tumors.

2. Diarylureas in Therapy or in Clinical Studies as Anticancer Agents

2.1. Sorafenib

Sorafenib (BAY-43-9006, Nexavar®, Figure 3) is an oral receptor TKI that determines the inhibition of Raf serine/threonine kinases and receptor tyrosine kinases (VEGF 1, 2, 3 and PDGF-β, FMS-like tyrosine kinase-3 (FLT-3), and c-KIT) that are components of signaling pathways controlling tumor growth and angiogenesis [31].

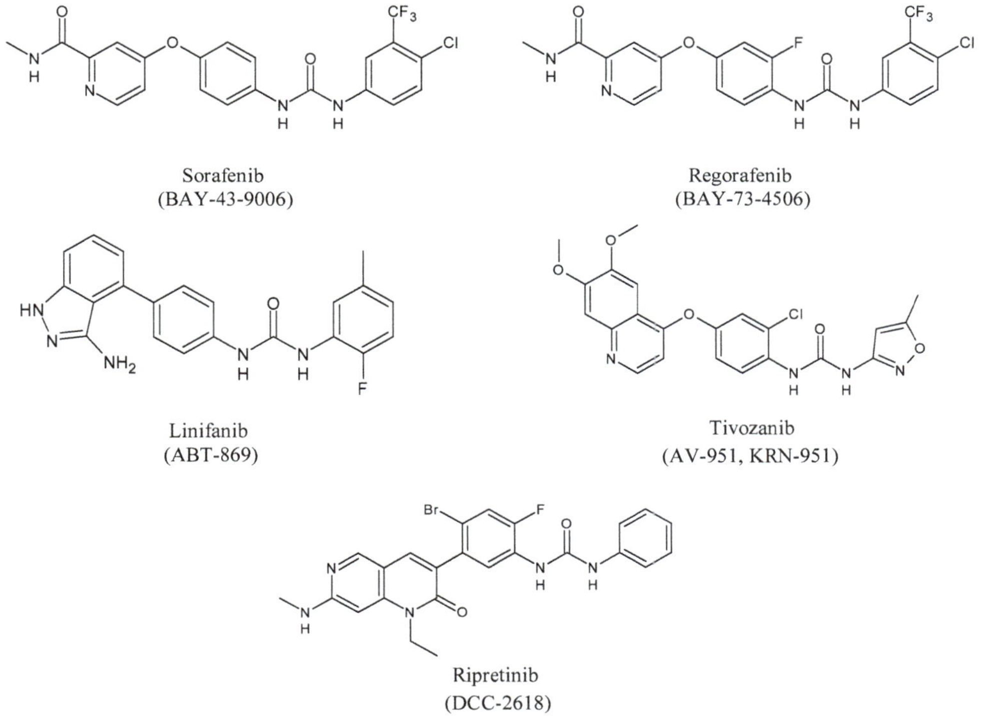

Figure 3. Structures of diarylureas.

Sorafenib inhibits the kinase activity of C-RAF and B-RAF (wild type and V600E mutant) showing IC_{50} = 6.22 and 38 nM, respectively. This compound is considered the most important drug in the late stage of injury for advanced stages of HCC [32] which is the second cause of cancer-related mortality all over the world [33]. For more than ten years, sorafenib has been the sole systemic treatment for advanced HCC [34]. However, some advanced HCC patients do not respond to therapy with sorafenib. Thus, combination studies of sorafenib with other drugs have been studied. Combined treatment with interferon-lambda 3 (IFN-λ3) and sorafenib show an effect of synergism in suppressing HCC cancer growth and in the promotion of cell apoptosis in vitro and in vivo [35]. A more recent study demonstrated that combination immunotherapy of sorafenib with atezolizumab, an immune checkpoint inhibitor (ICI) that target the programmed-cell death-1 receptor/ligand (PD-1/PD-L1) pathway, and bevacizumab, an anti-VEGF mAb, is superior to sorafenib alone as the first-line therapy of advanced HCC [36]. Moreover, given that liver function is essential for a correct prognosis, a precise rating for the safe prescription and clinical development of ICI in HCC is required. Recently, the albumin-bilirubin (ALBI) grade was used as an alternative biomarker for the prognosis [37]. The efficacy of sorafenib is limited by several factors as systematic tolerance and the poor solubility in water. Furthermore, the hydrophobicity of sorafenib is responsible of its low bioavailability as it decreases the absorption by the gastrointestinal tract. Nexavar® (Bayer Healthcare Pharmaceuticals–Onyx Pharmaceuticals) is used as tablets containing sorafenib tosylate to slightly improve the solubility. In order to increase the solubility and bioavailability of the drug, sorafenib-loaded lipid-based nanosuspensions were used [38]. The administration of sorafenib to the target cells could ameliorate patient survival and reduce the further proliferation of the tumor [39]. Thus, a drug delivery system for sorafenib has been recently studied in order to help the administration of therapies in malignant cells and raise its clinical efficacy [40]. Recently, the treatment of cancers of the gastrointestinal tract during the COVID-19 pandemic [41] has been studied: coronavirus-adapted institutional recommendations have been formulated. Sorafenib was recommended in the first-line setting only in patients with disease subtype Eastern Cooperative Oncology Group performance status (ECOG PS) 0 or 1 and Child-Pughscore A hepatocellular carcinoma [42]. In spite of its selectivity, sorafenib can determine adverse effects, such as severe respiratory and liver failure, fatigue, stomatitis, hand-foot syndrome, diarrhea, and myelosuppression, thus posing a challenge for oncologists [43]. The therapy with sorafenib might be done ad hoc to increase the therapeutic effects and reducing adverse effects [44]. Sorafenib is also studied for iodine-resistant advanced thyroid carcinoma [45]. Finally, it is important to note that sorafenib is also involved in cytoskeleton alteration that leads to cancer cells death by apoptosis. Wang et al. [46] reported that the treatment of Hep3B and PLC/PRF/5 human hepatoma cells with sorafenib induces a drastic loss of actin fibers and the redistribution of F-actin around the cell nuclei. This effect is due to the regulation of protein kinases and phosphatases that ends with cofilin dephosphorylation, which is an actin-binding factor necessary for the reorganization of actin. Chen et al. [47] demonstrated that the ability of sorafenib in inducing human prostate cancer cell line PC-3 apoptosis, through cytoskeleton destabilization, increased if combined with zinc exposure, suggesting that zinc may sensitize prostate cancer cells to sorafenib treatment. D'Alessandro et al. [48] demonstrated a synergistic effect on HCC cells migration using combined doses of sorafenib and/or vitamin K1 with insulin like growth factor I receptor (IGF1-R) antagonists enhancing the reduction and reorganization of F-actin, probably through the modulation of MAPK cascade.

2.2. Regorafenib

Regorafenib (BAY 73-4506, Stivarga®, Figure 3) is the fluorinated analogue of sorafenib. It is an orally active diphenylurea multikinase inhibitor that targets stromal (PDGFR-β, FGFR-1), angiogenic (VEGFR1-3, TIE-2), and oncogenic receptor tyrosine kinases (c-KIT, RET, and RAF-1) [49]. It is the first multi-targeting kinase inhibitor which was approved by FDA in 2012 for the treatment of mCRC patients in refractory to standard chemotherapy [50]. In addition, regorafenib treatment determined an important amelioration in progression-free survival (PFS) in comparison with placebo in patients with metastatic GISTs after standard treatments; thus, it has also received FDA-approval for this indication since 2013. Then, in 2017, FDA approved regorafenib as a therapy for patients with advanced HCC [51]. Regorafenib showed a significant amelioration of PFS and overall survival (OS) in comparison with placebo. Regorafenib has been described in phase II clinical trials in different tumors, including RCC, soft-tissue sarcoma (STS), second- and third-line treatments for medullary thyroid cancer. However, several post-marketing observational studies, after the treatment of mCRC patients showed extensive data of toxicities (CORRELATE, REBECCA, RECORA, Japanese post-marketing study) [52–54]. It was found that adverse reactions due to regorafenib frequently occurred in the initial stages of treatment, mostly in the first cycle [55]. Thus, in patients with mCRC during the first cycle of regorafenib, the use of a dose-escalation strategy treatment was suggested [56]. This strategy was then supported by a multicenter, open-label, phase II study [57]. Recently, during the COVID-19 outbreak, regorafenib has been considered as a therapy for gastrointestinal cancer [42]. A drug-delivery system has been studied for regorafenib, too [40]. The pivotal RESORCE (NCT01774344) phase III trial studied regorafenib therapy in patients with HCC who were tolerant to sorafenib, but who had progressed during sorafenib treatment [58]. Regorafenib has been demonstrated to inhibit glioblastoma multiforme (GBM) growth through PSAT1-mediated autophagy arrest [59], and to have beneficial effect in Alzheimer's disease (AD) and formation of dendritic spine in vitro and in vivo [60].

2.3. Linifanib

Linifanib (ABT-869, Abbott Laboratories, Abbott Park, IL, USA, Figure 3) is an orally available TKI which targets VEGFR and PDGFR with relevant specificity and low off-target inhibition. Linifanib can also inhibit FLT-3 [61]. It does not show significant activity against representative cytosolic tyrosine and serine/threonine kinases [62]. Linifanib is a colony-stimulating factor-1 receptor (CSF-1R) inhibitor through the inhibition of the phosphorylation of CSF-1R tyrosine kinase in transfected cells [63]. It is used as a therapy for non-small cell lung carcinoma (NSCLC), liver cancer, breast cancer, colorectal cancer [45]. Preclinical and early clinical trials showed interesting activity in various human neoplasms with a satisfactory profile of toxicity. Linifanib competes with ATP in the binding site domain of tyrosine kinase, thus it prevents downstream signaling [64]. Phase II trial studies show that linifanib is useful for the treatment of patients with advanced, refractory colorectal cancer that expresses k-Ras mutations [65]. In an open-label phase II trial linifanib showed interesting clinical activity, as monotherapy, in patients with advanced HCC [66]. Linifanib versus sorafenib was studied in terms of efficacy and tolerability. Linifanib and sorafenib showed similar OS in advanced HCC. Linifanib did not meet predefined superiority and non-inferiority OS boundaries; thus, the study did not reach the primary end point. Secondary end points, time to progression (TTP), and objective response rate (ORR), favored Linifanib; safety results favored sorafenib [67]. Although linifanib is currently examined in HCC clinical trials, it has not yet been studied in preclinical and clinical studies for gastric cancer. 5-Fluorouracil (5-FU) and cisplatin represent the first-line chemotherapy for patients with gastric cancer and the combined use with linifanib inhibits synergistically the viability of some gastric cancer cell lines and led to remarkable suppression of VEGF-induced angiogenesis in vitro and in vivo [68]. Linifanib has demonstrated to be also useful in the treatment of anaplastic thyroid cancer (ATC), that is considered the most aggressive form of thyroid cancer. The synergistic use of linifanib and irinotecan significantly increased the survival of ATC-affected mice. These observations have been made by using an orthotopic in vivo model that better recapitulates features of human tumors than the more simplistic subcutaneous xenograft models, suggesting a potential role of this co-treatment in ATC patient's treatment [59]. Finally, linifanib was demonstrated to interfere with adipocyte browning. It suppresses STAT3 signaling pathway, thus leading to the enhancement of adipocyte browning and inhibition of adipogenesis. Linifanib's blocking browning effect was demonstrated as the phosphorylation of STAT3 was reduced by linifanib and the STAT3 activator SD19, as well [27].

2.4. Tivozanib

Tivozanib (AV-951, KRN-951, FOTIVDA®, Figure 3), used as the hydrochloride monohydrate salt, is a bioavailable inhibitor of angiogenesis which targets VEGFR tyrosine kinases with high antitumor activity. It is a VEGF-TKI specific for VEGFR1–3, showing an inhibitor effect at nanomolar concentrations, with IC50 values of 30 nM, 6.5 nM, and 15 nM for VEGFR1, 2 and 3, respectively. The compound is unique in that it is highly specific for VEGFR1–3, and presents minimal residual effects on c-KIT and PDGFR-β [70]. It presents a long half-life, too [71]. It has shown considerable efficacy for the treatment of advanced RCC over the past decade. In August 2017, tivozanib was approved by the EMEA as a first-line therapy for patients with advanced RCC and those who are VEGFR and mTOR pathway inhibitor-naïve following disease progression after previous therapy with cytokines for advanced RCC. Tivozanib was compared with sorafenib in a phase III trial for patients with metastatic RCC. Tivozanib improved PFS, but not OS, and showed a differentiated safety profile, in comparison with sorafenib, as initial targeted therapy for metastatic RCC [72]. Preclinical data and phase III trials of tivozanib in RCC, TIVO-1, and TIVO-3 have been recently summarized. Given the agent's excellent tolerability profile it is appropriate for those patients with heavily pretreated disease that could exhibit clinical deterioration. Currently, the standard therapy is represented by nivolumab and ipilimumab, followed by cabozantinib. Tivozanib may represent a third-line treatment after failure of these agents [73]. The results of phase III TIVO-3 trial (American Society of Clinical Oncology Virtual Scientific Program, 2020) showed that tivozanib significantly improved PFS, compared with sorafenib, in patients with highly relapsed or refractory metastatic

RCC [74–77]. Tivozanib activity has been also investigated in hepatocellular carcinoma in association with durvalumab [78] and in recurrent, platinum-resistant ovarian cancer, fallopian tube cancer, and primary peritoneal cancer [79,80]. Tivozanib has been studied in phase I and II clinical trials as monotherapy and in combination with other drugs for the treatment of STS [81], glioblastoma [82], breast [83], and colorectal cancers, and other advanced gastrointestinal cancers [84,85].

2.5. Ripretinib

Ripretinib (DCC-2618, QINLOCK^{TM}, Figure 3) is an oral inhibitor of tyrosine kinase that primarily inhibits KIT proto-oncogene receptor tyrosine kinase and platelet-derived growth factor receptor A (PDGFRA) kinase signaling. Ripretinib also inhibits other kinases, such as PDGFRB, TIE2, VEGFR2, and BRAF. It was designed for cancers and myeloproliferative neoplasms, especially GISTs. Ripretinib is a "switch-control" kinase inhibitor that forces the activation loop (or activation "switch") into the inactive conformation. In preclinical cancer models it has shown efficacy, and preliminary clinical data show that Ripretinib inhibits a broad range of KIT mutants in patients with drug-resistant GISTs [86]. The INVICTUS study demonstrated the efficacy and safety of ripretinib as the fourth-line treatment *versus* placebo in patients with advanced GISTs [87]. In May 2020, ripretinib received approval from the US FDA for the treatment of patients with advanced GISTs who had received previous treatment with more than two kinase inhibitors [88]. Ripretinib is being evaluated in an ongoing phase III study (INTRIGUE) as a second-line therapy in comparison with sunitinib after progressing on imatinib [89]. Recently, ripretinib is being investigated in clinical trials for systemic mastocytosis (SM) [90], and has been also proposed for the treatment of STS [91].

2.6. Mechanisms of Inhibition of Diarylureas

The proposed inhibitory mechanisms of diarylureas depend on their structure. Garuti et al. [5] reported the crystal structure of ^{V600E}B-RAF kinase domains in complex with sorafenib. The pyridyl ring is shown to occupy the ATP adenine-binding pocket and to interact with three amino acids residues. The trifluoromethyl phenyl, that is a lipophilic moiety, fits into a hydrophobic pocket. The urea moiety forms two hydrogen bonds with ^{V600E}B-RAF, one with the aspartate, and one with the glutamate residue. Recently, the 2D interaction of the co-crystallized sorafenib inside the active site of B-Raf has been reported [92]. Regorafenib differs from sorafenib only for a fluorine atom, thus its interactions are similar to those of sorafenib. Chen et al. (2017) proposed an alternative mechanism for colorectal cancer for regorafenib. It seems that it interacts with microRNA-21 (miR-21), an oncogenic miRNA which plays a crucial role in resisting programmed cell death in CRC cells. RNA–ligand docking, molecular dynamics simulation showed that regorafenib can directly bind to miR-21 pre-element [93]. Docking studies of linifanib with FLT3 were recently reported, evidencing that 3-amino-indazole interacts with the ATP-bind site [61]. Kajal et al. (2018) has reported the 2D co-crystal-binding conformation of VEGFR2-Tivozanib, in which tivozanib mimics the binding pattern of ATP [94]. Finally, studies on the mechanism of action of ripretinib have been recently reported [86].

3. Other Diarylureas

In this paragraph several studies on other diarylureas are described (Table 1). Babić et al. synthesized several diarylurea derivatives in order to study their cytostatic activity [95]. The compounds were tested on tumor cell lines: HCT 116 (colon carcinoma), SW 620 (colon carcinoma), MCF-7 (breast carcinoma), H460 (lung carcinoma), L1210 (murine leukemia), CEM (human lymphoma), and HeLa (cervix carcinoma). Compounds **1a–e** exerted the highest effect (IC$_{50}$ from 1 to 4.3 µM, with an average of 2.6 $\pm$ 1.6 µM) even though with low selectivity for the different tumor cell lines. The compounds were also cytostatic against primary human embryonic lung (HEL) fibroblast cells. Kapuriya et al. [11] studied a series of water-soluble N-mustard-benzene conjugates containing a urea linker. The urea linker was introduced in order to overcome the low solubility of compounds previously studied. The authors studied a series of water-soluble *N*-mustards, in which the phenyl *N*-mustard is linked to a benzene ring through a urea linker. In particular, the diarylurea **2** (BO-1055), as the hydrochloride salt, exhibited high in vitro cytotoxicity and therapeutic efficacy against various human tumor cell lines. It was demonstrated to possess potent therapeutic effect against several human solid tumor cell lines, including human breast cancer (MX-1), colon cancer (HCT-116), and prostate cancer (PC3), in xenograft model. The DNA repair capacity of compound BO-1055, named ureidomustin, was then studied. It was proposed for the treatment of tumors with deficient nucleotide excision repair (NER), homologous recombination (HR), and O^6-methylguanine-DNA methyltransferase (MGMT) DNA repair genes, or in synergy with other drugs in tumors in which DNA damage response has been repressed [96].BO-1055 was also proposed as a therapeutic agent for Ewing sarcoma and rhabdomyosarcoma given its potency and relative lack of toxicity against normal tissue [97]. It also showed a potent activity against B-cell lymphomas, as mantle cell lymphoma (MCL) and diffuse large B-cell lymphoma (DLBCL) [98].

In a following paper, the same research group indicated that compound **2** has a quite narrow therapeutic window; thus, following a bioisosteric approach, an inversion of the carboxamide functionality was addressed. Compound **3** was superior to compound **2** against colon cancer (HCT-116) and lung cancer (H460) cell lines, and displayed minor toxicity. Cotreatment of compound **3** and 5-fluorouracil suppressed the growth of HCT-116 xenografts. Moreover, compound **3** could induce DNA cross-linking and cell-cycle arrest at the G2/M phase. This compound was selected for early preclinical studies [99]. A series of diarylureas was studied for in vitro antiproliferative activities against HepG2, MGC-803, and A549 cancer cell lines [100]. Compound **4** displayed optimal antiproliferative activity against the three cell lines in comparison with sorafenib and gefitinib. Indeed, it induced A549 cells apoptosis through the cell cycle block at the G0/G1 phase, the increase of intracellular reactive oxygen species, and the reduction of mitochondrial membrane potential. This compound also influenced the Raf/MEK/ERK pathway. A series of diarylurea derivatives was studied for its cytotoxicity in vitro against H-460, HT-29, A549, and MDA-MB-231 cancer cell lines [101]. Some of them showed higher activity than sorafenib (IC_{50} between 0.089 and 5.46 µM). In particular, compound **5** was the most potent both in cellular (IC_{50} = 0.15, 0.089, 0.36, and 0.75 µM, respectively) and enzymatic assay (IC_{50} = 56 nM against EGFR). The antiproliferative activity of diarylureas bearing a 4-anilinoquinazoline group was evaluated via MTT assay against A431 and A549 cells [102]. Three compounds showed high antiproliferative activities and their inhibitory activity against EGFR-TK was evaluated. Compound **6** was a potent EGFR-TK inhibitor. This completely inhibited cancer growth in established nude mouse A549 xenograft model in vivo, at 50 mg/kg. Diarylureas were studied as LIM-kinase (Limk) inhibitors for their therapeutic potential against prostate cancers [103]. Limk is a serine-threonine protein kinase existing in two isoforms, LIM kinase 1 (Limk1) and LIM kinase 2 (Limk2). The inhibition of Limk1 activity in cancer prostate cells and tissues determines reduction of phosphorylated cofilin and cancer cells motility, thus reducing invasiveness of the tumor and evolution to metastasis. The substituted diarylurea **7**, at 1 µM, inhibited only Limk1 and STK16 with $\geq$80% inhibition. The use of Limk inhibitors has been also suggested to target the invasive machinery in GBM [104]. Recently, a diarylurea, N69B, was evaluated for its anticancer activity and its molecular mechanism was investigated [105]. The compound was shown to inhibit proliferation of murine and human cancer cells in vitro, and reduce tumor growth in mouse 4T1 breast tumor model in vivo. Compound N69B significantly increased protein levels of cathepsins, in particular cathepsin D, a lysosomal aspartyl protease with various biological functions. Several diarylureas bearing a coumarine moiety [106,107] have been recently tested for their in vitro antiproliferative activities against the H4IIE and HepG2 cancer cell lines, and has been proposed as a promising lead for further optimization [108]. Compound **8b** exhibited a higher inhibition of H4IIE cells compared to sorafenib. **8a** also showed a better inhibition against HepG2 cells than sorafenib. In particular, **8b** arrested cell cycle at the S phase and induced H4IIE cells apoptosis. A library of diarylureas has been designed and the in vitro antiproliferative activities was studied against HT-29 and A549 cancer cell lines. Compound **9** was the most active against HT-29 cells showing an IC_{50} value of 3.38 μM, compared to that of sorafenib (IC_{50} = 17.28 μM). It induced cell cycle arrest at G_0/G_1 phase, interfered with Raf/MEK/ERK signaling pathway, increased intracellular reactive oxygen species level, and led to HT-29 cells apoptosis [109]. The same research group studied a series of benzo[*b*]thiophene-diarylureas with potential anticancer effects, too. Compound **10** was the most active (IC_{50} = 5.91 and 14.64 μM on HT-29 and A549 cells, respectively). It induced apoptosis and cell cycle arrest at the G0/G1 phase on HT-29 cells, too [110]. Several diphenyl indazoles, containing diarylurea moieties, in the low micromolar range, inhibited cell viability of various cancer cell lines including murine metastatic breast cancer 4T1, murine glioma GL261-luc2, human triple negative breast cancer MDA-MB-231, human pancreatic cancer MIAPaCa-2, and human colorectal adenocarcinoma WiDr. The lead candidate **11** significantly reduced the tumor growth in aggressive stage IV breast cancer 4T1 syngraft model in vivo [111]. A series of diarylureas bearing a substituted thiadiazole as one of the two aryl moieties was studied against human chronic myeloid leukemia (CML) cell line K562. The diarylurea **12** exhibited the least cytotoxicity and higher biological activity (IC_{50} = 0.038 μM). It also displayed good induced-apoptosis effect for human CML cell line K562; its effect seems to happen via a significant reduction of protein phosphorylation of PI3K/AKT signal pathway by human phospho-kinase array analysis [112]. Forchlorfenuron (FCF; N-(2-Chloro-4-pyridyl)-N′-phenylurea) is a small synthetic diarylurea currently used in agriculture as a plant fertilizer that increases fruit size because of its potent cytokinin activity. FCF inhibits proliferation, anchorage-independent growth, migration, and invasion of cancer cell lines in various cancer types, such as prostate, mesothelioma, lung, colon, breast, ovary, and cervix [113]. FCF was also found to be effective in a mouse model, in which tumor growth was inhibited. FCF treatment caused the suppression of HIF-1α and HER2, both of them playing a crucial role in cancer cell survival [114]. Recently, several FCF analogues (UR214-1, UR214-7, and UR214-9) were demonstrated to be more effective in decreasing viability and proliferation in both ovarian and endometrial cancer cell lines, and suppress HER2 expression at a concentration lower than that of FCF. Moreover, FCF and its analogues were found to decrease the expression of human epididymis protein 4 (HE4), which is commonly upregulated in ovarian and endometrial cancers [115]. Diarylurea PQ401 is a small molecule that behaves as an inhibitor of IGF-1R signaling. It is also able to prevent breast cancer cells growth in in vivo mouse models [116]. It has also shown anti-cancer properties in glioma by inducing cellular apoptosis in U87MG cells, thus reducing cell viability and proliferation and attenuating cell mobility in vitro. Moreover, through a mouse xenograft model, PQ401 administration led to the suppression of glioma tumor growth in vivo in mice [117]. Recently, PQ401 potential as a putative chemotherapy drug in osteosarcoma cells has been investigated. PQ401 effectively suppressed osteosarcoma cell growth, migration, and colony formation in vitro, as well as induced apoptosis in vitro. PQ401 inhibited U2OS cell viability almost as effective as *cis*platin. PQ401 can significantly cause U2OS cell apoptosis and clonogenesis at the IC_{50} concentration with the blockade of IGF1-R phosphorylation and related downstream signaling [118]. The diphenyl urea-derivative DUD was designed on the basis of a docking study for the optimization of a natural product, taspine. The anti-metastatic potential of DUD for NSCLC was studied in vitro. DUD inhibited A549 cells migration by reversing EMT via Wnt/β-catenin and PI3K/Akt signaling, thus it has been suggested as a potential therapy for NSCLC treatment [119]. Several fluorinated diarylureas were studied as activators of adenosine monophosphate-activated kinase (AMPK). Compound FND-4b determined the induction of phosphorylated AMPK and the decrease in markers of cell proliferation, as cyclin D1, in all CRC cell lines. Apoptosis was also increased in CRC cells treated with FND-4b [120]. Thidiazuron (TDZ, 1-phenyl-3-(1,2,3-thiadiazol-5-yl) urea) is a synthetic plant hormone which has been widely used as herbicide, pesticide, and as growth regulator in plant tissue culture [121]. Given its cytotoxic effect on HeLa (human cervical carcinoma) cell lines, it has been recently proposed as a potential agent to act against cervical cancer cells. It has also suggested to have a role on apoptosis in cancer cells through DNA damage. Furthermore, the activity of TDZ as anticancer was tested against Hela cells by mitochondrial dysfunction, DNA damage, in silico caspase-3 inhibition, and some gene expression [122]. This observation has been recently confirmed by radiolabeling TDZ with 99mTc. The in silico study supported the ability of 99mTc-TDZ complex to bind caspase-3 protein that is overexpressed in cancers, suggesting that 99mTc-TDZ might be a potential agent for diagnosis of solid tumors, such as the cervix cancer [123].

Table 1. Structures of compounds described in the literature.

Structure	Compd	Ref
a: R = cyclophenyl b: R = cyclohexyl c: R = cyclohexylmethyl d: R = benzyl e: R = phenylethyl	**1a–e**	[95]
	2 (BO-1055)	[11]
	3	[99]
	4	[100]
	5	[101]
	6	[102]
	7	[103]
	N69B	[105]

Table 1. Cont.

Structure	Compd	Ref
a: $R = NO_2$ **b:** $R = CF_3$	**8a,b**	[106–108]
	9	[109]
	10	[110]
	11	[111]
	12	[112]
	Forchlorfenuron (FCF)	[113,114]
UR214-1: $R^1 = R^4 = H$; $R^2 = S\text{-}CF_3$; $R^3 = Cl$ **UR214-7:** $R^1 = F$; $R^2 = CF_3$; $R^3 = R^4 = H$ **UR214-9:** $R^1 = F$; $R^2 = CF_3$; $R^3 = H$; $R^4 = Cl$	UR214-1 UR214-7 UR214-9	[115]
	PQ401	[116–118]
	DUD	[119]
	FND-4b	[120]
	Thidiazuron (TDZ)	[121–123]

4. Summary

Diarylureas are considered a privileged structure in medicinal chemistry, particularly for anticancer drugs. Sorafenib is a neovascular blocker that prevents the formation of new blood vessels, followed by the growth of cancer tissue, through multiple kinase inhibitors that target angiogenesis. It is approved for the treatment of advanced inoperable HCC and advanced RCC. An alternative for the treatment of these tumors under study may be represented by tivozanib. Regorafenib is an effective therapy for patients with advanced GSTIs or mCRC. Linifanib may represent a promising therapeutic agent for human gastric cancer, NSCLC, liver cancer, breast cancer, colorectal cancer. Ripretinib is addressed to GISTs. In this paper, we report an overview of the development and application of these drugs. The current treatment trends in oncology have shifted to immunotherapy combinations with ICI, as anti-PD-L1-directed monoclonal antibodies. Pending improved understanding of HCC, RCC, GSTIs, and mCRC tumorigenesis, it would be very interesting to evaluate the combination of various treatment modalities. Diarylurea combined with a checkpoint inhibitor could be a promising treatment strategy to be deeply investigated in the future. Moreover, this review encompasses the recent advances in scientific literature in the broad area of diarylureas as anticancer agents. The newly synthesized compounds of this class, that are now in phase of study, may represent promising small molecules able to unseat or help the already known existing drugs.

Funding: This research received no external funding.

Conflicts of Interest: The authors declare no conflict of interest.

Abbreviations

AD	Alzheimer's disease
AMPK	adenosine monophosphate-activated kinase
ATC	Anaplastic thyroid cancer
CML	chronic myeloid leukemia
CSF-1R	colony stimulating factor-1 receptor
DLBCL	diffuse large B-cell lymphoma
ECOG PS	Eastern Cooperative Oncology Group performance status
EMEA	European Medicinal Agency
FCF	Forchlorfenuron
FDA	Food and Drug Administration
FLT-3	FMS-like tyrosine kinase-3
5-FU	5-Fluorouracil
GBM	glioblastoma multiform
GIST	gastrointestinal stromal tumors
HCC	hepatocellular carcinoma
HE4	human epididymis protein 4
HEL	human embryonic lung
HR	homologues recombination
ICI	Immune checkpoint inhibitors
IFN-λ3	Interferon-lambda 3
IGF1-R	insulin like growth factor I receptor
Limk	LIM-kinase
MCL	mantle cell lymphoma
mCRC	metastatic colorectal cancer
MGMT	O^6-methylguanine-DNA methyltransferase
NER	nucleotide excision repair
NSCLC	non-small cell lung carcinoma
ORR	objective response rate
OS	overall survival
PD-1/PD-L1	programmed-cell death-1 receptor/ligand
PDGFR	platelet derived growth factor receptor

PFS	progression-free survival
PKIs	protein kinase inhibitors
RCC	renal cell carcinoma
RTKs	receptor tyrosine kinases
SM	systemic mastocytosis
STS	soft-tissue sarcoma
TDZ	Thidiazuron
TKIs	tyrosine protein kinases
TTP	time to progression
VEGFR-2	vascular endothelial growth factor receptor 2

References

1. Ghosh, A.K.; Brindisi, M. Urea derivatives in modern drug discovery and medicinal chemistry. *J. Med. Chem.* **2019**, *63*, 2751–2788. [CrossRef] [PubMed]

2. López-Muñoz, F.; Ucha-Udabe, R.; Alamo, C. The history of barbiturates a century after their clinical introduction. *Neuropsychiatr. Dis. Treat.* **2005**, *1*, 329–343. [PubMed]

3. Jagtap, A.D.; Kondekar, N.B.; Sadani, A.A.; Chern, J.W. Ureas: Applications in Drug Design. *Curr. Med. Chem.* **2017**, *24*, 622–651. [CrossRef] [PubMed]

4. Asif, M. Short Notes on Diaryl Ureas Derivatives. *J. Adv. Res. BioChem. Pharmacol.* **2018**, *1*, 38–41.

5. Garuti, L.; Roberti, M.; Bottegoni, G.; Ferraro, M. Diaryl urea: A privileged structure in anticancer agents. *Curr. Med. Chem.* **2016**, *23*, 1528–1548. [CrossRef]

6. Ceramella, J.; Mariconda, A.; Rosano, C.; Iacopetta, D.; Caruso, A.; Longo, P.; Sinicropi, M.S.; Saturnino, C. α–ω Alkenyl-bis-S-guanidine thiourea dihydrobromide affects HeLa cell growth hampering tubulin polymerization. *ChemMedChem* **2020**. [CrossRef]

7. Wu, Y.C.; Ren, X.Y.; Rao, G.W. Research Progress of Diphenyl Urea Derivatives as Anticancer Agents and Synthetic Methodologies. *Mini Rev. Org. Chem.* **2019**, *16*, 617–630. [CrossRef]

8. Rizza, P.; Pellegrino, M.; Caruso, A.; Iacopetta, D.; Sinicropi, M.S.; Rault, S.; Lancelot, J.C.; El-Kashef, H.; Lesnard, A.; Rochais, C.; et al. 3-(Dipropylamino)-5-hydroxybenzofuro [2, 3-f] quinazolin-1 (2H)-one (DPA-HBFQ-1) plays an inhibitory role on breast cancer cell growth and progression. *Eur. J. Med. Chem.* **2016**, *107*, 275–287. [CrossRef]

9. Saturnino, C.; Barone, I.; Iacopetta, D.; Mariconda, A.; Sinicropi, M.S.; Rosano, C.; Campana, A.; Catalano, S.; Longo, P.; Andò, S. N-heterocyclic carbene complexes of silver and gold as novel tools against breast cancer progression. *Fut. Med. Chem.* **2016**, *8*, 2213–2229. [CrossRef]

10. Iacopetta, D.; Grande, F.; Caruso, A.; Mordocco, R.A.; Plutino, M.R.; Scrivano, L.; Ceramella, J.; Miuà, N.; Saturnino, C.; Puoci, F.; et al. New insights for the use of quercetin analogs in cancer treatment. *Fut. Med. Chem.* **2017**, *9*, 2011–2028. [CrossRef]

11. Kapuriya, N.; Kakadiya, R.; Dong, H.; Kumar, A.; Lee, P.-C.; Zhang, X.; Chou, T.-C.; Lee, T.-C.; Chen, C.-H.; Lam, K.; et al. Design, synthesis, and biological evaluation of novel water-soluble N-mustards as potential anticancer agents. *Bioorg. Med. Chem.* **2011**, *19*, 471–485. [CrossRef] [PubMed]

12. El-Nassan, H.B. Recent progress in the identification of BRAF inhibitors as anticancer agents. *Eur. J. Med. Chem.* **2014**, *72*, 170–205. [CrossRef] [PubMed]

13. Ravez, S.; Barczyk, A.; Six, P.; Cagnon, A.; Garofalo, A.; Goossens, L.; Depreux, P. Inhibition of tumor cell growth and angiogenesis by 7-aminoalkoxy-4-aryloxy-quinazoline ureas, a novel series of multi-tyrosine kinase inhibitors. *Eur. J. Med. Chem.* **2014**, *79*, 360–381. [CrossRef] [PubMed]

14. Wang, C.; Gao, H.; Dong, J.; Zhang, Y.; Su, P.; Shi, Y.; Zhang, J. Biphenyl derivatives incorporating urea unit as novel VEGFR-2 inhibitors: Design, synthesis and biological evaluation. *Bioorg. Med. Chem.* **2014**, *22*, 277–284. [CrossRef] [PubMed]

15. Lu, Y.Y.; Zhao, C.R.; Wang, R.Q.; Li, W.-B.; Qu, X.-J. A novel anticancer diarylurea derivative HL-40 as a multi-kinases inhibitor with good pharmacokinetics in Wistar rats. *Biomed. Pharmacother.* **2015**, *69*, 255–259. [CrossRef]

16. Curtin, M.L.; Frey, R.R.; Heyman, H.R.; Soni, N.B.; Marcotte, P.A.; Pease, L.J.; Glaser, K.B.; Magoc, T.J.; Tapang, P.; Albert, D.H.; et al. Thienopyridine ureas as dual inhibitors of the VEGF and Aurora kinase families. *Bioorg. Med. Chem. Lett.* **2012**, *22*, 3208–3212. [CrossRef]

17. Ceramella, J.; Iacopetta, D.; Barbarossa, A.; Caruso, A.; Fedora, G.; Bonomo, M.G.; Mariconda, A.; Longo, P.; Saturnino, C.; Sinicropi, M.S. Carbazole derivatives as kinase-targeting inhibitors for cancer treatment. *Mini Rev. Med. Chem.* **2020**, *20*, 444–465. [CrossRef]

18. Tella, S.H.; Kommalapati, A.; Mahipal, A. Systemic therapy for advanced hepatocellular carcinoma: Targeted therapies. *Chin. Clin. Oncol.* **2020**. [CrossRef]

19. Escudier, B.; Worden, F.; Kudo, M. Sorafenib: Key lessons from over 10 years of experience. *Exp. Rev. Anticanc. Ther.* **2019**, *19*, 177–189. [CrossRef]

20. Mazzocca, A.; Napolitano, A.; Silletta, M.; Spalato Ceruso, M.; Santini, D.; Tonini, G.; Vincenzi, B. New frontiers in the medical management of gastrointestinal stromal tumours. *Ther. Adv. Med. Oncol.* **2019**, *11*. [CrossRef]

21. Strumberg, D.; Scheulen, M.E.; Schultheis, B.; Richly, H.; Frost, A.; Büchert, M.; Christensen, O.; Jeffers, M.; Heinig, R.; Boix, O.; et al. Regorafenib (BAY 73-4506) in advanced colorectal cancer: A phase I study. *Br. J. Canc.* **2012**, *106*, 1722–1727. [CrossRef] [PubMed]

22. Llovet, J.M.; Ricci, S.; Mazzaferro, V.; Hilgard, P.; Gane, E.; Blanc, J.F.; De Oliveira, A.C.; Santoro, A.; Raoul, J.-L.; Forner, A.; et al. Sorafenib in advanced hepatocellular carcinoma. *N. Engl. J. Med.* **2008**, *359*, 378–390. [CrossRef] [PubMed]

23. Nagai, H.; Mukozu, T.; Ogino, Y.U.; Matsui, D.; Matsui, T.; Wakui, N.; Momiyama, K.; Igarashi, Y.; Sumino, Y.; Higai, K. Sorafenib and hepatic arterial infusion chemotherapy for advanced hepatocellular carcinoma with portal vein tumor thrombus. *Anticancer Res.* **2015**, *35*, 2269–2277. [PubMed]

24. Sirohi, B.; Philip, D.S.; Shrikhande, S.V. Regorafenib in gastrointestinal stromal tumors. *Future Oncol.* **2014**, *10*, 1581–1587. [CrossRef] [PubMed]

25. Scrivano, L.; Parisi, O.I.; Iacopetta, D.; Ruffo, M.; Ceramella, J.; Sinicropi, M.S.; Puoci, F. Molecularly imprinted hydrogels for sustained release of sunitinib in breast cancer therapy. *Pol. Adv. Technol.* **2019**, *30*, 743–748. [CrossRef]

26. Mossenta, M.; Busato, D.; Baboci, L.; Di Cintio, F.; Toffoli, G.; Dal Bo, M. New insight into therapies targeting angiogenesis in hepatocellular carcinoma. *Cancers* **2019**, *11*, 1086. [CrossRef]

27. Zhao, S.; Chu, Y.; Zhang, Y.; Zhou, Y.; Jiang, Z.; Wang, Z.; Mao, L.; Li, K.; Sun, W.; Li, P.; et al. Linifanib exerts dual anti-obesity effect by regulating adipocyte browning and formation. *Life Sci.* **2019**, *222*, 117–124. [CrossRef]

28. Jacob, A.; Shook, J.; Hutson, T.E. Tivozanib, a highly potent and selective inhibitor of VEGF receptor tyrosine kinases, for the treatment of metastatic renal cell carcinoma. *Future Oncol.* **2020**, *16*, 2147–2164. [CrossRef]

29. Bournez, C.; Carles, F.; Peyrat, G.; Aci-Sèche, S.; Bourg, S.; Meyer, C.; Bonnet, P. Comparative assessment of protein kinase inhibitors in public databases and in PKIDB. *Molecules* **2020**, *25*, 3226. [CrossRef]

30. Nemunaitis, J.; Bauer, S.; Blay, J.Y.; Choucair, K.; Gelderblom, H.; George, S.; Schöffski, P.; von Mehren, M.; Zalcberg, J.; Achour, H.; et al. Intrigue: Phase III study of ripretinib versus sunitinib in advanced gastrointestinal stromal tumor after imatinib. *Future Oncol.* **2020**, *16*, 4251–4264. [CrossRef]

31. Iyer, R.; Fetterly, G.; Lugade, A.; Thanavala, Y. Sorafenib: A clinical and pharmacologic review. *Exp. Opin. Pharmacother.* **2010**, *11*, 1943–1955. [CrossRef] [PubMed]

32. Raoul, J.L.; Kudo, M.; Finn, R.S.; Edeline, J.; Reig, M.; Galle, P.R. Systemic therapy for intermediate and advanced hepatocellular carcinoma: Sorafenib and beyond. *Canc. Treat. Rev.* **2018**, *68*, 16–24. [CrossRef] [PubMed]

33. Qiu, X.; Li, M.; Wu, L.; Xin, Y.; Mu, S.; Li, T.; Song, K. Severe Fatigue is an Important Factor in the Prognosis of Patients with Advanced Hepatocellular Carcinoma Treated with Sorafenib. *Cancer Manag. Res.* **2020**, *12*, 7983. [CrossRef] [PubMed]

34. Rich, N.E.; Yopp, A.C.; Singal, A.G. Medical Management of Hepatocellular Carcinoma. *J. Oncol. Pract.* **2017**, *13*, 356–364. [CrossRef] [PubMed]

35. Yan, Y.; Wang, L.; He, J.; Liu, P.; Lv, X.; Zhang, Y.; Xu, X.; Zhang, L.; Zhang, Y. Synergy with interferon-lambda 3 and sorafenib suppresses hepatocellular carcinoma proliferation. *Biomed. Pharmacother.* **2017**, *88*, 395–402. [CrossRef] [PubMed]

36. Cheng, A.-L.; Qin, S.; Ikeda, M.; Galle, P.; Ducreux, M.; Zhu, A.; Kim, T.-Y.; Kudo, M.; Breder, V.; Merle, P.; et al. LBA3IMbrave150: Efficacy and safety results from a ph III study evaluating atezolizumab (atezo) + bevacizumab (bev) vs sorafenib (Sor) as first treatment (tx) for patients (pts) with unresectable hepatocellular carcinoma (HCC). *Ann. Oncol.* **2019**, *30*, ix186–ix187. [CrossRef]

37. Pinato, D.J.; Kaneko, T.; Saeed, A.; Pressiani, T.; Kaseb, A.; Wang, Y.; Szafron, D.; Jun, T.; Dharmapuri, S.; Naqash, A.R.; et al. Immunotherapy in hepatocellular cancer patients with mild to severe liver dysfunction: Adjunctive role of the ALBI grade. *Cancers* **2020**, *12*, 1862. [CrossRef]

38. Zhang, N.; Zhang, B.; Gong, X.; Wang, T.; Liu, Y.; Yang, S. In vivo biodistribution, biocompatibility, and efficacy of sorafenib-loaded lipid-based nanosuspensions evaluated experimentally in cancer. *Int. J. Nanomed.* **2016**, *11*, 2329–2343. [CrossRef]

39. Guo, Y.; Zhong, T.; Duan, X.-C.; Zhang, S.; Yao, X.; Yin, Y.-F.; Huang, D.; Ren, W.; Zhang, Q.; Zhang, X. Improving anti-tumor activity of sorafenib tosylate by lipid- and polymer-coated nanomatrix. *Drug Deliv.* **2017**, *24*, 270–277. [CrossRef]

40. Srimathi, U.; Nagarajan, V.; Chandiramouli, R. Investigation on graphdiyne nanosheet in adsorption of sorafenib and regorafenib drugs: A DFT approach. *J. Mol. Liq.* **2019**, *277*, 776–785. [CrossRef]

41. Catalano, A. COVID-19: Could Irisin Become the Handyman Myokine of the 21st Century? *Coronaviruses* **2020**, *1*, 32–41. [CrossRef]

42. Pietrantonio, F.; Morano, F.; Niger, M.; Corallo, S.; Antista, M.; Raimondi, A.; Prisciandaro, M.; Pagani, F.; Prinzi, N.; Nichetti, F.; et al. Systemic treatment of patients with gastrointestinal cancers during the COVID-19 outbreak: COVID-19-adapted recommendations of the national cancer institute of Milan. *Clin. Colorect. Canc.* **2020**, *19*, 156–164. [CrossRef] [PubMed]

43. Granito, A.; Marinelli, S.; Negrini, G.; Menetti, S.; Benevento, F.; Bolondi, L. Prognostic significance of adverse events in patients with hepatocellular carcinoma treated with sorafenib. *Ther. Adv. Gastroenterol.* **2016**, *9*, 240–249. [CrossRef] [PubMed]

44. Tovoli, F.; Ielasi, L.; Casadei-Gardini, A.; Granito, A.; Foschi, F.G.; Rovesti, G.; Negrini, G.; Orsi, G.; Renzulli, M.; Piscaglia, F. Management of adverse events with tailored sorafenib dosing prolongs survival of hepatocellular carcinoma patients. *J. Hepatol.* **2019**, *71*, 1175–1183. [CrossRef]

45. Borriello, A.; Caldarelli, I.; Bencivenga, D.; Stampone, E.; Perrotta, S.; Oliva, A.; Della Ragione, F. Tyrosine kinase inhibitors and mesenchymal stromal cells: Effects on self-renewal, commitment and functions. *Oncotarget* **2017**, *8*, 5540. [CrossRef]

46. Wang, Z.; Wang, M.; Carr, B.I. Involvement of receptor tyrosine phosphatase DEP-1 mediated PI3K-cofilin signaling pathway in Sorafenib-induced cytoskeletal rearrangement in hepatoma cells. *J. Cell. Physiol.* **2010**, *224*, 559–565. [CrossRef]

47. Chen, X.; Che, X.; Wang, J.; Chen, F.; Wang, X.; Zhang, Z.; Fan, B.; Yang, D.; Song, X. Zinc sensitizes prostate cancer cells to sorafenib and regulates the expression of Livin. *Acta Biochim Biophys Sin.* **2013**, *45*, 353–358. [CrossRef]

48. D'Alessandro, R.; Refolo, M.G.; Lippolis, C.; Carella, N.; Messa, C.; Cavallini, A.; Carr, B.I. Strong enhancement by IGF1-R antagonists of hepatocellular carcinoma cell migration inhibition by Sorafenib and/or vitamin K1. *Cell. Oncol.* **2018**, *41*, 283–296. [CrossRef]

49. Ettrich, T.J.; Seufferlein, T. Regorafenib. In *Small Molecules in Oncology*; Springer: Cham, Switzerland; Berlin/Heidelberg, Germany, 2018; pp. 45–56.

50. Arai, H.; Battaglin, F.; Wang, J.; Lo, J.H.; Soni, S.; Zhang, W.; Lenz, H.J. Molecular insight of regorafenib treatment for colorectal cancer. *Canc. Treat. Rev.* **2019**, *81*, 101912. [CrossRef]

51. Pelosof, L.; Lemery, S.; Casak, S.; Jiang, X.; Rodriguez, L.; Pierre, V.; Bi, Y.; Liu, J.; Zirkelbach, J.F.; Patel, A.; et al. Benefit-risk summary of regorafenib for the treatment of patients with advanced hepatocellular carcinoma that has progressed on sorafenib. *Oncologist* **2018**, *23*, 496–500. [CrossRef]

52. O'Connor, J.M.; Öhler, L.; Scheithauer, W.; Metges, J.P.; Dourthe, L.M.; de Groot, J.W.; Thaler, J.; Yeh, K.H.; Lin, J.K.; Falcone, A.; et al. Real-world dosing of regorafenib in metastatic colorectal cancer (mCRC): Interim analysis from the prospective, observational CORRELATE study. *Liver* **2017**, *260*, 52. [CrossRef]

53. Komatsu, Y.; Muro, K.; Yamaguchi, K.; Satoh, T.; Uetake, H.; Yoshino, T.; Nishida, T.; Takikawa, H.; Kato, T.; Chosa, M.; et al. Safety and efficacy of regorafenib post-marketing surveillance (PMS) in Japanese patients with metastatic colorectal cancer (mCRC). *J. Clin. Oncol.* **2017**, *35*, 721. [CrossRef]

54. Adenis, A.; de la Fouchardiere, C.; Paule, B.; Burtin, P.; Tougeron, D.; Wallet, J.; Dourthe, L.M.; Etienne, P.L.; Mineur, L.; Clisant, S.; et al. Survival, safety, and prognostic factors for outcome with Regorafenib in patients with metastatic colorectal cancer refractory to standard therapies: Results from a multicenter study (REBECCA) nested within a compassionate use program. *BMC Cancer* **2016**, *16*, 412.

55. Grothey, A.; George, S.; Van Cutsem, E.; Blay, J.Y.; Sobrero, A.; Demetri, G.D. Optimizing treatment outcomes with regorafenib: Personalized dosing and other strategies to support patient care. *Oncologist* **2014**, *19*, 669–680. [CrossRef] [PubMed]

56. NCCN Clinical Practice Guidelines in Oncology (NCCN Guidelines). Colon Cancer. Version 4.2018 National Comprehensive Cancer Network. Available online: https://www.nccn.org/professionals/physician_gls/pdf/colon.pdf (accessed on 12 November 2020).

57. Bekaii-Saab, T.S.; Ou, F.S.; Ahn, D.H.; Boland, P.M.; Ciombor, K.K.; Heying, E.N.; Dockter, T.J.; Jacobs, N.L.; Pasche, B.C.; Cleary, J.M.; et al. Regorafenib dose-optimisation in patients with refractory metastatic colorectal cancer (ReDOS): A randomised, multicentre, open-label, phase 2 study. *Lancet Oncol.* **2019**, *20*, 1070–1082. [CrossRef]

58. Bruix, J.; Qin, S.; Merle, P.; Granito, A.; Huang, Y.H.; Bodoky, G.; Pracht, M.; Yokosuka, O.; Rosmorduc, O.; Breder, V.; et al. Regorafenib for patients with hepatocellular carcinoma who progressed on sorafenib treatment (RESORCE): A randomised, double-blind, placebo-controlled, phase 3 trial. *Lancet* **2017**, *389*, 56–66. [CrossRef]

59. Jiang, J.; Zhang, L.; Chen, H.; Lei, Y.; Zhang, T.; Wang, Y.; Jin, P.; Lan, J.; Zhou, L.; Huang, Z.; et al. Regorafenib induces lethal autophagy arrest by stabilizing PSAT1 in glioblastoma. *Autophagy* **2020**, *16*, 106–122. [CrossRef]

60. Han, K.M.; Kang, R.J.; Jeon, H.; Lee, H.; Lee, J.S.; Park, H.H.; Jeon, S.G.; Suk, K.; Seo, J.; Hoe, H.S. Regorafenib regulates AD pathology, neuroinflammation, and dendritic spinogenesis in cells and a mouse model of AD. *Cells* **2020**, *9*, 1655. [CrossRef]

61. Shi, Z.H.; Liu, F.T.; Tian, H.Z.; Zhang, Y.M.; Li, N.G.; LU, T. Design, synthesis and structure-activity relationship of diaryl-ureas with novel isoxazol [3,4-*b*]pyridine-3-amino-structure as multi-target inhibitors against receptor tyrosine kinase. *Bioorg. Med. Chem.* **2018**, *26*, 4735–4744. [CrossRef]

62. Zhou, J.; Goh, B.C.; Albert, D.H.; Chen, C.S. ABT-869, a promising multi-targeted tyrosine kinase inhibitor: From bench to bedside. *J. Hematol. Oncol.* **2009**, *2*, 33. [CrossRef]

63. Guo, J.; Marcotte, P.A.; McCall, J.O.; Dai, Y.; Pease, L.J.; Michaelides, M.R.; Davidsen, S.K.; Glaser, K.B. Inhibition of phosphorylation of the colonystimulating factor-1 receptor (c-Fms) tyrosine kinase in transfected cells by ABT-869 and other tyrosine kinase inhibitors. *Mol. Cancer Ther.* **2006**, *5*, 1007–1013. [CrossRef] [PubMed]

64. Aversa, C.; Leone, F.; Zucchini, G.; Serini, G.; Geuna, E.; Milani, A.; Valdembri, D.; Martinello, R.; Montemurro, F. Linifanib: Current status and future potential in cancer therapy. *Exp. Rev. Anticancer Ther.* **2015**, *15*, 677–687. [CrossRef] [PubMed]

65. Lin, W.H.; Hsu, J.T.A.; Hsieh, S.Y.; Chen, C.T.; Song, J.S.; Yen, S.C.; Hsu, T.; Chen, C.H.; Chou, L.H.; Yang, Y.N.; et al. Discovery of 3-phenyl-1*H*-5-pyrazolylamine derivatives containing a urea pharmacophore as potent and efficacious inhibitors of FMS-like tyrosine kinase-3 (FLT3). *Bioorg. Med. Chem.* **2013**, *21*, 2856–2867. [CrossRef] [PubMed]

66. Toh, H.C.; Chen, P.J.; Carr, B.I.; Knox, J.J.; Gill, S.; Ansell, P.; McKeegan, E.M.; Dowell, B.; Pedersen, M.; Qin, Q.; et al. Phase 2 trial of linifanib (ABT-869) in patients with unresectable or metastatic hepatocellular carcinoma. *Cancer* **2013**, *119*, 380–387. [CrossRef]

67. Cainap, C.; Qin, S.; Huang, W.T.; Chung, I.J.; Pan, H.; Cheng, Y.; Kudo, M.; Kang, Y.K.; Chen, P.J.; Toh, H.C.; et al. Linifanib versus Sorafenib in patients with advanced hepatocellular carcinoma: Results of a randomized phase III trial. *J. Clin. Oncol.* **2015**, *33*, 172–180. [CrossRef]

68. Chen, J.; Guo, J.; Chen, Z.; Wang, J.; Liu, M.; Pang, X. Linifanib (ABT-869) potentiates the efficacy of chemotherapeutic agents through the suppression of receptor tyrosine kinase-mediated akt/mtor signaling pathways in gastric cancer. *Sci. Rep.* **2016**, *6*, 1–11. [CrossRef]

69. Banchi, M.; Orlandi, P.; Gentile, D. Synergistic activity of linifanib and irinotecan increases the survival of mice bearing orthotopically implanted human anaplastic thyroid cancer. *Am. J. Cancer Res.* **2020**, *10*, 2120–2127.

70. Nakamura, K.; Taguchi, E.; Miura, T.; Yamamoto, A.; Takahashi, K.; Bichat, F.; Guilbaud, N.; Hasegawa, K.; Kubo, K.; Fujiwara, Y.; et al. KRN951, a highly potent inhibitor of vascular endothelial growth factor receptor tyrosine kinases, has antitumor activities and affects functional vascular properties. *Cancer Res.* **2006**, *66*, 9134–9142. [CrossRef]

71. Momeny, M.; Moghaddaskho, F.; Gortany, N.K.; Yousefi, H.; Sabourinejad, Z.; Zarrinrad, G.; Mirshahvaladi, S.; Eyvani, H.; Barghi, F.; Ahmadinia, L. Blockade of vascular endothelial growth factor receptors by tivozanib has potential anti-tumour effects on human glioblastoma cells. *Sci. Rep.* **2017**, *7*, 44075. [CrossRef]

72. Motzer, R.J.; Nosov, D.; Eisen, T.; Bondarenko, I.; Lesovoy, V.; Lipatov, O.; Tomczak, P.; Lyulko, O.; Alyasova, A.; Harza, M.; et al. Tivozanib versus sorafenib as initial targeted therapy for patients with metastatic renal cell carcinoma: Results from a phase III trial. *J. Clin. Oncol.* **2013**, *31*, 3791. [CrossRef]

73. Salgia, N.J.; Zengin, Z.B.; Pal, S.K. Tivozanib in renal cell carcinoma: A new approach to previously treated disease. *Ther. Adv. Med. Oncol.* **2020**, *12*. [CrossRef] [PubMed]

74. Wright, K.M. Final data analysis supports tivozanib as superior treatment for patients with RCC. *Oncology (Williston Park, NY)* **2020**, *34*, 257.

75. Rini, B.I.; Pal, S.K.; Escudier, B.J.; Atkins, M.B.; Hutson, T.E.; Porta, C.; Verzoni, E.; Needle, M.N.; McDermott, D.F. Tivozanib versus sorafenib in patients with advanced renal cell carcinoma (TIVO-3): A phase 3, multicentre, randomised, controlled, open-label study. *Lancet Oncol.* **2020**, *21*, 95–104. [CrossRef]

76. Pal, S.K.; Escudier, B.J.; Atkins, M.B. Final overall survival results from a phase 3 study to compare tivozanib to sorafenib as third-or fourth-line therapy in subjects with metastatic renal cell carcinoma. *Eur. Urol.* **2020**, *78*, 783–785. [CrossRef]

77. Westerman, M.E.; Wood, C.G. Editorial Commentary: Tivozanib versus sorafenib in patients with advanced renal cell carcinoma (TIVO-3): A phase 3, multicentre, randomised, controlled, open-label study. *Ann. Transl. Med.* **2020**, *8*, 1037. [CrossRef]

78. Iyer, R.V.; Li, D.; Dayyani, F.; Needle, M.N.; Abrams, T.A. A phase Ib/II, open-label study of tivozanib in combination with durvalumab in subjects with untreated advanced hepatocellular carcinoma. *J. Clin. Oncol.* **2020**, *38* (Suppl. 16599). [CrossRef]

79. Swetzig, W.M.; Lurain, J.R.; Berry, E.; Pineda, M.J.; Shahabi, S.; Perry, L.; Neubauer, N.L.; Nieves-Neira, W.; Schink, J.C.; Schiller, A.; et al. Efficacy and safety of tivozanib in recurrent, platinum resistant ovarian, fallopian tube or primary peritoneal cancer. *J. Clin. Oncol.* **2019**, *37* (Suppl. 5538). [CrossRef]

80. Momeny, M.; Sabourinejad, Z.; Zuzzinrad, G.; Maghaddaskho, F.; Eyvani, H.; Yousefi, H.; Mirshahvaladi, S.; Poursani, E.M.; Barghi, F.; Poursheikhani, A.; et al. Anti-tumour activity of tivozanib, a pan-inhibitor of VEGF receptors, in therapy-resistant ovarian carcinoma cells. *Sci. Rep.* **2017**, *7*, 1–13. [CrossRef]

81. Martin-Liberal, J.; Pérez, E.; García Del Muro, X. Investigational therapies in phase II clinical trials for the treatment of soft tissue sarcoma. *Exp. Opin. Invest. Drugs* **2019**, *28*, 39–50. [CrossRef]

82. Kalpathy-Cramer, J.; Chandra, V.; Da, X.; Ou, Y.; Emblem, K.E.; Muzikansky, A.; Cai, X.; Douw, L.; Evans, J.G.; Dietrich, J.; et al. Phase II study of tivozanib, an oral VEGFR inhibitor, in patients with recurrent glioblastoma. *J. Neurooncol.* **2017**, *131*, 603–610. [CrossRef]

83. Jamil, M.O.; Hathaway, A.; Mehta, A. Tivozanib: Status of development. *Curr. Oncol. Rep.* **2015**, *17*, 24. [CrossRef] [PubMed]

84. Oldenhuis, C.N.; Loos, W.J.; Esteves, B.; van Doorn, L.; Cotreau, M.M.; Strahs, A.L.; den Hollander, M.W.; Gietema, J.A.; de Vries, E.G.E.; Eskens, F.A.L.M. A phase Ib study of the VEGF receptor tyrosine kinase inhibitor tivozanib and modified FOLFOX-6 in patients with advanced gastrointestinal malignancies. *Clin. Colorectal Cancer* **2015**, *14*, 18–24.e1. [CrossRef] [PubMed]

85. Benson, A.B.; Kiss, I.; Bridgewater, J.; Eskens, F.A.; Sasse, C.; Vossen, S.; Chen, J.; Van Sant, C.; Ball, H.A.; Keating, A.; et al. BATON-CRC: A phase II randomized trial comparing tivozanib plus mFOLFOX6 with bevacizumab plus mFOLFOX6 in stage IV metastatic colorectal cancer. *Clin. Cancer Res.* **2016**, *22*, 5058–5067. [CrossRef] [PubMed]

86. Smith, B.D.; Kaufman, M.D.; Lu, W.P.; Gupta, A.; Leary, C.B.; Wise, S.C.; Rutkoski, T.J.; Ahn, Y.M.; Ahn, G.; Bulfer, S.L.; et al. Ripretinib (DCC-2618) is a switch control kinase inhibitor of a broad spectrum of oncogenic and drug-resistant KIT and PDGFRA variants. *Cancer Cell* **2019**, *35*, 738–751. [CrossRef]

87. Blay, J.Y.; Serrano, C.; Heinrich, M.C.; Zalcberg, J.; Bauer, S.; Gelderblom, H.; Schöffski, P.; Jones, R.L.; Attia, S.; D'Amato, G.; et al. Ripretinib in patients with advanced gastrointestinal stromal tumours (INVICTUS): A double-blind, randomised, placebo-controlled, phase 3 trial. *Lancet Oncol.* **2020**, *21*, e415. [CrossRef]

88. US Food and Drug Administration. QINLOCK Prescribing Information. 2020. Available online: https://www.accessdata.fda.gov/drugsatfda_docs/label/2020/213973s000lbl.pdf (accessed on 12 November 2020).

89. Khoshnood, A. Gastrointestinal stromal tumor—A review of clinical studies. *J. Oncol. Pharm. Pract.* **2019**, *25*, 1473–1485. [CrossRef]

90. Jawhar, M.; Gotlib, J.; Reiter, A. Tyrosine kinase inhibitors in systemic mastocytosis. In *Mastocytosis*; Springer: Cham, Switzerland, 2020; pp. 257–265.

91. Haddox, C.L.; Riedel, R.F. Individualizing systemic therapy for advanced soft tissue sarcomas based on tumor histology and biology. *Exp. Rev. Anticancer Ther.* **2020**, *20*, 5–8. [CrossRef]

92. Thabit, M.G.; Mostafa, A.S.; Selim, K.B.; Elsayed, M.A.; Nasr, M.N. Design, synthesis and molecular modeling of phenyl dihydropyridazinone derivatives as B-Raf inhibitors with anticancer activity. *Bioorg. Chem.* **2020**, *103*, 104148. [CrossRef]

93. Chen, X.; Xie, B.; Cao, L.; Zhu, F.; Zhu, B.; Lv, H.; Fan, X.; Han, L.; Bie, L.; Cao, X.; et al. Direct binding of microRNA-21 pre-element with Regorafenib: An alternative mechanism for anti-colorectal cancer chemotherapy? *J. Mol. Graph. Model.* **2017**, *73*, 48–53. [CrossRef]

94. Kajal, K.; Panda, A.K.; Bhat, J.; Chakraborty, D.; Bose, S.; Bhattacharjee, P.; Sarkar, T.; Chatterjee, S.; Kar, S.K.; Sa, G. Andrographolide binds to ATP-binding pocket of VEGFR2 to impede VEGFA-mediated tumor-angiogenesis. *Sci. Rep.* **2019**, *9*, 1–10. [CrossRef]

95. Babić, Ž.; Crkvenčić, M.; Rajić, Z.; Mikecin, A.M.; Kralj, M.; Balzarini, J.; Petrova, M.; Vanderleyden, J.; Zorc, B. New sorafenib derivatives: Synthesis, antiproliferative activity against tumour cell lines and antimetabolic evaluation. *Molecules* **2012**, *17*, 1124–1137. [CrossRef] [PubMed]

96. Kuo, C.Y.; Chou, W.C.; Wu, C.C.; Wong, T.S.; Kakadiya, R.; Lee, T.C.; Su, T.L.; Wang, H.C. Repairing of N-mustard derivative BO-1055 induced DNA damage requires NER, HR, and MGMT-dependent DNA repair mechanisms. *Oncotarget* **2015**, *6*, 25770. [CrossRef] [PubMed]

97. Ambati, S.R.; Shieh, J.H.; Pera, B.; Lopes, E.C.; Chaudhry, A.; Wong, E.W.; Saxena, A.; Su, T.L.; Moore, M.A. BO-1055, a novel DNA cross-linking agent with remarkable low myelotoxicity shows potent activity in sarcoma models. *Oncotarget* **2016**, *7*, 43062. [CrossRef] [PubMed]

98. Lopes, E.C.; Correa, F.; Peguero, E.; Ambati, S.R.; Shieh, J.H.; Su, T.L.; Moore, M.A.S. Pre-Clinical Evaluation of a Novel DNA Crosslinking Agent, BO-1055 in B-Cell Lymphoma. *Blood* **2014**, *124*, 5483. [CrossRef]

99. Tala, S.D.; Ou, T.H.; Lin, Y.W.; Tala, K.S.; Chao, S.H.; Wu, M.H.; Tsai, T.H.; Kakadiya, R.; Suman, S.; Chen, C.H.; et al. Design and synthesis of potent antitumor water-soluble phenyl N-mustard-benzenealkylamide conjugates via a bioisostere approach. *Eur. J. Med. Chem.* **2014**, *76*, 155–169. [CrossRef] [PubMed]

100. Chen, J.N.; Wang, X.F.; Li, T.; Wu, D.W.; Fu, X.B.; Zhang, G.J.; Shen, X.C.; Wang, H.S. Design, synthesis, and biological evaluation of novel quinazolinyl-diaryl urea derivatives as potential anticancer agents. *Eur. J. Med. Chem.* **2016**, *107*, 12–25. [CrossRef]

101. Jiang, N.; Bu, Y.; Wang, Y.; Nie, M.; Zhang, D.; Zhai, X. Design, synthesis and structure-activity relationships of novel diaryl urea derivatives as potential EGFR inhibitors. *Molecules* **2016**, *21*, 1572. [CrossRef]

102. Zuo, S.J.; Zhang, S.; Mao, S.; Xie, X.X.; Xiao, X.; Xin, M.H.; Xuan, W.; He, Y.Y.; Cao, Y.X.; Zhang, S.Q. Combination of 4-anilinoquinazoline, arylurea and tertiary amine moiety to discover novel anticancer agents. *Bioorg. Med. Chem.* **2016**, *24*, 179–190. [CrossRef]

103. Yin, Y.; Zheng, K.; Eid, N.; Howard, S.; Jeong, J.H.; Yi, F.; Guo, J.; Park, C.M.; Bibian, M.; Wu, W.; et al. Bis-aryl urea derivatives as potent and selective LIM kinase (Limk) inhibitors. *J. Med. Chem.* **2015**, *58*, 1846–1861. [CrossRef]

104. Park, J.B.; Agnihotri, S.; Golbourn, B.; Bertrand, K.C.; Luck, A.; Sabha, N.; Smith, C.A.; Byron, S.; Zadeh, G.; Croul, S.; et al. Transcriptional profiling of GBM invasion genes identifies effective inhibitors of the LIM kinase-Cofilin pathway. *Oncotarget* **2014**, *5*, 9382. [CrossRef]

105. Wu, J.; Huang, Y.; Xie, Q.; Zhang, J.; Zhan, Z. A novel bis-aryl urea compound inhibits tumor proliferation via cathepsin D-associated apoptosis. *Anticancer Drugs* **2020**, *31*, 500–506. [CrossRef] [PubMed]

106. Sinicropi, M.S.; Caruso, A.; Conforti, F.; Marrelli, M.; El Kashef, H.; Lancelot, J.C.; Rault, S.; Statti, G.A.; Menichini, F. Synthesis, inhibition of NO production and antiproliferative activities of some indole derivatives. *J. Enzyme Inhib. Med. Chem.* **2009**, *24*, 1148–1153. [CrossRef] [PubMed]

107. Iacopetta, D.; Catalano, A.; Ceramella, J.; Barbarossa, A.; Carocci, A.; Fazio, A.; La Torre, C.; Caruso, A.; Ponassi, M.; Rosano, C.; et al. Synthesis, anticancer and antioxidant properties of new indole and pyranoindole derivatives. *Bioorg. Chem.* **2020**, *105*, 104440. [CrossRef] [PubMed]

108. Kurt, B.Z.; Kandas, N.O.; Dag, A.; Sonmez, F.; Kucukislamoglu, M. Synthesis and biological evaluation of novel coumarin-chalcone derivatives containing urea moiety as potential anticancer agents. *Arab. J. Chem.* **2020**, *13*, 1120–1129. [CrossRef]

109. Azimian, F.; Hamzeh-Mivehroud, M.; Mojarrad, J.S.; Hemmati, S.; Dastmalchi, S. Synthesis and biological evaluation of diaryl urea derivatives designed as potential anticarcinoma agents using de novo structure-based lead optimization approach. *Eur. J. Med. Chem.* **2020**, *201*, 112461. [CrossRef] [PubMed]

110. Zarei, O.; Azimian, F.; Hamzeh-Mivehroud, M.; Shahbazi Mojarrad, J.; Hemmati, S.; Dastmalchi, S. Design, synthesis, and biological evaluation of novel benzo[b]thiophene-diaryl urea derivatives as potential anticancer agents. *Med. Chem. Res.* **2020**, 1–11. [CrossRef]

111. Solano, L.N.; Nelson, G.L.; Ronayne, C.T.; Jonnalagadda, S.; Jonnalagadda, S.K.; Kottke, K.; Chitren, R.; Johnson, J.L.; Pandey, M.K.; Jonnalagadda, S.C.; et al. Synthesis, in vitro, and in vivo evaluation of novel N-phenylindazolyl diarylureas as potential anti-cancer agents. *Sci. Rep.* **2020**, *10*, 1–10. [CrossRef]

112. Li, W.; Chu, J.; Fan, T.; Zhang, W.; Yao, M.; Ning, Z.; Wang, M.; Sun, J.; Zhao, X.; Wen, A. Design and synthesis of novel 1-phenyl-3-(5-(pyrimidin-4-ylthio)-1,3,4-thiadiazol-2-yl) urea derivatives with potent anti-CML activity throughout PI3K/AKT signaling pathway. *Bioorg. Med. Chem. Lett.* **2019**, *29*, 1831–1835. [CrossRef]

113. Blum, W.; Henzi, T.; Pecze, L.; Diep, K.L.; Bochet, C.G.; Schwaller, B. The phytohormone forchlorfenuron decreases viability and proliferation of malignant mesothelioma cells in vitro and in vivo. *Oncotarget* **2019**, *10*, 6944–6956. [CrossRef]

114. Marcus, E.A.; Tokhtaeva, E.; Turdikulova, S.; Capri, J.; Whitelegge, J.P.; Scott, D.R.; Sachs, G.; Berditchevski, F.; Vagin, O. Septin oligomerization regulates persistent expression of ErbB2/HER2 in gastric cancer cells. *Biochem. J.* **2016**, *473*, 1703–1718. [CrossRef]

115. Kim, K.K.; Singh, R.K.; Khazan, N.; Kodza, A.; Singh, N.A.; Jones, A.; Sivagnanalingam, U.; Towner, M.; Itamochi, H.; Turner, R.; et al. Development of potent forchlorfenuron analogs and their cytotoxic effect in cancer cell lines. *Sci. Rep.* **2020**, *10*, 1–8. [CrossRef] [PubMed]
116. Gable, K.L.; Maddux, B.A.; Penaranda, C.; Zavodovskaya, M.; Campbell, M.J.; Lobo, M.; Robinson, L.; Schow, S.; Kerner, J.A.; Goldfine, I.D.; et al. Diarylureas are small-molecule inhibitors of insulin-like growth factor I receptor signaling and breast cancer cell growth. *Mol. Cancer Ther.* **2006**, *5*, 1079–1086. [CrossRef] [PubMed]
117. Zhou, X.; Zhao, X.; Li, X.; Ping, G.; Pei, S.; Chen, M.; Wang, Z.; Zhou, W.; Jin, B. PQ401, an IGF-1R inhibitor, induces apoptosis and inhibits growth, proliferation and migration of glioma cells. *J. Chemother.* **2016**, *28*, 44–49. [CrossRef] [PubMed]
118. Qi, B.; Zhang, R.; Sun, R.; Guo, M.; Zhang, M.; Wei, G.; Zhang, L.; Yu, S.; Huang, H. IGF-1R inhibitor PQ401 inhibits osteosarcoma cell proliferation, migration and colony formation. *Int. J. Clin. Exp. Pathol.* **2019**, *12*, 1589–1598.
119. Dai, B.; Fan, M.; Yu, R.; Su, Q.; Wang, B.; Yang, T.; Liu, F.; Zhang, Y. Novel diphenyl urea derivative serves as an inhibitor on human lung cancer cell migration by disrupting EMT via Wnt/β-catenin and PI3K/Akt signaling. *Toxicol. In Vitro* **2020**, *69*, 105000. [CrossRef]
120. Sinner, H.F.; Johnson, J.; Rychahou, P.G.; Watt, D.S.; Zaytseva, Y.Y.; Liu, C.; Evers, B.M. Novel chemotherapeutic agent, FND-4b, activates AMPK and inhibits colorectal cancer cell proliferation. *PLoS ONE* **2019**, *14*, e0224253. [CrossRef]
121. Parmar, V.R.; Jasrai, Y.T. Effect of thidiazuron (TDZ) on in vitro propagation of valuable medicinal plant: *Uraria picta* (Jacq.) Desv. ex DC. *J. Agricult. Res. (03681157)* **2015**, *53*, 513–521.
122. Enkhtaivan, G.; Kim, D.H.; Pandurangan, M. Cytotoxic effect of TDZ on human cervical cancer cells. *J. Photochem. Photobiol. B Biol.* **2017**, *173*, 493–498. [CrossRef]
123. Shamsel-Din, H.A.; Gizawy, M.A.; Abdelaziz, G. Molecular docking and preliminary bioevaluation of 99m Tc-Thiadiazuron as a novel potential agent for cervical cancer imaging. *J. Radioanal. Nucl. Chem.* **2020**, *326*, 1375–1381. [CrossRef]

applied sciences

Review

Schiff Bases: Interesting Scaffolds with Promising Antitumoral Properties

Domenico Iacopetta [1], Jessica Ceramella [1], Alessia Catalano [2,*], Carmela Saturnino [3,*], Maria Grazia Bonomo [3], Carlo Franchini [2] and Maria Stefania Sinicropi [1]

[1] Department of Pharmacy, Health and Nutritional Sciences, University of Calabria, 87036 Arcavacata di Rende, Italy; domenico.iacopetta@unical.it (D.I.); jessicaceramella@gmail.com (J.C.); s.sinicropi@unical.it (M.S.S.)

[2] Department of Pharmacy-Drug Sciences, University of Bari "Aldo Moro", 70126 Bari, Italy; carlo.franchini@uniba.it

[3] Department of Science, University of Basilicata, 85100 Potenza, Italy; mariagrazia.bonomo@unibas.it

[*] Correspondence: alessia.catalano@uniba.it (A.C.); carmela.saturnino@unibas.it (C.S.); Tel.: +39-0805442731 (A.C.); +39-097126442 (C.S.)

Abstract: Schiff bases, named after Hugo Schiff, are highly reactive organic compounds broadly used as pigments and dyes, catalysts, intermediates in organic synthesis, and polymer stabilizers. Lots of Schiff bases are described in the literature for various biological activities, including antimalarial, antibacterial, antifungal, anti-inflammatory, and antiviral. Schiff bases are also known for their ability to form complexes with several metals. Very often, complexes of Schiff bases with metals and Schiff bases alone have demonstrated interesting antitumor activity. Given the innumerable vastness of data regarding antitumor activity of all these compounds, we focused our attention on mono- and bis-Schiff bases alone as antitumor agents. We will highlight the most significant examples of compounds belonging to this class reported in the literature.

Keywords: Schiff bases; antitumor agents; apoptosis; antiproliferative activity; imines

check for **updates**

Citation: Iacopetta, D.; Ceramella, J.; Catalano, A.; Saturnino, C.; Bonomo, M.G.; Franchini, C.; Sinicropi, M.S. Schiff Bases: Interesting Scaffolds with Promising Antitumoral Properties. *Appl. Sci.* **2021**, *11*, 1877. https://doi.org/10.3390/app11041877

Academic Editor: Fethi Bedioui

Received: 2 February 2021
Accepted: 17 February 2021
Published: 20 February 2021

Publisher's Note: MDPI stays neutral with regard to jurisdictional claims in published maps and institutional affiliations.

Copyright: © 2021 by the authors. Licensee MDPI, Basel, Switzerland. This article is an open access article distributed under the terms and conditions of the Creative Commons Attribution (CC BY) license (https://creativecommons.org/licenses/by/4.0/).

1. Introduction

Since their discovery by the German chemist Hugo Schiff [1], Schiff bases (imines), scaffolds with high chemical reactivity, and their metal complexes have been very well known for catalysis in various synthetic processes and for their biological properties. In therapy, Schiff bases and their metal complexes have been reported to manifest a wide range of biological activities [2,3] such as antimicrobial [4], ureases inhibitory [5], anti-inflammatory [6,7], anti-ulcerogenic [8], antioxidant [9–11], pesticidal, cytotoxic, and anticancer [12] including DNA damage [13–15]. Schiff bases have been also successfully used in scientific studies [16] as highly efficient and selective sensing materials for optical, electrochemical [17,18], and membrane sensors [19]. Zinc-Schiff bases have been proposed as carrier vehicles for the delivery of zinc to prostate cells. Indeed, the use of the membrane-penetrating peptide Novicidin connected to zinc-Schiff base has been studied as a therapeutic approach for prostate cancer [20]. Schiff base ligands, as some other organic small molecules [21], have received great attention from researchers thanks to their easy preparation and ability to form complexes with almost all metals, due to the electron-donating nitrogen in their base structure [22–24]. Several metal complexes, in which the metal is coordinated to various ligands, are able not only to stabilize the metal but also to modify its chemical and pharmaceutical properties and are receiving attention in medicinal chemistry [25–30]. The general structure of a Schiff base is shown in Figure 1, R^1, R^2 and R^3 being an alkyl or aryl moiety.

Appl. Sci. **2021**, *11*, 1877. https://doi.org/10.3390/app11041877 — https://www.mdpi.com/journal/applsci

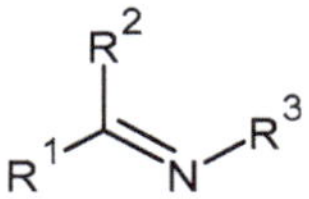

Figure 1. General structure of a Schiff base.

Schiff bases are particularly interesting in the field of antitumor agents [31–34] as many other small organic molecules (for instance, diarylureas [35], indoles [36,37], carbazoles [38], phthalimides [39], and so on [40,41]). The most salient and recent data on Schiff bases will be, herein, reviewed.

For instance, in a recent study, the introduction of Schiff bases in the N-phenylcarbazole/triphenylamine modified half-sandwiched iridium(III) compounds determined an enhancement of antitumor activity of about 13 times that of the clinical cisplatin [42]. This review focused on studies of the last decades on mono- and bis-Schiff bases as antiproliferative agents, paying attention particularly on Schiff bases showing high activity (concentration which kills or inhibits cell viability by 50% (IC_{50}) in the range of micromolar to nanomolar).

2. Schiff Bases as Antiproliferative Agents

2.1. Mono-Schiff Bases

Vicini et al. (2003) [43] studied a series of Schiff bases and tested their antiproliferative activity against a panel of human cell lines derived from hematological and solid tumors. The most interesting compounds were **1–3** (Table 1). All of them inhibited the growth of leukemia cell lines, with IC_{50} values ranging between 1.5 and 7 µM against human CD4$^+$ lymphocytes (MT-4), human CD4$^+$ acute T-lymphoblastic leukemia (CCRF-CEM), human splenic B-lymphoblastoid cells (WIL-2NS), and human acute B-lymphoblastic leukemia (CCRF-SB). The 2-Amino-6-mercaptopurine was used as reference drug (IC_{50} between 0.1 and 0.5 µM). Particularly, compound **3** was also active against solid tumor-derived cell lines' skin melanoma and breast adenocarcinoma cells (IC_{50} = 6 and 10 nM) against human skin melanoma SK-MEL-28 and human breast adenocarcinoma MCF-7 cell lines, respectively. The values for 2-amino-6-mercaptopurine were 5 and 4 µM, respectively. Zhou et al. (2007) [44] studied several imines bearing thiazole and triazole moieties and evaluated their antiproliferative activities against leukemia, stomach, and larynx cancer cell lines. The 2,4-dinitro substituted Schiff base **4** displayed high activity against HL-60, BGC-823 and Hep-2 cell lines, showing percentage inhibition of 91.97, 98.49, and 91.16%, respectively. Abdel-Hafez et al. (2009) [45] studied several Schiff bases as derivatives of xanthotoxin and evaluated their antitumor activities against cervical carcinoma (HeLa) and breast carcinoma (MCF 7) cell lines. The Schiff base **5** was inactive against MCF-7 cell line but was the most interesting against HeLa, showing an IC_{50} value of 7.2 µM and a percent viability of 70% (xanthotoxin, 7.6 µM and 62%, respectively). Kraicheva et al. (2009) [46] studied three Schiff bases and evaluated their antiproliferative activity, using the 3-[4,5-dimethylthiazol-2-yl]-2,5-diphenyltratrazolium bromide (MTT) assay, against four human leukemic cell lines, viz., LAMA-84 (peripheral chronic myeloid leukemia cells), K-562 (non-adherent chronic myelogenous leukemia cells of the erythroleukemia type), HL-60 (acute promyelocytic leukemia cells) and its multi-drug-resistant sub-line HL-60/Dox (multi-drug resistant acute myeloblastic leukemia cell line), characterized by the overexpression of MRP-1 protein (ABC-C1). Compound **6** showed antiproliferative activity (IC_{50} = 39.9 µM, 29.9 µM, and 68.6 µM against LAMA-84, K-562, and HL-60/Dox, respectively), while compound **7** was less active (IC_{50} = 251.9 µM, 212.9 µM, and 226.1 µM against LAMA-84, K-562, and HL-60/Dox, respectively). Both the investigated compounds were identified as capable of evoking the distinctly marked lower cytotoxic effects (with the IC_{50} values over 400 µM) against the sensitive leukemic cell line HL-60 in a preliminary antitumor screening.

Nawaz et al. (2009) [47] studied Schiff bases with ferrocene addition and evaluated their antitumor, antioxidant, and DNA-protecting activities. Antitumor activity was evaluated by Potato disc tumor induction assay using *Agrobacterium tumefaciens* (At-10) to induce tumors on potato discs, that is, a prescreen assay and its results were in accordance with other commonly used in vitro antitumor assays. All the tested compounds inhibited tumor production for treatment of 1000, 100, and 10 µg/mL concentration at $p < 0.05$ (vincristine, used as positive control, showed 100% tumor inhibition at all concentrations tested). The inhibition was observed in a dose-dependent manner with the highest inhibition at 1000 µg/mL concentration. Moreover, the highest tumor growth inhibition of 71% was observed with ferrocene containing Schiff base **8**, followed by **9** with 58% inhibition at 1000 µg/mL. IC_{50} values were 20 and 563 µg/mL versus 0.003 µg/mL of vincristine. Zaheer et al. (2010) [48] studied several Schiff bases and tested their cytotoxic activity by the brine shrimp lethality assay. Medium lethal concentration (LD_{50}) values for compounds **10** and **11** were 292.95 and 18.22 ppm, respectively.

Cheng et al. (2010) [49] studied eight Schiff bases and evaluated their antiproliferative effects on human hepatoma HepG2 cells by sulforhodamine B assay. Compounds **12** and **13** were comparable to positive control, etoposide, showing IC_{50} values of 5.6 and 6.8 µM, respectively, versus 4.1 µM of etoposide. Jesmin et al. (2010) [50] studied two Schiff bases, PHP [*N*-(1-phenyl-2-hydroxy-2-phenylethylidine)-2-hydroxylphenylimine, **14**) and HHP [*N*-(2-hydroxybenzylidine)-2-hydroxylphenylimine, **15**) as anticancer agents acting on Ehrlich ascites carcinoma (EAC) cells in Swiss albino mice. All compounds were more active than standard anticancer drug, bleomycin, in improving the life span, lowering tumor weight, and inhibiting the tumor cell growth of EAC cell-bearing mice. The toxicity of the tested compounds was evaluated by measuring LD_{50} values that were of 16 and 15.5 mg/kg for **14** and **15**, respectively. The maximum percentage cell growth inhibition of 93% was observed with **15** with dose loading of 2 mg/kg. Etaiw et al. (2011) [51] studied a Schiff base derived from 2-aminobenzothiazole and 2-thiophenecarboxaldehyde (**16**) for its antiproliferative activity against five human cancer cell lines (cervical carcinoma, HeLa; breast carcinoma, MCF-7; liver carcinoma, HepG2; colon carcinoma, HCT-116; and larynx carcinoma, HEP2). Compound **16** showed activity against HeLa cancer cells (IC_{50} = T0.186 µM). Moreover, its complexes with Cu(II), Fe(III), and Ni(II) showed a higher activity. Hranjec et al. (2011) [52] prepared series of 14 imines and studied the suppression of proliferation of different human cancer cell lines (HeLa (cervical carcinoma), SW620 (colorectal adenocarcinoma, metastatic), MiaPaCa-2 (pancreatic carcinoma), MCF-7 (breast epithelial adenocarcinoma, metastatic)) and their cytotoxic activity on normal human fibroblasts (WI38 normal diploid human fibroblasts) using the MTT assay. Compounds **17** and **18** exerted a strong non-specific antiproliferative effect on all cell lines tested and a concentration-dependent effect on HeLa and MCF-7 cell lines at micromolar concentrations (IC_{50} = 4.73 and 3.24 µM on HeLa and 9.23 and 15.27 µM on MCF-7). However, they were also highly cytotoxic on human fibroblasts. Shaker et al. (2011) [53] synthesized surfactants containing Schiff bases with hydrocarbon chains of different lengths (from C12 to C18). In vitro anticancer cytotoxic activity of these compounds was investigated using EAC as a model system of mice cell tumor at different concentrations (25, 50, and 100%) against liver carcinoma (HepG2), breast carcinoma (MCF-7), and colon carcinoma (HCT-116) cell lines. Compound **19**, bearing a C14 hydrocarbon chain, caused the death of 95% of EAC cell at the highest concentration. The IC_{50} values for compound **19** at different concentrations ranged from 1 to 10 mg/mL. It showed high activity in in vitro system on the tumor cell lines investigated and the highest cytotoxic effect on HepG2, HCT-116, and MCF-7, respectively, and SBC12 surfactant-affected tumor tissue at very low concentrations at values lower than their critical micelle concentration (cmc) values.

Kraicheva et al. (2012) [54] synthesized two anthracene-containing Schiff bases, 9-anthrylidene-p-toluidine (**20**) and 9-anthrylidene-furfurylamine (**21**), and tested their anticancer activities in vitro on a panel of human epithelial cancer cell lines (cell lines from ductal carcinoma of the breast with low and high metastatic potential, MCF-7 and MDA-MB-231, respectively; colostrum-derived myoepithelial cells, expressing polyoma virus large T-antigen, HBL-100 line; bladder carcinoma, 647-V; hepatocellular carcinoma, HepG2;

colon carcinoma, HT-29; cervical carcinoma, HeLa). Compounds **20** and **21** showed high cytotoxic activity toward colon carcinoma HT-29 cell line (IC_{50} = 0.08 and 0.20 mg/mL versus 0.58 mg/mL of doxorubicin). The authors also performed their safety testing, both in vitro (Neutral Red Uptake Assay, 3T3 NRU test) and *in vivo* on ICR mice for genotoxicity and antiproliferative activity. Both compounds were shown not to induce clearly expressed dose-effect clastogenic activities, in contrast to the alkylating agent Mitomycin C. Bae et al. (2012) [55] synthesized new Schiff bases and evaluated their anti-melanogenesis activity, in murine B16F10 melanoma cells, through the inhibition of tyrosinase. Compound **12** exhibited the most potent and non-competitive inhibition on mushroom tyrosinase, even better than the kojic acid used as positive reference (IC_{50} value of 17.22 μM versus 51.11 μM of the kojic acid). This compound decreased the melanin production stimulated by the alpha-melanocyte-stimulating hormone and inhibited murine tyrosinase activity in a dose-dependent manner. Sondhi et al. (2012) [56] synthesized several mono-Schiff bases and bis-Schiff bases and studied their anticancer activities against five human cancer cell lines (lung, NCI H-522; ovary, PA1; breast T47D; colon, HCT-15; liver, HepG2) by MTT assay. The percentage growth (PG) inhibition of cancer cell lines was determined at a concentration of 1×10^{-5} M. The most active mono-Schiff base against lung cancer cells (49% versus 59% of actinomycin-D) was compound **22**. The other active bis-Schiff bases are reported in the next paragraph. Klimczak et al. (2013) [57] studied several small molecules bearing an imine moiety and studied their activity against four esophageal cancer cell lines. Compound **23** was the most active of the series, showing IC_{50} values of 50.12, 158.49, and 111.2 μM against KYSE 150, KYSE 30 and KYSE 270.

Hafez et al. (2013) [58] studied several Schiff bases as antiproliferative agents against various cell lines by using the MTT assay. The most interesting compounds, more active than standard drugs, were **24–26**. In particular, **24** was active on ovarian carcinoma (SK OV-3) cell line (IC_{50} = 0.44 μM versus 4.16 μM of doxorubicin), whereas compound **25** showed good activity on leukemia (U937) (IC_{50} = 0.09 nM versus 4.45 of doxorubicin), neuroblastoma (GOTO and NB-1) (IC_{50} = 0.45 nM and 0.64 nM, respectively, versus IC_{50} = 4.73 nM and IC_{50} = 5.15 nM, respectively, of doxorubicin), and fibrosarcoma (HT1080) cell lines (IC_{50} = 0.54 nM versus 1.16 nM of tamoxifen). Finally, compound **26** was active on cervical carcinoma (KB) (IC_{50} = 0.54 μM versus 4.46 μM of fluorouracil), CNS (SF-268) (IC_{50} = 0.30 nM versus 7.68 nM of cytarabine), leukemia (K-562) (IC_{50} = 0.43 nM versus 6.66 of doxorubicin), liver (HepG2) (IC_{50} = 0.09 nM versus 1.31 nM of tamoxifen), and non-small cell lung (NCI H460) cancer cell lines (IC_{50} = 6.60 nM versus 2.13 nM of gencitabine hydrochloride). Hassan et al. (2015) [59] synthesized several imines and evaluated their cytotoxicity against four human cancer cell lines (colon HCT-116, lung A549, breast MCF-7, and liver HepG2) according to Sulforhodamine-B stain (SRB) assay. Compound **27** was the only compound to show slight activity against liver HepG2 (IC_{50} = 6.20 μg/mL) and breast MCF-7 (IC_{50} = 7 μg/mL) cells in comparison with the standard drug, doxorubicin (IC_{50} = 4.20 and 4.70 μg/mL, respectively).

Zhao et al. (2013) [60] studied a series of Schiff bases and evaluated the in vitro antiproliferative activities against human breast cancer cell MCF-7 and mouse lymphocyte leukemia cell L1210 by the WST-8 ([2-(2-methoxy-4-nitrophenyl)-3-(4-nitrophenyl)-5-(2,4-disulfophenyl)-2*H*-tetrazolium monosodium salt]) assay as a substitute for the most classic MTT assay. The lead compound 2-phenyl-4-carboxyl-1,3-selenazole (PCS) was taken as a comparison. Compounds **28–30** were the most potent compounds against MCF-7 (IC_{50} = 4.02, 7.55 and 8.51 μM, respectively, versus 16.56 μM of PCS). Compound **31** was the most active against L1210 (IC_{50} = 38.73 μM versus 60.11 μM of PCS). Noureen et al. (2013) [61] reported a study on Schiff bases and evaluation of their antioxidant, antitumor, and anti-inflammatory potentials. The antitumor activity was assessed by the potato disc anti-tumor assay. Compounds **32** and **33** were the most active, showing IC_{50} values of 0.15 and 8.03 μg/mL, respectively, versus 0.003 μg/mL of vincristine, used as reference drug.

Zhang et al. (2014) [62] synthesized a series of Schiff bases and evaluated the in vitro antitumor activity against three human tumor cell lines (human liver SMMC-7721, human breast MCF-7, and human lung A549) using the WST-8 assay and 5-Fluorouracil (5-FU) as a positive control. Compound **34** was the most active against SMMC-7721 cells (IC_{50} = 2.84 µM versus 5-FU, IC_{50} = 5.62 µM), whereas compounds **35** and **36** showed significant antiproliferative activity against MCF-7 cells (IC_{50} = 4.56 and 4.25 µM, respectively, versus 14.26 µM of 5-FU). Finally, the most interesting compounds against A549 cells were **37** and **36** (IC_{50} = 4.11 and 4.13 µM, respectively, versus 8.13 µM of 5-FU). Gupta et al. (2015) [63] synthesized 13 Schiff bases and studied their potential as Hsp 90ATPase inhibitors by malachite green assay and antiproliferative activity against PC3 prostate cancer cell lines by MTT assay. Compound **38** showed a high effect toward PC3 cells with an IC_{50} of 4.83 µM (versus 2.45 µM of geldanamycin), followed by compounds **39** and **40** (IC_{50} = 7.43 µM and 7.15 µM, respectively), which were the other promising anticancer molecules among the newly synthesized compounds. In malachite green assay for Hsp90 ATPase suppression, none of the molecules demonstrated IC_{50} values in nanomolar range. Only compounds **41** and **42** showed the maximum inhibitory potential, with an IC_{50} value of 0.02 µM. In conclusion, the authors identified the compound **38**, showing sub micro-molar target affinity and good cellular potency, as the lead molecule for preclinical evaluation in animals and development of Hsp90 inhibitors as anticancer agents. Abd-Elzaher et al. (2016) [64] synthesized and studied a Schiff base ligand (**43**) and its complexes with metal ions. Compound **43** was tested for its anticancer activity against different human tumor cell lines (liver HepG2, breast MCF-7, and colorectal HCT116) and doxorubicin was used as a reference drug. It showed IC_{50} = 9.22, 10.00, and 9.50 µM against the tree cell lines, respectively (IC_{50} values for doxorubicin were 4.20, 4.40, and 5.25 µM, respectively).

Sabbah et al. (2018) [65] described the design, synthesis, and biological evaluation of new phenylimino-1,2-diphenylethanol derivatives in human colon carcinoma (HCT-116), breast adenocarcinoma (MCF-7), and breast carcinoma (T47D) cell lines. Among the tested compounds, the authors evidenced a selectivity toward the adopted cells lines, indicating that the highest inhibitory activity toward the MCF-7 and T47D cells was obtained under the imine **44** treatment (IC_{50} values of 0.024 and 0.034 M, respectively). Moreover, they suggested that this different selectivity could depend on the difference forms of the phosphatidylinositol 3-kinases (PI3Ks) present in the adopted cell models. This hypothesis was proven by the means of in silico and in vitro studies, indicating that the phosphoinositide 3-kinase α (PI3K α) is one of the targets of the compound **44**, which influences the fundamental PI3K/Akt signaling pathway [66] leading, ultimately, to cancer cell apoptosis. At the same time, compound **44** reduces the expression of the Vascular Endothelial Growth factor (VEGF) in MCF-7 cells, suggesting a role in inhibiting the angiogenesis process. However, no evidence about the effects on normal cell lines has been reported. Hassan et al. (2018) [67] described a series of Schiff bases and evaluated their antiproliferative activities against HepG2 (liver) and MCF-7 (breast) cell lines using the MTT assay. The majority of prepared Schiff bases displayed better antitumor activity than doxorubicin. Compounds **45** and **46** were the most interesting of the series. Compound **45** was the most active against HepG2 cell line compared to doxorubicin (IC_{50} = 66.3 µM versus 80.9 µM), while compound **46** showed high activity against MCF-7 (IC_{50} = 60.8 µM versus 65.6 µM of doxorubicin). They were also demonstrated to induce apoptosis in HepG2 and MCF-7, increasing the caspase-3 levels. Hassanin et al. (2018) [68] reported a series of Schiff bases bearing a pyranoquinolinone moiety. They were evaluated for topoisomerase IIβ (TOP2B) inhibitory activity [69,70] and cytotoxicity against breast cancer cell line (MCF-7). The compounds **47–49** displayed a significant TOP2B cytotoxicity compared to the reference doxorubicin (IC_{50} = 0.042, 0.83, and 0.6 µM versus IC_{50} = 1.17 µM of doxorubicin).

Several Schiff bases derived from 2-aminobenzothiazole were reported by Saipriya et al. (2018) [71], who performed in vitro MTT assay on HeLa cell lines to validate the cytotoxic activity against cervical cancer cells. Compound **50** showed high activity with an IC_{50} value of 2.517 µg/mL (cisplatin: IC_{50} = 17.2 µg/mL). Uddin et al. (2019) [72] studied a series of Schiff bases and evaluated their cytotoxicity against cancer cell lines (HeLa and

MCF-7) and a normal cell line (BHK-21) by means of the MTT assay. Compounds **51** and **53** showed a slight cytotoxic activity against HeLa (IC_{50} = 56.7 and 20.8 μM, respectively, versus 5.13 μM of carboplatin) and BHK-21 cells (IC_{50} = 32.2 and 60.2 μM, respectively). The mechanism of action for the active compound L5 was deepened, studying the pro-apoptotic mechanism by fluorescence microscopy, cell cycle analysis, caspase-9 and -3 activity, reactive oxygen species (ROS) production, and DNA binding. Compound **52** exhibited disintegrated cell membranes and condensed cellular protein, probably due to the lipids' and proteins' oxidation, suggesting that it could be a potent drug against cancer. Several Schiff bases of tetrahydrocurcumin have been recently reported by Mahal et al. (2019) [73] as potential anticancer agents. The in vitro anticancer activity was evaluated against three human cancer cell lines: human epithelial lung carcinoma (A549) and cervical cancer (HeLa) and human breast adenocarcinoma (MCF-7) cells. Most compounds exhibited moderate to good anticancer activity against all three tested cell lines and were significantly more active than tetrahydrocurcumin. The most interesting was compound **53** (IC_{50} = 11.9, 12.7, 4.8 μM, against the three cell lines considered, respectively). Erturk et al. (2020) [74] synthesized and studied two Schiff bases (**54** and **55**) for different biological activities, among them the antitumor one, against MCF-7 human breast cancer cell line. The IC_{50} values were 6.70, 2.20, and <0.1 mM for **54**; 1.00, 0.30, and 0.14 mM for **55** for 24, 48, and 72 h, respectively. The higher activity of compound containing 10-chloroanthracene **55** than that containing 8-hydroxyquinoline **54** was in agreement with theoretical calculations obtained by various spectroscopic analyses and single-crystal X-ray diffraction and Hirshfeld surface analysis and fingerprint plots of the two compounds. Suyambulingam et al. (2020) [75] synthesized two Schiff bases (**56** and **57**) and evaluated their antiproliferative activity against MCF-7 cells, obtaining IC_{50} values of 80.19 μM for compound **56** and 44.12 μM for compound **57** (doxorubicin: IC_{50} = 2.05 μM). Molecular docking studies were also carried out against six different active sites [76,77]. Mishra et al. (2020) [78] studied several Schiff bases containing a benzothiazole nucleus and studied the DNA binding interaction with pBR322 plasmid DNA by means of electrophoretic mobility shift assay [79]. The anticancer study was performed using the MTT assay. Imine **58** showed 85.82% inhibition of MCF-7 cancer cell lines at a concentration of 200 μg/mL. It was less toxic to normal cells at the concentration required to produce the anticancer effect (IC_{50} = 973 μg/mL).

Table 1. Mono-Schiff bases.

1	IC_{50} = 1.5–7 μM (MT-4, CCRF-CEM, WIL-2NS, CCRF-SB).	Vicini et al., 2003 [43]
2	IC_{50} = 1.5–7 μM (MT-4, CCRF-CEM, WIL-2NS, CCRF-SB).	Vicini et al., 2003 [43]
3	IC_{50} = 1.5–7 μM (MT-4, CCRF-CEM, WIL-2NS, CCRF-SB). IC_{50} = 6 μM (SK-MEL-28). IC_{50} = 10 μM (MCF-7).	Vicini et al., 2003 [43]

Table 1. *Cont.*

Structure	Activity	Reference
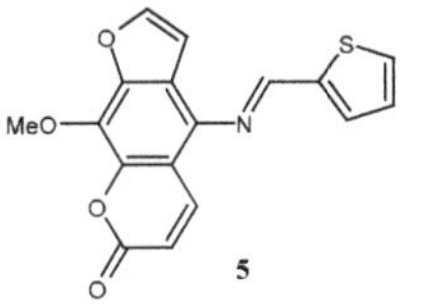 **5**	IC$_{50}$ = 7.2 μM (HeLa).	Abdel-Hafez et al., 2009 [45]
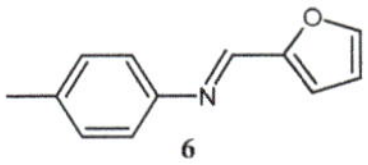 **6**	IC$_{50}$ = 39.9 μM (LAMA-84), 29.9 μM (K-562), 68.6 μM (HL-60/Dox). IC$_{50}$ > 400 μM (HL-60).	Kraicheva et al., 2009 [46]
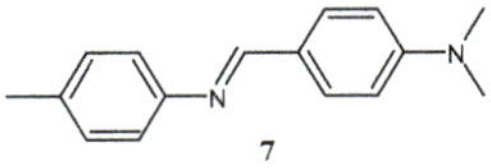 **7**	IC$_{50}$ = 251.9 μM (LAMA-84), 212.9 μM (K-562), 226.1 μM (HL-60/Dox). IC$_{50}$ > 400 μM (HL-60).	Kraicheva et al., 2009 [46]
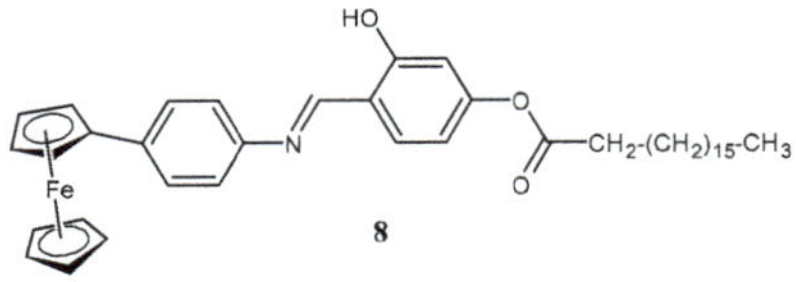 **8**	Tumor growth inhibition = 71%; IC$_{50}$ = 20 μg/mL (Potato disc tumor induction assay using *A. tumefaciens* (At-10).	Nawaz et al., 2009 [47]
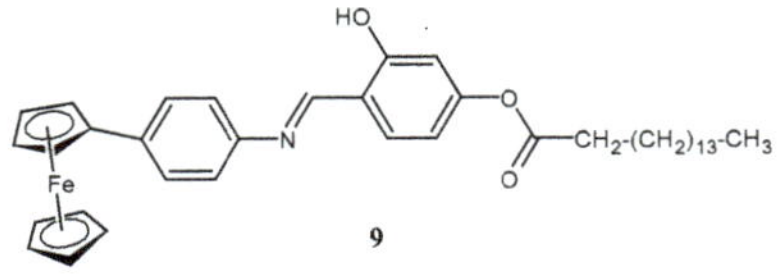 **9**	Tumor growth inhibition = 58%; IC$_{50}$ = 563 μg/mL (Potato disc tumor induction assay using *A. tumefaciens* (At-10).	Nawaz et al., 2009 [47]
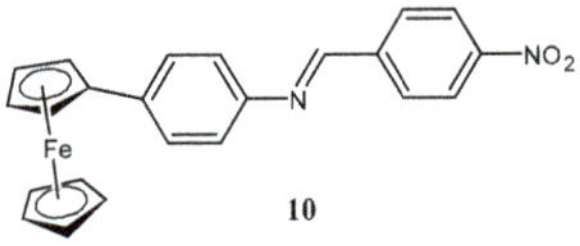 **10**	LD$_{50}$ = 292.95 ppm (brine shrimp lethality assay).	Zaheer et al., 2010 [48]
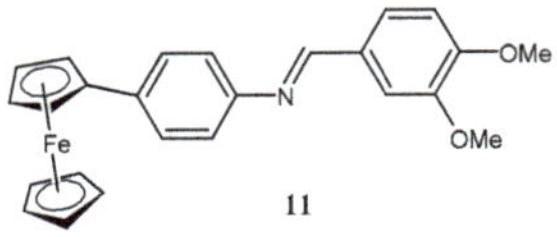 **11**	LD$_{50}$ = 18.22 ppm (brine shrimp lethality assay).	Zaheer et al., 2010 [48]
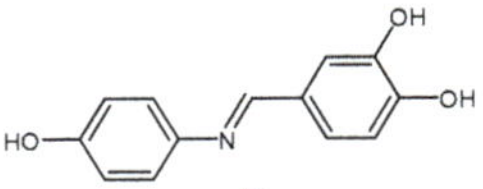 **12**	IC$_{50}$ = 5.6 μM (HepG2). IC$_{50}$ = 17.22 μM (murine B16F10 melanoma cells).	Cheng et al., 2010 [49] Bae et al., 2012 [55]
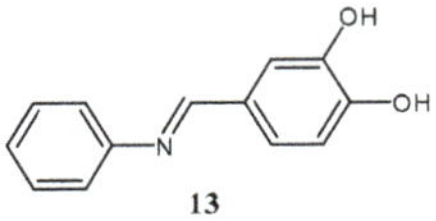 **13**	IC$_{50}$ = 6.8 μM (HepG2).	Cheng et al., 2010 [49]

Table 1. *Cont.*

Structure	Activity	Reference
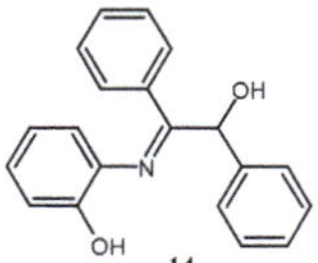 **14** (PHP)	LD_{50} = 16 mg/kg.	Jesmin et al., 2010 [50]
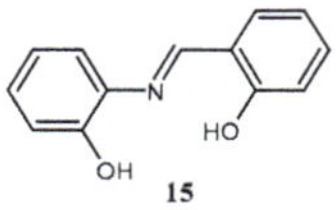 **15** (HHP)	LD_{50} = 15.5 mg/kg; maximum percentage cell growth inhibition = 93% (EAC).	Jesmin et al., 2010 [50]
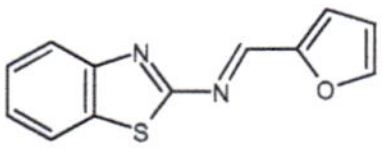 **16**	IC_{50} = 0.186 µM (HeLa).	Etaiw et al., 2011 [51]
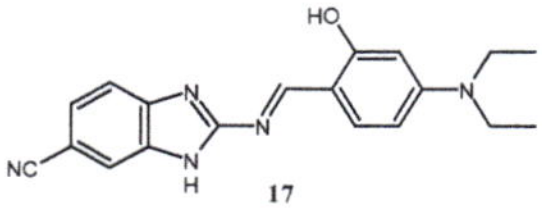 **17**	IC_{50} = 4.73 (HeLa), 9.23 (MCF-7, metastatic).	Hranjec et al., 2011 [52]
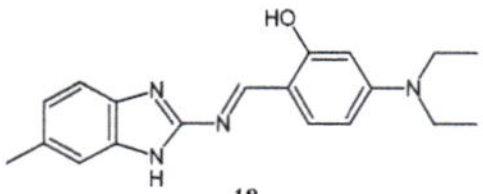 **18**	IC_{50} = 3.24 µM (HeLa), 15.27 µM (MCF-7, metastatic).	Hranjec et al., 2011 [52]
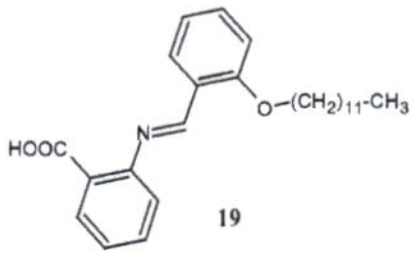 **19**	IC_{50} = 1–10 mg/mL (HepG2, HCT-116 and MCF-7).	Shaker et al., 2011 [53]
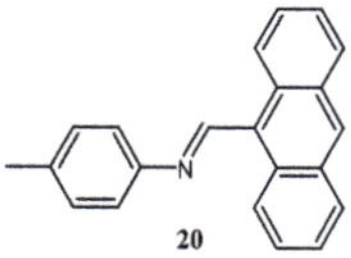 **20**	IC_{50} = 0.08 mg/mL (HT-29 cell line).	Kraicheva et al., 2012 [54]
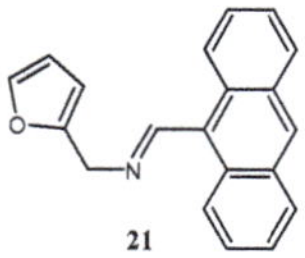 **21**	IC_{50} = 0.20 mg/mL (HT-29 cell line).	Kraicheva et al., 2012 [54]
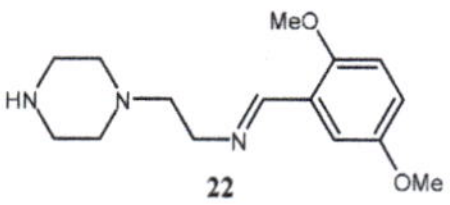 **22**	PG inhibition = 49% (at 1×10^{-5} M against NCI H-522).	Sondhi et al., 2012 [56]

Table 1. *Cont.*

Structure	Activity	Reference
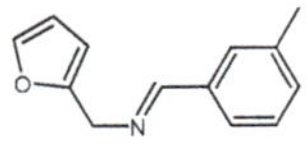 **23**	IC_{50} = 50.12 µM (KYSE 150), 158.49 µM (KYSE30), 111.2 µM (KYSE 270).	Klimczak et al., 2013 [57]
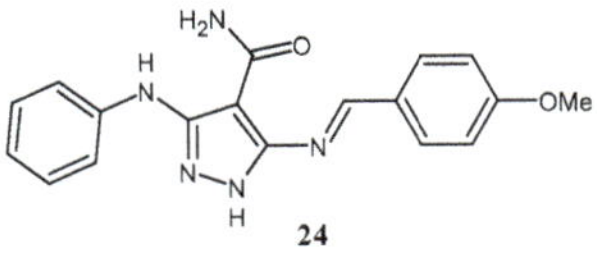 **24**	IC_{50} = 0.44 µM (SK OV-3).	Hafez et al., 2013 [58]
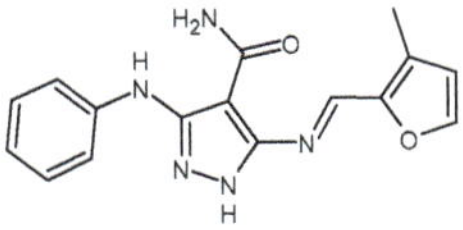 **25**	IC_{50} = 0.09 nM (U937), IC_{50} = 0.45 nM (GOTO), IC_{50} = 0.64 nM (NB-1), IC_{50} = 0.54 nM (HT1080).	Hafez et al., 2013 [58]
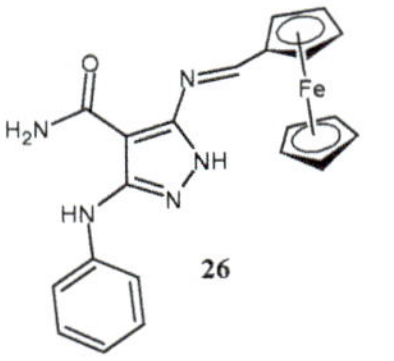 **26**	IC_{50} = 0.54 µM (KB), IC_{50} = 0.30 nM (CNS, SF-268), IC_{50} = 0.43 nM (K-562), IC_{50} = 0.09 nM (HepG2), IC_{50} = 6.60 nM (NCI H460).	Hafez et al., 2013 [58]
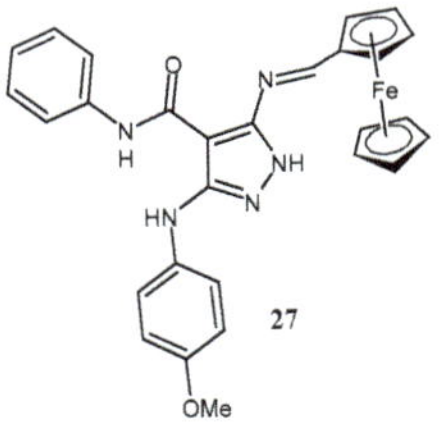 **27**	IC_{50} = 6.20 µg/mL (HepG2), IC_{50} = 7 µg/mL (MCF-7).	Hassan et al., 2015 [59]
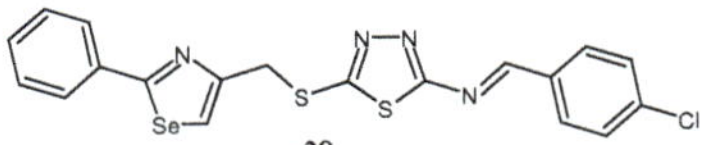 **28**	IC_{50} = 4.02 µM (MCF-7).	Zhao et al., 2013 [60]
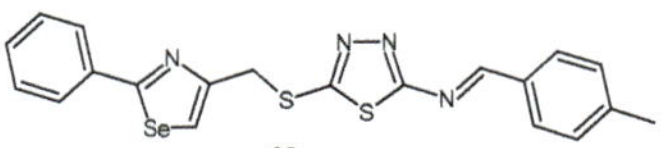 **29**	IC_{50} = 7.55 µM (MCF-7).	Zhao et al., 2013 [60]
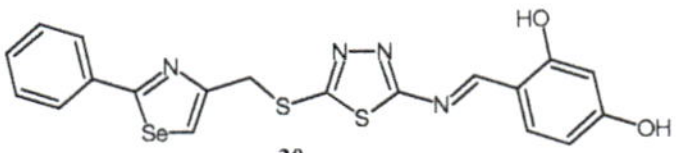 **30**	IC_{50} = 8.51 µM (MCF-7).	Zhao et al., 2013 [60]
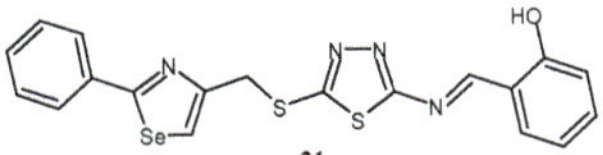 **31**	IC_{50} = 38.73 µM (L1210).	Zhao et al., 2013 [60]
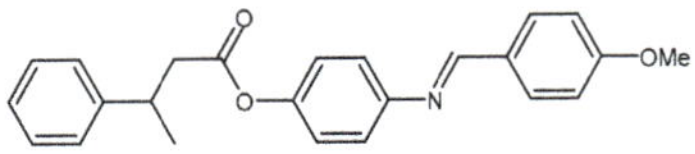 **32**	IC_{50} = 0.15 µg/mL (potato disc).	Noureen et al., 2013 [61]

Table 1. *Cont.*

Structure	Activity	Reference
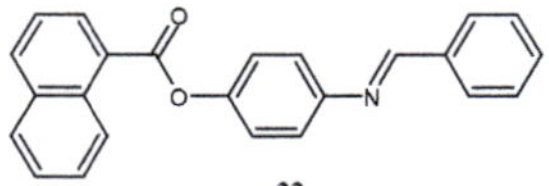 **33**	IC$_{50}$ = 8.03 µg/mL (potato disc).	Noureen et al., 2013 [61]
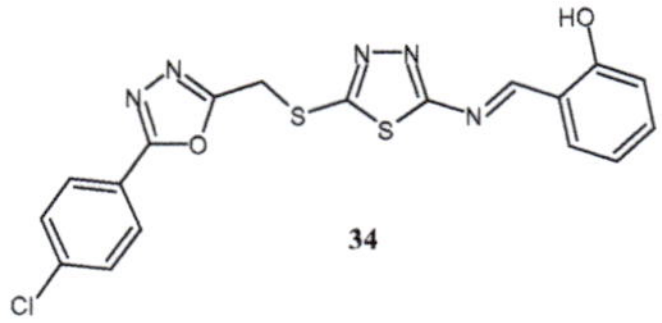 **34**	IC$_{50}$ = 2.84 µM (SMMC-7721).	Zhang et al., 2014 [62]
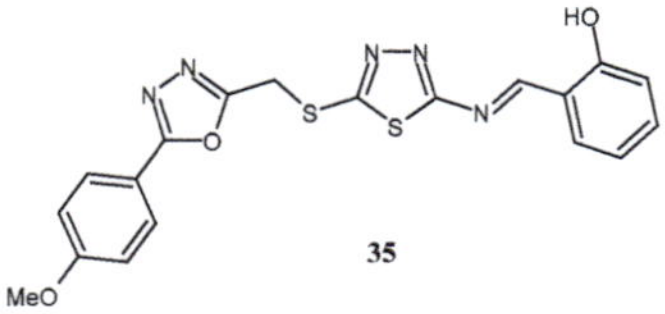 **35**	IC$_{50}$ = 4.56 µM (MCF-7).	Zhang et al., 2014 [62]
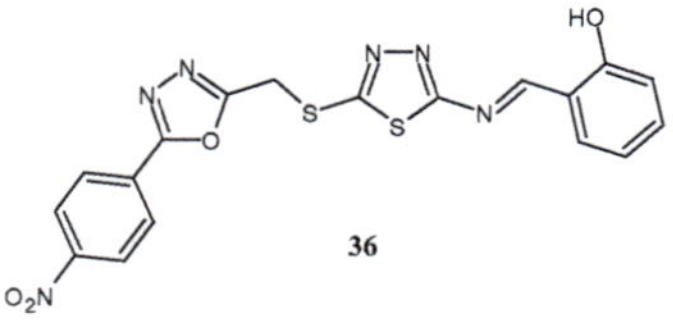 **36**	IC$_{50}$ = 4.25 µM (MCF-7), IC$_{50}$ = 4.13 µM (A549).	Zhang et al., 2014 [62]
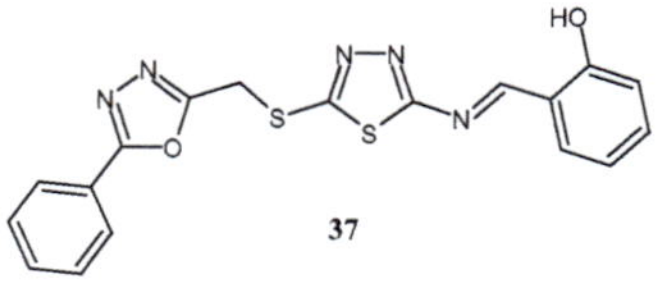 **37**	IC$_{50}$ = 4.11 µM (A549).	Zhang et al., 2014 [62]
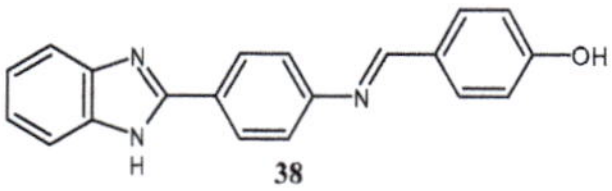 **38**	IC$_{50}$ = 4.83 µM (PC3).	Gupta et al., 2015 [63]
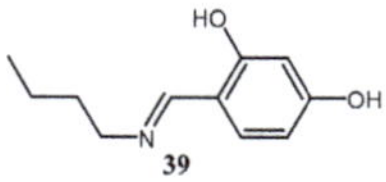 **39**	IC$_{50}$ = 7.43 µM (PC3).	Gupta et al., 2015 [63]
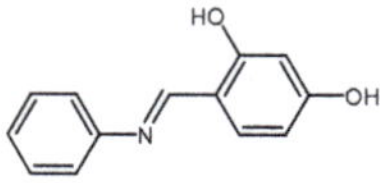 **40**	IC$_{50}$ = 7.15 µM (PC3).	Gupta et al., 2015 [63]
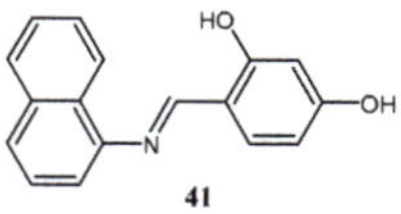 **41**	IC$_{50}$ value of 0.02 µM (Hsp90 ATPase suppression).	Gupta et al., 2015 [63]

Table 1. *Cont.*

Structure	Activity	Reference
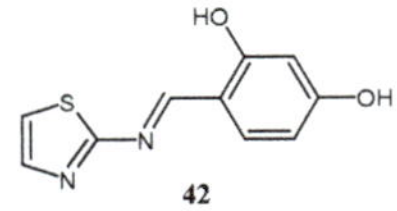 **42**	IC$_{50}$ = 0.02 μM (Hsp90 ATPase suppression).	Gupta et al., 2015 [63]
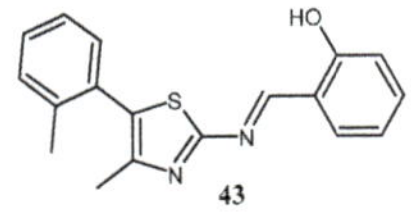 **43**	IC$_{50}$ = 9.22 (HepG2), IC$_{50}$ = 10.00 μM (MCF-7), IC$_{50}$ = 9.50 μM (HCT116).	Abd-Elzaher et al., 2016 [64]
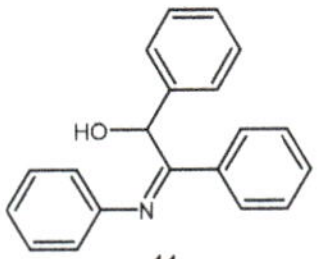 **44**	IC$_{50}$ = 0.024 M (MCF-7), IC$_{50}$ = 0.034 M (T47D).	Sabbah et al., 2018 [65]
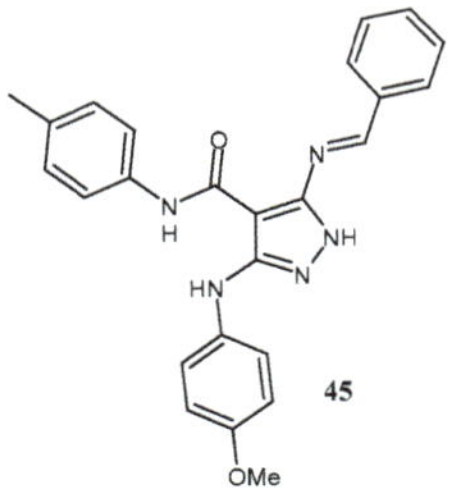 **45**	IC$_{50}$ = 66.3 μM (HepG2).	Hassan et al., 2018 [67]
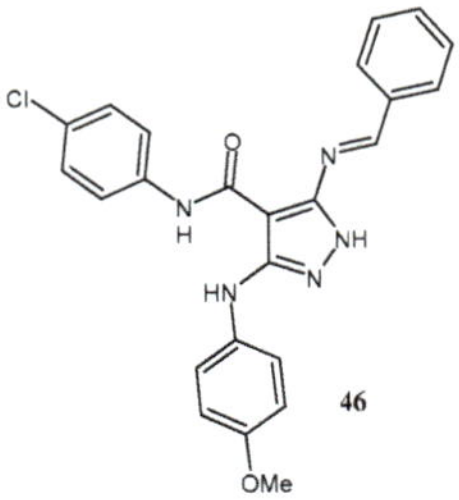 **46**	IC$_{50}$ = 60.8 μM (MCF-7).	Hassan et al., 2018 [67]
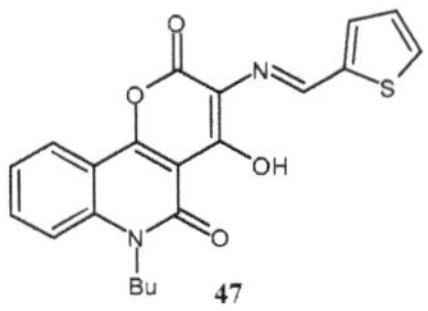 **47**	IC$_{50}$ = 0.042 μM (MCF-7).	Hassanin et al., 2018 [68]
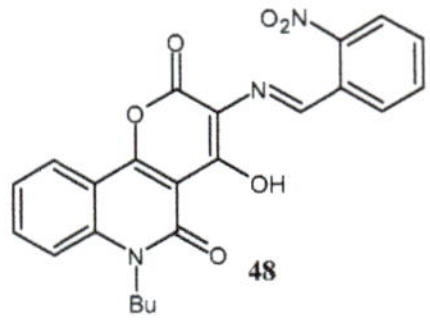 **48**	IC$_{50}$ = 0.83 μM (MCF-7).	Hassanin et al., 2018 [68]

Table 1. *Cont.*

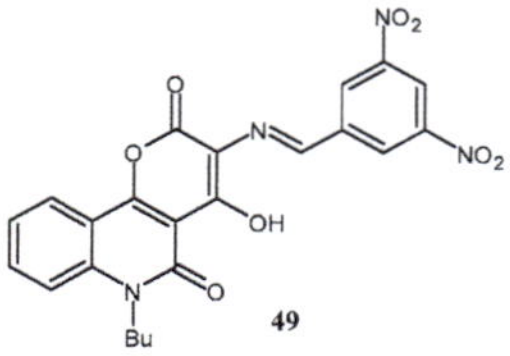 **49**	IC$_{50}$ = 0.6 μM (MCF-7).	Hassanin et al., 2018 [68]
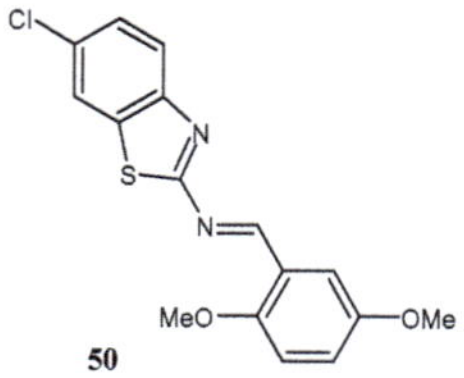 **50**	IC$_{50}$ = 2.517 μg/mL (HeLa).	Saipriya et al., 2018 [71]
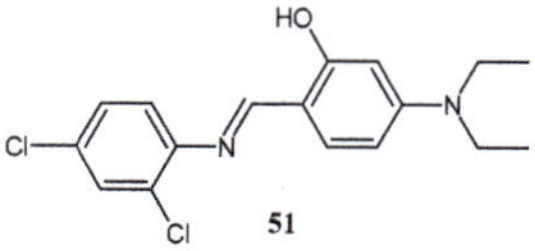 **51**	IC$_{50}$ = 56.7 μM (HeLa), IC$_{50}$ = 32.2 μM (BHK-21).	Uddin et al., 2019 [72]
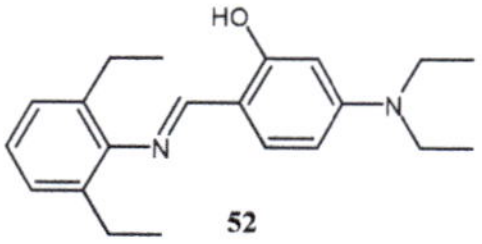 **52**	IC$_{50}$ = 20.8 μM (HeLa), IC$_{50}$ = 60.2 μM (BHK-21).	Uddin et al., 2019 [72]
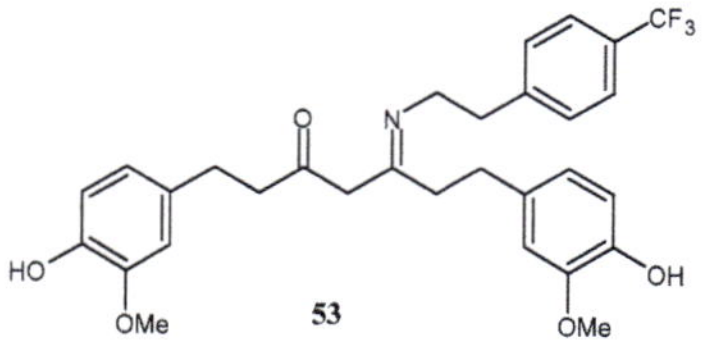 **53**	IC$_{50}$ = 11.9 μM (A549), IC$_{50}$ = 12.7 μM (HeLa), IC$_{50}$ = 4.8 μM (MCF-7).	Mahal et al., 2019 [73]
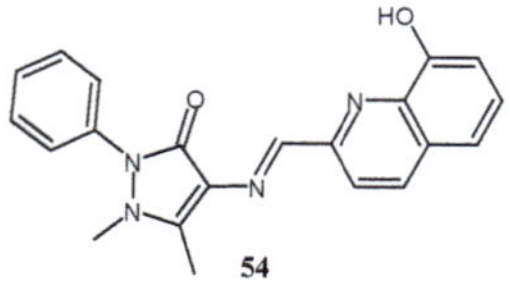 **54**	IC$_{50}$ = 6.70, 2.20, and < 0.1 mM (for 24, 48 and 72 h against MCF-7).	Erturk et al., 2020 [74]
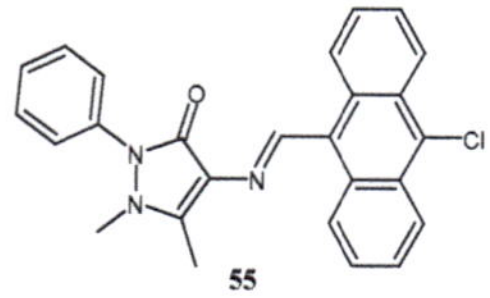 **55**	IC$_{50}$ = 1.00, 0.30 and 0.14 mM (for 24, 48 and 72 h against MCF-7).	Erturk et al., 2020 [74]
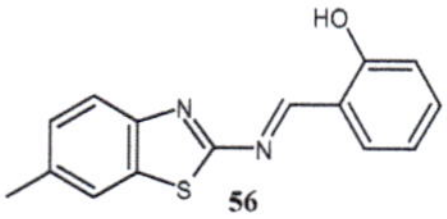 **56**	IC$_{50}$ = 80.19 μM (MCF-7).	Suyambulingam et al., 2020 [75]

Table 1. *Cont.*

57	IC_{50} = 44.12 µM (MCF-7).	Suyambulingam et al., 2020 [75]
58	Percentage inhibition = 85.82% (at 200 µg/mL against MCF-7).	Mishra et al., 2020 [78]

2.2. Bis-Schiff Bases

Desai et al. (2001) [80] reported a study on 10 Schiff bases and evaluated their antiproliferative activity by measuring their effect on the (PG) of 57 different cancer cell lines, including lung, colon, central nervous system (CNS), ovarian, renal, prostrate, melanoma, leukemia, and breast cancer. Compounds **59–62** (Table 2) showed activity against different cell lines. Particularly, compounds **59** and **60** were effective on leukemia (SR and MOLT-4) and colon (COLO-205), compound **62** on leukemia (SR and MOLT-4), CNS (SF-539) and melanoma (SK-MEL-28 and UACC-257). Against some of the abovementioned cell lines, other compounds have also been reported to be active, particularly, compound **61** on CNS, melanoma, and breast and compound **59** on leukemia, colon, and breast. Compound **60** was the most effective of imines analogues on leukemia. Padhye et al. (2009) [81] studied several Schiff bases and copper complexes as proteasome inhibitors and apoptosis inducers in human colon cancer HCT-116 cells, in comparison to curcumin. Among the synthesized compounds, the authors demonstrated that the fluorine-substituted curcumin analogs were superior to the curcumin against HCT-116 and BxPC-3 pancreatic cancer cells, probably because of the higher metabolic stability allowed by the fluoro substituents. Compound **63** showed approximately 80%, 60%, and 60% proteasome inhibition at 10, 20, and 30 µM (curcumin: 27%, 47%, and 64% at 10, 20, and 30 µM, respectively). In the study by Sondhi et al. (2012) [56], described in the paragraph below, the activity of mono-Schiff bases and bis-Schiff bases was evaluated. Compound **64** showed activity against ovary (PA1) cancer cells (62% versus 93% of actinomycin-D) and **65** against both breast (T47D) and ovary (PA1) (41% and 53%, respectively, versus 21% and 93% of actinomycin-D, respectively) cancer cells.

Shokrollahi et al. (2020) [82] recently studied four tetrahydrobenzothiazole-based Schiff bases and tested their cytotoxic activity against the human breast cancer (MCF-7) and hepatocellular liver carcinoma (HepG2) cell lines by MTT assay. The compounds showed cytotoxic activity against both cell lines in a concentration-dependent manner. Compound **66** was the most active against MCF-7 (IC_{50} = 7.75 and 34.52 µM, at 24 and 48 h, respectively) and HepG2 (IC_{50} = 3.01 and 1.29 µM, at 24 and 48 h, respectively). Morsy et al. (2021) [83] recently reported some selected bis-Schiff bases studied for their in vitro antiproliferative activity toward three human carcinoma HepG2 (liver), MCF-7 (breast), and RPE-1 (normal retina pigmented epithelium) cell lines using MTT assay. The results showed that compound **67** was found to be the active candidate against HepG2 (IC_{50} = 84.2 µM versus 25.3 µM of doxorubicin) and MCF-7 cells (IC_{50} = 99.4 µM versus 20.9 µM of doxorubicin), while compound **68** was found to be the most active of the series against RPE-1 cells (IC_{50} = 127.7 µM versus 19.1 µM of doxorubicin).

Table 2. Bis-Schiff bases.

Structure	Activity	Reference
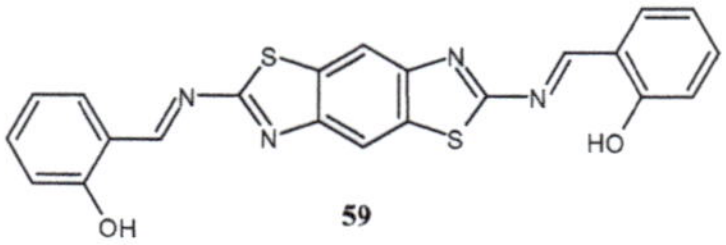 **59**	effective on leukemia (SR and MOLT-4), colon (COLO-205) and breast. IC_{50} values not reported.	Desai et al., 2001 [80]
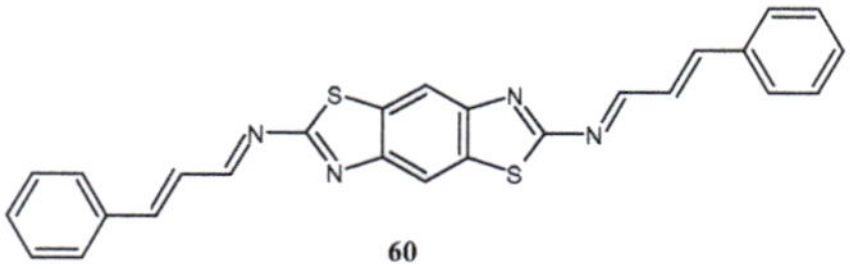 **60**	effective on leukemia (SR and MOLT-4), colon (COLO-205) and leukemia. IC_{50} values not reported.	Desai et al., 2001 [80]
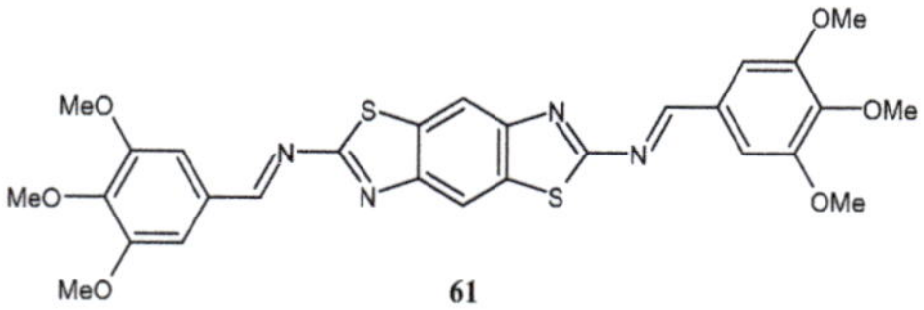 **61**	effective on CNS (SF-539), melanoma (SK-MEL-28 and UACC-257) and breast. IC_{50} values not reported.	Desai et al., 2001 [80]
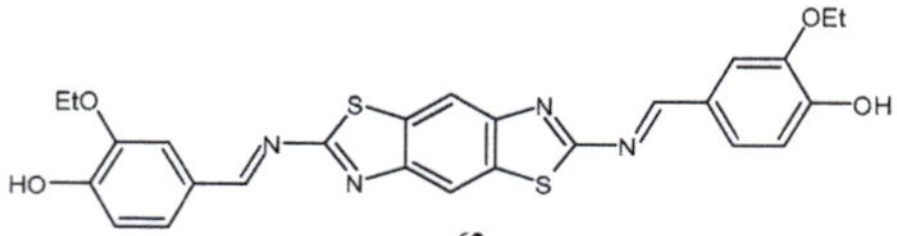 **62**	effective on leukemia (SR and MOLT-4), CNS (SF-539), melanoma (SK-MEL-28 and UACC-257). IC_{50} values not reported.	Desai et al., 2001 [80]
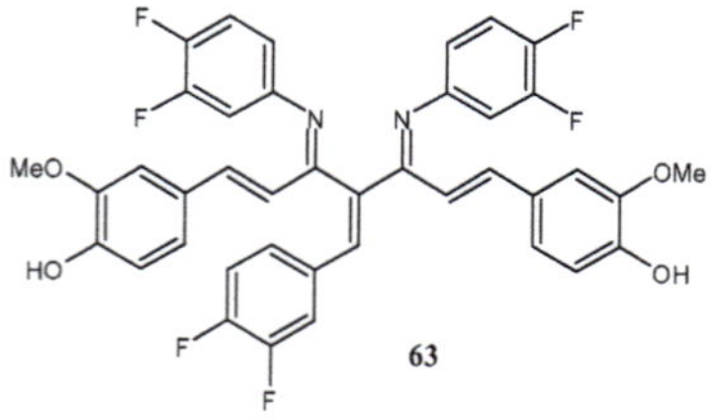 **63**	proteasome inhibition: about 80%, 60% and 60% at 10, 20 and 30 μM.	Padhye et al., 2009 [81]
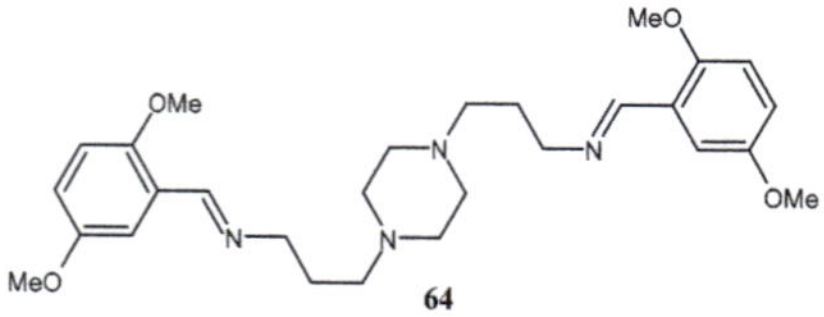 **64**	PG inhibition: 62% (ovary, PA1).	Sondhi et al., 2012 [56]
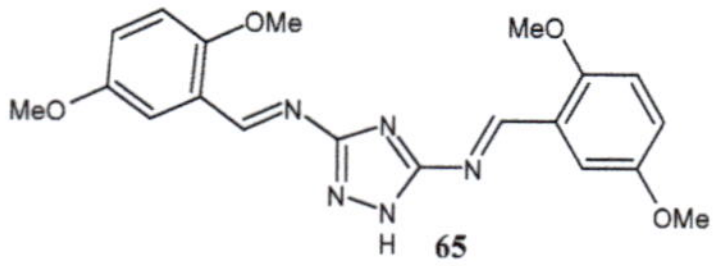 **65**	PG inhibition: 41% (T47D) and 53% (PA1).	Sondhi et al., 2012 [56]
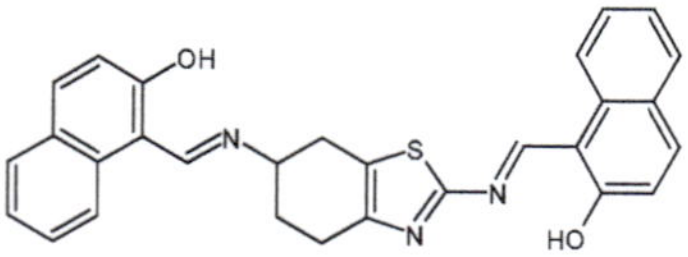 **66**	$IC_{50} = 7.75$ and 34.52 μM, at 24 and 48 h, against MCF-7), $IC_{50} = 3.01$ and 1.29 μM, at 24 and 48 h, against HepG2).	Shrollaki et al., 2020 [82]

Table 2. *Cont.*

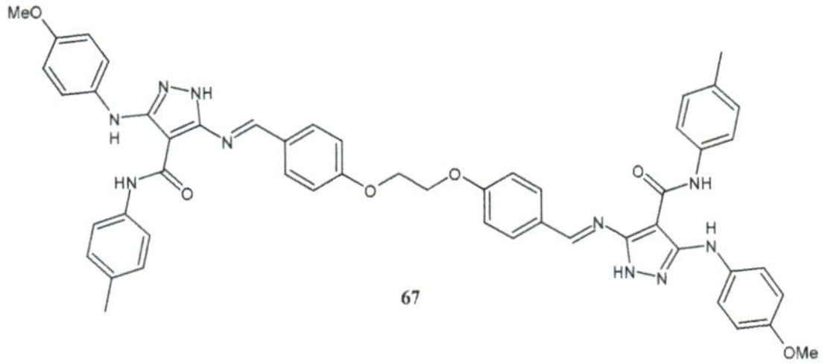	IC$_{50}$ = 84.2 μM (H₂pG2), IC$_{50}$ = 99.4 μM (MCF-7).	Morsy et al., 2021 [83]

67

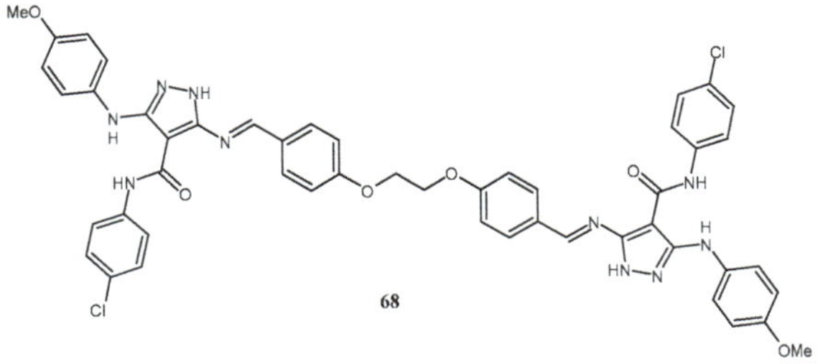	IC$_{50}$ = 127.7 μM (RPE-1).	Morsy et al., 2021 [83]

68

3. Summary

Schiff bases have long attracted researchers due to their chemical reactivity and to the broad range of pharmacological activities that they exert as such or complexed with metals, including antibacterial, antifungal, anti-inflammatory, antioxidant, and anticancer. They are also employed as versatile tools in several applications such as fluorescent turn-on/turn-off sensors for the determination of diverse analytes. Their easy preparation and capability of forming complexes with almost all metals make them interesting compounds in medicinal chemistry. Recently, several organic compounds bearing Schiff base structure or their complexes with metals were used as effective drugs against cancer. Metal complexes' actions are multiple, depending on the broad range of coordination numbers, geometries, and kinetic properties and, after the worldwide use of cisplatin, different papers reported the importance of Schiff bases' anticancer actions in metal complexes. The reviewed paper indicated that the use of these compounds offered better anticancer properties with respect to the reference molecules, viz., cisplatin, doxorubicin, and vincristine, for instance, both in in vitro and in vivo. It is worthy to note that these compounds displayed the anticancer effects against a very broad variety of tumor cell models, solid or liquid, without hampering, in the most cases, the growth of the normal cells used as control. Furthermore, Schiff bases may target different intracellular regulator enzymes, together with the already known interactions with nuclear DNA, producing cancer cells' death by apoptosis. Finally, different evidence about their ability to modulate the intracellular redox equilibrium, strongly associated with tumor prevention, onset, and progression, have been revealed, confirming the multiple actions exerted by these molecules. In this review, studies regarding mono- and bis-Schiff bases with potent antitumor activity on several cell lines were reviewed. In the future, the study of structure–activity relationships of Schiff bases against cancer cells may help in synthesizing new and effective antitumor agents derived by modification of the already studied imines.

Funding: No financial support.

Institutional Review Board Statement: Not applicable.

Informed Consent Statement: Not applicable.

Conflicts of Interest: The authors declare no conflict of interest.

Abbreviations

647-V	bladder carcinoma cell lines
A549	lung cancer cell lines
B16F10	murine melanoma cells
BGC-823	stomach cancer cell lines
BHK-21	normal cell line
CCRF-CEM	human CD4$^+$ acute T-lymphoblastic leukemia cells
CCRF-SB	human acute B-lymphoblastic leukemia cells
SF-268	central nervous system (CNS) cancer cell lines
SF-539	colon cell lines
COLO-205	colon cell lines
EAC	Ehrlich Ascites Carcinoma
5-FU	5-Fluorouracil
GOTO	neuroblastoma cancer cell lines
HBL-100	colostrum derived myoepithelial cells, expressing polyoma virus large T-antigen line
HCT-15	colon cancer cells
HCT-116	human colon cancer cells lines
HeLa	cervical carcinoma cell lines
Hep-2	larynx cancer cell lines
HepG2	human hepatoma cell lines
HHP	N-(2-hydroxy benzylidine)-2-hydroxyl phenyl imine
HL-60	leukemia cell lines
HL-60/Dox	multi-drug resistant acute myeloblastic leukemia cell line
HT1080	fibrosarcoma cell lines
IC$_{50}$	concentration which kills or inhibits cell viability by 50%
K-562	non-adherent chronic myelogenous leukemia cells of the erythroleukemia type
KB	cervical carcinoma cell lines
KYSE 30	esophageal cancer cell lines
KYSE 150	esophageal cancer cell lines
KYSE 270	esophageal cancer cell lines
L1210	mouse lymphocyte leukemia cells
LAMA-84	peripheral chronic myeloid leukemia cells
LD$_{50}$	medium lethal concentration
MCF-7	human breast adenocarcinoma cell lines (low metastatic potential)
MDA-MB-231	human breast adenocarcinoma cell lines (high metastatic potential)
MiaPaCa-2	pancreatic carcinoma
MOLT-4	leukemia cell lines
MT-4	human CD4$^+$ lymphocytes
MTT	3-[4,5-dimethylthiazol-2-yl]-2,5-diphenyltratrazolium bromide
NB-1	neuroblastoma cancer cell lines
NCI H460	non-small cell lung cancer cell lines
NCI H-522	lung cancer cell lines
NRU	Neutral Red Uptake
PA1	ovary cancer cell lines
PC3	prostate cancer cell lines
PCS	2-phenyl-4-carboxyl-1,3-selenazole
PG	percentage growth
PHP	N-(1-phenyl, 2-hydroxy-2-phenyl ethylidine)-2-hydroxyl phenyl imine
ROS	reactive oxygen species
RPE-1	normal retina pigmented epithelium
SK-MEL-28	melanoma cell lines
SK OV-3	ovarian carcinoma
SMMC-7721	human liver cancer cell lines
SRB	Sulforhodamine-B stain
SW620	colorectal adenocarcinoma, metastatic

T47D	breast cancer cell lines
U937	leukemia cell lines
UACC-257	melanoma cell lines
WIL-2NS	human splenic B-lymphoblastoid cells
WST-8	2-(2-methoxy-4-nitrophenyl)-3-(4-nitrophenyl)-5-(2,4-disulfophenyl)-2*H*-tetrazolium monosodium salt

References

1. Qin, W.; Long, S.; Panunzio, M.; Biondi, S. Schiff bases: A short survey on an evergreen chemistry tool. *Molecules* **2013**, *18*, 12264–12289. [CrossRef] [PubMed]
2. Hameed, A.; al-Rashida, M.; Uroos, M.; Ali, S.A.; Khan, K.M. Schiff bases in medicinal chemistry: A patent review (2010–2015). *Expert Opin. Ther. Pat.* **2017**, *27*, 63–79. [CrossRef] [PubMed]
3. Murtaza, G.; Mumtaz, A.; Khan, F.A.; Ahmad, S.; Azhar, S.; Najam-Ul-Haq, M.; Atif, M.; Khan, S.A.; Maalik, A.; Alam, F.; et al. Recent pharmacological advancements in schiff bases: A review. *Acta Pol. Pharm.* **2014**, *71*, 531–535. [PubMed]
4. da Silva, C.M.; da Silva, D.L.; Modolo, L.V.; Alves, R.B.; de Resende, M.A.; Martins, C.V.; de Fatima, A. Schiff bases: A short review of their antimicrobial activities. *J. Adv. Res.* **2011**, *2*, 1–8. [CrossRef]
5. de Fátima, Â.; de Paula Pereira, C.; Olímpio, C.R.; de Freitas Oliveira, B.G.; Franco, L.L.; da Silva, P.H. Schiff bases and their metal complexes as urease inhibitors–a brief review. *J. Adv. Res.* **2018**, *13*, 113–126. [CrossRef]
6. Kajal, A.; Bala, S.; Kamboj, S.; Sharma, N.; Saini, V. Schiff bases: A versatile pharmacophore. *J. Catal.* **2013**, *2013*, 893512. [CrossRef]
7. Pandey, A.; Dewangan, D.; Verma, S.; Mishra, A.; Dubey, R.D. Synthesis of Schiff bases of 2-amino-5-aryl-1, 3,4-thiadiazole and its analgesic, anti-inflammatory, antibacterial and antitubercular activity. *Int. J. Chem. Tech. Res.* **2011**, *3*, 178–184.
8. Alafeefy, A.M.; Bakht, M.A.; Ganaie, M.A.; Ansarie, M.N.; El-Sayed, N.N.; Awaad, A.S. Synthesis, analgesic, anti-inflammatory and anti-ulcerogenic activities of certain novel Schiff's bases as fenamate isosteres. *Bioorg. Med. Chem. Lett.* **2015**, *25*, 179–183. [CrossRef]
9. Kumar, M.; Padmini, T.; Ponnuvel, K. Synthesis, characterization and antioxidant activities of Schiff bases are of cholesterol. *J. Saudi Chem. Soc.* **2017**, *21*, S322–S328. [CrossRef]
10. Teran, R.; Guevara, R.; Mora, J.; Dobronski, L.; Barreiro-Costa, O.; Beske, T.; Pérez-Barrera, J.; Araya-Maturana, R.; Rojas-Silva, P.; Poveda, A.; et al. Characterization of antimicrobial, antioxidant, and leishmanicidal activities of schiff base derivatives of 4-aminoantipyrine. *Molecules* **2019**, *24*, 2696. [CrossRef]
11. Shah, S.S.; Shah, D.; Khan, I.; Ahmad, S.; Ali, U.; ur Rahman, A. Synthesis and Antioxidant Activities of Schiff Bases and Their Complexes: An Updated Review. *Biointerf. Res. Appl. Chem.* **2020**, *10*, 6936–6963.
12. Zehra, S.; Shavez Khan, M.; Ahmad, I.; Arjmand, F. New tailored substituted benzothiazole Schiff base Cu(II)/Zn(II) antitumor drug entities: Effect of substituents on DNA binding profile, antimicrobial and cytotoxic activity. *J. Biomol. Struct. Dynam.* **2019**, *37*, 1863–1879. [CrossRef]
13. Utreja, D.V.; Singh, S.; Kaur, M. Schiff bases and their metal complexes as anti-cancer agents: A review. *Curr. Bioact. Compd.* **2015**, *11*, 215–230. [CrossRef]
14. Sztanke, K.; Maziarka, A.; Osinka, A.; Sztanke, M. An insight into synthetic Schiff bases revealing antiproliferative activities in vitro. *Bioorg. Med. Chem.* **2013**, *21*, 3648–3666. [CrossRef]
15. Demirci, S.; Doğan, A.; Başak, N.; Telci, D.; Dede, B.; Orhan, C.; Tuzcu, M.; Şahin, K.; Şahin, N.; Özercan, I.H.; et al. A Schiff base derivative for effective treatment of diethylnitrosamine-induced liver cancer in vivo. *Anti-Cancer Drugs* **2015**, *26*, 555–564. [CrossRef] [PubMed]
16. Berhanu, A.L.; Mohiuddin, I.; Malik, A.K.; Aulakh, J.S.; Kumar, V.; Kim, K.-H. A review of the applications of Schiff bases as optical chemical sensors. *TrAC Trends Anal. Chem.* **2019**, *116*, 74–91. [CrossRef]
17. Nourifard, F.; Payehghadr, M. Conductometric studies and application of new Schiff base ligand as carbon paste electrode modifier for mercury and cadmium determination. *Int. J. Environ. Anal. Chem.* **2016**, *96*, 552–567. [CrossRef]
18. Rahimi, M.; Amini, A.; Behmadi, H. Novel symmetric Schiff-base benzobisthiazole-salicylidene derivative with fluorescence turn-on behavior for detecting Pb^{2+} ion. *J. Photochem. Photobiol. A Chem.* **2020**, *388*, 112190. [CrossRef]
19. Gupta, V.K.; Singh, A.K.; Gupta, B. A cerium(III) selective polyvinyl chloride membrane sensor based on a Schiff base complex of N,N′-bis[2-(salicylideneamino)ethyl]ethane-1,2-diamine. *Anal. Chim. Acta* **2006**, *575*, 198–204. [CrossRef] [PubMed]
20. Milosavljevic, V.; Haddad, Y.; Merlos Rodrigo, M.A.; Moulick, A.; Polanska, H.; Hynek, D.; Hegel, Z.; Kopel, P.; Adam, V. The zinc-schiff base-novicidin complex as a potential prostate cancer therapy. *PLoS ONE* **2016**, *11*, e0163983. [CrossRef]
21. Catalano, A.; Iacopetta, D.; Pellegrino, M.; Aquaro, S.; Franchini, C.; Sinicropi, M.S. Diarylureas: Repositioning from Antitumor to Antimicrobials or Multi-Target Agents against New Pandemics. *Antibiotics* **2021**, *10*, 92. [CrossRef] [PubMed]
22. Abu-Dief, A.M.; Mohamed, I.M. A review on versatile applications of transition metal complexes incorporating Schiff bases. *Beni-Suef Univ. J. Basic Appl. Sci.* **2015**, *4*, 119–133.
23. Ghosh, P.; Dey, S.; Ara, M.; Karim, K.; Islam, A.B. A review on synthesis and versatile applications of some selected Schiff bases with their transition metal complexes. *Egypt. J. Chem.* **2020**, *63*, 5–6. [CrossRef]
24. Sakthivel, A.; Jeyasubramanian, K.; Thangagiri, B.; Raja, J.D. Recent advances in Schiff base metal complexes derived from 4-amoniantipyrine derivatives and their potential applications. *J. Mol. Struct.* **2020**, *1222*, 128885. [CrossRef]

25. Saturnino, C.; Barone, I.; Iacopetta, D.; Mariconda, A.; Sinicropi, M.S.; Rosano, C.; Campana, A.; Catalano, S.; Longo, P.; Ando, S. N-heterocyclic carbene complexes of silver and gold as novel tools against breast cancer progression. *Future Med. Chem.* **2016**, *8*, 2213–2229. [CrossRef]

26. Chimento, A.; Saturnino, C.; Iacopetta, D.; Mazzotta, R.; Caruso, A.; Plutino, M.R.; Mariconda, A.; Ramunno, A.; Sinicropi, M.S.; Pezzi, V.; et al. Inhibition of human topoisomerase I and II and anti-proliferative effects on MCF-7 cells by new titanocene complexes. *Bioorg. Med. Chem.* **2015**, *23*, 7302–7312. [CrossRef]

27. Iacopetta, D.; Mariconda, A.; Saturnino, C.; Caruso, A.; Palma, G.; Ceramella, J.; Muia, N.; Perri, M.; Sinicropi, M.S.; Caroleo, M.C.; et al. Novel gold and silver carbene complexes exert antitumor effects triggering the reactive oxygen species dependent intrinsic apoptotic pathway. *ChemMedChem* **2017**, *12*, 2054–2065. [CrossRef]

28. Sirignano, E.; Saturnino, C.; Botta, A.; Sinicropi, M.S.; Caruso, A.; Pisano, A.; Lappano, R.; Maggiolini, M.; Longo, P. Synthesis, characterization and cytotoxic activity on breast cancer cells of new half-titanocene derivatives. *Bioorg. Med. Chem. Lett.* **2013**, *23*, 3458–3462. [CrossRef]

29. Ceramella, J.; Mariconda, A.; Iacopetta, D.; Saturnino, C.; Barbarossa, A.; Caruso, A.; Rosano, C.; Sinicropi, M.S.; Longo, P. From coins to cancer therapy: Gold, silver and copper complexes targeting human topoisomerases. *Bioorg. Med. Chem. Lett.* **2020**, *30*, 126905. [CrossRef] [PubMed]

30. Iacopetta, D.; Rosano, C.; Sirignano, M.; Mariconda, A.; Ceramella, J.; Ponassi, M.; Saturnino, C.; Sinicropi, M.S.; Longo, P. Is the way to fight cancer paved with gold? Metal-based carbene complexes with multiple and fascinating biological features. *Pharmaceuticals* **2020**, *13*, 91. [CrossRef]

31. Matela, G. Schiff Bases and Complexes: A Review on Anti-Cancer Activity. *Anti-Cancer Agents Med. Chem. Former. Curr. Med. Chem. Anti-Cancer Agents* **2020**, *20*, 1908–1917. [CrossRef]

32. Hodnett, E.M.; Mooney, P.D. Antitumor activities of some Schiff bases. *J. Med. Chem.* **1970**, *13*, 786. [CrossRef]

33. Hodnett, E.M.; Dunn, W.J. Structure-antitumor activity correlation of some Schiff bases. *J. Med. Chem.* **1970**, *13*, 768–770. [CrossRef]

34. Iacopetta, D.; Lappano, R.; Mariconda, A.; Ceramella, J.; Sinicropi, M.S.; Saturnino, C.; Talia, M.; Cirillo, F.; Martinelli, F.; Puoci, F. Newly synthesized imino-derivatives analogues of resveratrol exert inhibitory effects in breast tumor cells. *Int. J. Mol. Sci.* **2020**, *21*, 7797. [CrossRef]

35. Catalano, A.; Iacopetta, D.; Sinicropi, M.S.; Franchini, C. Diarylureas as Antitumor Agents. *Appl. Sci.* **2021**, *11*, 374. [CrossRef]

36. Sinicropi, M.S.; Caruso, A.; Conforti, F.; Marrelli, M.; El Kashef, H.; Lancelot, J.C.; Rault, S.; Statti, G.A.; Menichini, F. Synthesis, inhibition of no production and antiproliferative activities of some indole derivatives. *J. Enzym. Inhib. Med. Chem.* **2009**, *24*, 1148–1153. [CrossRef] [PubMed]

37. Iacopetta, D.; Catalano, A.; Ceramella, J.; Barbarossa, A.; Carocci, A.; Fazio, A.; La Torre, C.; Caruso, A.; Ponassi, M.; Rosano, C.; et al. Synthesis, anticancer and antioxidant properties of new indole and pyranoindole derivatives. *Bioorg. Chem.* **2020**, *105*, 104440. [CrossRef] [PubMed]

38. Sinicropi, M.S.; Iacopetta, D.; Rosano, C.; Randino, R.; Caruso, A.; Saturnino, C.; Muia, N.; Ceramella, J.; Puoci, F.; Rodriquez, M.; et al. N-thioalkylcarbazoles derivatives as new anti-proliferative agents: Synthesis, characterisation and molecular mechanism evaluation. *J. Enzym. Inhib. Med. Chem.* **2018**, *33*, 434–444. [CrossRef]

39. Iacopetta, D.; Carocci, A.; Sinicropi, M.S.; Catalano, A.; Lentini, G.; Ceramella, J.; Curcio, R.; Caroleo, M.C. Old Drug Scaffold, New Activity: Thalidomide-Correlated Compounds Exert Different Effects on Breast Cancer Cell Growth and Progression. *ChemMedChem* **2017**, *12*, 381–389. [CrossRef]

40. Filosa, R.; Peduto, A.; de Caprariis, P.; Saturnino, C.; Festa, M.; Petrella, A.; Pau, A.; Pinna, G.A.; Colla, P.L.; Busonera, B.; et al. Synthesis and antiproliferative properties of N3/8-disubstituted-3,8-diazabicyclo[3.2.1]octane analogues of 3,8-bis[2-(3,4,5-trimethoxyphenyl)pyridin-4-yl]methyl-piperazine. *Eur. J. Med. Chem.* **2007**, *42*, 293–306. [CrossRef] [PubMed]

41. Ceramella, J.; Caruso, A.; Occhiuzzi, M.A.; Iacopetta, D.; Barbarossa, A.; Rizzuti, B.; Dallemagne, P.; Rault, S.; El-Kashef, H.; Saturnino, C.; et al. Benzothienoquinazolinones as new multi-target scaffolds: Dual inhibition of human Topoisomerase I and tubulin polymerization. *Eur. J. Med. Chem.* **2019**, *181*, 111583. [CrossRef] [PubMed]

42. Chen, S.; Liu, X.; Ge, X.; Wang, Q.; Xie, Y.; Hao, Y.; Zhang, Y.; Zhang, L.; Shang, W.; Liu, Z. Lysosome-targeted iridium(III) compounds with pyridine-triphenylamine Schiff base ligands: Syntheses, antitumor applications and mechanisms. *Inorg. Chem. Front.* **2020**, *7*, 91–100. [CrossRef]

43. Vicini, P.; Geronikaki, A.; Incerti, M.; Busonera, B.; Poni, G.; Kabras, C.A.; Colla, P.L. Synthesis and biological evaluation of benzo[*d*]isothiazole, benzothiazole and thiazole Schiff bases. *Bioorg. Med. Chem.* **2003**, *11*, 4785–4789. [CrossRef]

44. Zhou, X.; Shao, L.; Jin, Z.; Liu, J.-B.; Dai, H.; Fang, J.-X. Synthesis and antitumor activity evaluation of some Schiff bases derived from 2-aminothiazole derivatives. *Heteroat. Chem.* **2007**, *18*, 55–59. [CrossRef]

45. Abdel-Hafez, O.M.; Amin, K.M.; Abdel-Latif, N.A.; Mohamed, T.K.; Ahmed, E.Y.; Maher, T. Synthesis and antitumor activity of some new xanthotoxin derivatives. *Eur. J. Med. Chem.* **2009**, *44*, 2967–2974. [CrossRef] [PubMed]

46. Kraicheva, I.; Bogomilova, A.; Tsacheva, I.; Momekov, G.; Troev, K. Synthesis, NMR characterization and *in vitro* antitumor evaluation of new aminophosphonic acid diesters. *Eur. J. Med. Chem.* **2009**, *44*, 3363–3367. [CrossRef] [PubMed]

47. Nawaz, H.; Akhter, Z.; Yameen, S.; Siddiqi, H.M.; Mirza, B.; Rifat, A. Synthesis and biological evaluations of some Schiff-base esters of ferrocenyl aniline and simple aniline. *J. Organomet. Chem.* **2009**, *694*, 2198–2203. [CrossRef]

48. Zaheer, M.; Shah, A.; Akhter, Z.; Qureshi, R.; Mirza, B.; Tauseef, M.; Bolte, M. Synthesis, characterization, electrochemistry and evaluation of biological activities of some ferrocenyl Schiff bases. *Appl. Organomet. Chem.* **2011**, *25*, 61–69. [CrossRef]

49. Cheng, L.; Tang, J.; Luo, H.; Jin, X.; Dai, F.; Yang, J.; Qian, Y.; Li, X.; Zhou, B. Antioxidant and antiproliferative activities of hydroxyl-substituted Schiff bases. *Bioorg. Med. Chem. Lett.* **2010**, *20*, 2417–2420. [CrossRef]

50. Jesmin, M.; Ali, M.M.; Khanam, J.A. Antitumour activities of some Schiff bases derived from benzoin, salicylaldehyde, aminophenol and 2,4-dinitrophenyl hydrazine. *Thai. J. Pharm. Sci.* **2010**, *34*, 20–31.

51. Etaiw, S.E.; Abd El-Aziz, D.M.; Abd El-Zaher, E.H.; Ali, E.A. Synthesis, spectral, antimicrobial and antitumor assessment of Schiff base derived from 2-aminobenzothiazole and its transition metal complexes. *Spectrochim. Acta A Mol. Biomol. Spectrosc.* **2011**, *79*, 1331–1337. [CrossRef]

52. Hranjec, M.; Starčević, K.; Pavelić, S.K.; Lučin, P.; Pavelić, K.; Zamola, G.K. Synthesis, spectroscopic characterization and antiproliferative evaluation in vitro of novel Schiff bases related to benzimidazoles. *Eur. J. Med. Chem.* **2011**, *46*, 2274–2279. [CrossRef]

53. Shaker, N.O.; Abd El-Salam, F.H.; El-Sadek, B.M.; Kandeel, E.M.; Baker, S.A. Anionic Schiff base amphiphiles: Synthesis, surface, biocidal and antitumor activities. *J. Am. Sci.* **2011**, *7*, 427–436.

54. Kraicheva, I.; Tsacheva, I.; Vodenicharova, E.; Tashev, E.; Tosheva, T.; Kril, A.; Topashka-Ancheva, M.; Iliev, I.; Gerasimova, T.; Troev, K. Synthesis, antiproliferative activity and genotoxicity of novel anthracene-containing aminophosphonates and a new anthracene-derived Schiff base. *Bioorg. Med. Chem.* **2012**, *20*, 117–124. [CrossRef]

55. Bae, S.J.; Ha, Y.M.; Park, Y.J.; Park, J.Y.; Song, Y.M.; Ha, T.K.; Chun, P.; Moon, H.; Chung, H.Y. Design, synthesis, and evaluation of (*E*)-*N*-substituted benzylidene–aniline derivatives as tyrosinase inhibitors. *Eur. J. Med. Chem.* **2012**, *57*, 383–390. [CrossRef]

56. Sondhi, S.M.; Arya, S.; Rani, R.; Kumar, N.; Roy, P. Synthesis, anti-inflammatory and anticancer activity evaluation of some mono- and bis-Schiff's bases. *Med. Chem. Res.* **2012**, *21*, 3620–3628. [CrossRef]

57. Klimczak, A.A.; Kuropatwa, A.; Lewkowski, J.; Szemraj, J. Synthesis of new N-arylamino (2-furyl) methylphosphonic acid diesters, and in vitro evaluation of their cytotoxicity against esophageal cancer cells. *Med. Chem. Res.* **2013**, *22*, 852–860. [CrossRef]

58. Hafez, T.S.; Osman, S.A.; Yosef, H.A.; Abd El-All, A.S.; Hassan, A.S.; El-Sawy, A.A.; Abdallah, M.M.; Youns, M. Synthesis, structural elucidation and in vitro antitumor activities of some pyrazolopyrimidines and Schiff Bases derived from 5-amino-3-(arylamino)-1*H*-pyrazole-4-carboxamides. *Sci. Pharm.* **2013**, *81*, 339–357. [CrossRef]

59. Hassan, A.S.; Hafez, T.S.; Osman, S.A.; Ali, M.M. Synthesis and in vitro cytotoxic activity of novel pyrazolo[1,5-a]pyrimidines and related Schiff bases. *Turk. J. Chem.* **2015**, *39*, 1102–1113. [CrossRef]

60. Zhao, H.C.; Shi, Y.P.; Liu, Y.M.; Li, C.W.; Xuan, L.N.; Wang, P.; Zhang, K.; Chen, B.Q. Synthesis and antitumor-evaluation of 1, 3-selenazole-containing 1, 3, 4-thiadiazole derivatives. *Bioorg. Med. Chem. Lett.* **2013**, *23*, 6577–6579. [CrossRef] [PubMed]

61. Noureen, A.; Saleem, S.; Fatima, T.; Siddiqi, H.M.; Mirza, B. Synthesis, characterization, biological evaluation and QSAR of some Schiff base esters: Promising new antitumor, antioxidant and anti-inflammatory agents. *Pak. J. Pharm. Sci.* **2013**, *26*, 113–124.

62. Zhang, K.; Wang, P.; Xuan, L.-N.; Fu, X.-Y.; Jing, F.; Li, S.; Liu, Y.-M.; Chen, B.-Q. Synthesis and antitumor activities of novel hybrid molecules containing 1,3,4-oxadiazole and 1,3,4-thiadiazole bearing Schiff base moiety. *Bioorg. Med. Chem. Lett.* **2014**, *24*, 5154–5156. [CrossRef] [PubMed]

63. Gupta, S.D.; Revathi, B.; Mazaira, G.I.; Galigniana, M.D.; Subrahmanyam, C.V.; Gowrishankar, N.L.; Raghavendra, N.M. 2,4-dihydroxy benzaldehyde derived Schiff bases as small molecule Hsp90 inhibitors: Rational identification of a new anticancer lead. *Bioorg. Chem.* **2015**, *59*, 97–105. [CrossRef]

64. Abd-Elzaher, M.M.; Labib, A.A.; Mousa, H.A.; Moustafa, S.A.; Ali, M.M.; El-Rashedy, A.A. Synthesis, anticancer activity and molecular docking study of Schiff base complexes containing thiazole moiety. *Beni Suef Univ. J. Basic. Appl. Sci.* **2016**, *5*, 85–96. [CrossRef]

65. Sabbah, D.A.; Al-Tarawneh, F.; Talib, W.H.; Sweidan, K.; Bardaweel, S.K.; Al-Shalabi, E.; Zhong, H.A.; Abu Sheikha, G.; Abu Khalaf, R.; Mubarak, M.S. Benzoin schiff bases: Design, synthesis, and biological evaluation as potential antitumor agents. *Med. Chem.* **2018**, *14*, 695–708. [CrossRef] [PubMed]

66. Perri, A.M.; Agosti, V.; Olivo, E.; Concolino, A.; De Angelis, M.T.; Tammè, L.; Fiumara, C.V.; Cuda, G.; Scumaci, D. Histone proteomics reveals novel post-translational modifications in breast cancer. *Aging* **2019**, *11*, 11722–11755. [CrossRef] [PubMed]

67. Hassan, A.S.; Awad, H.M.; Magd-El-Din, A.A.; Hafez, T.S. Synthesis and in vitro antitumor evaluation of novel Schiff bases. *Med. Chem. Res.* **2018**, *27*, 915–927. [CrossRef]

68. Hassanin, H.M.; Serya, R.A.; Abd Elmoneam, W.R.; Mostafa, M.A. Synthesis and molecular docking studies of some novel Schiff bases incorporating 6-butylquinolinedione moiety as potential topoisomerase IIβ inhibitors. *R. Soc. Open Sci.* **2018**, *5*, 172407. [CrossRef]

69. Saturnino, C.; Caruso, A.; Iacopetta, D.; Rosano, C.; Ceramella, J.; Muia, N.; Mariconda, A.; Bonomo, M.G.; Ponassi, M.; Rosace, G.; et al. Inhibition of human topoisomerase II by *N,N,N*-trimethylethanammonium iodide alkylcarbazole derivatives. *ChemMedChem* **2018**, *13*, 2635–2643. [CrossRef]

70. Iacopetta, D.; Rosano, C.; Puoci, F.; Parisi, O.I.; Saturnino, C.; Caruso, A.; Longo, P.; Ceramella, J.; Malzert-Freon, A.; Dallemagne, P.; et al. Multifaceted properties of 1,4-dimethylcarbazoles: Focus on trimethoxybenzamide and trimethoxyphenylurea derivatives as novel human topoisomerase II inhibitors. *Eur. J. Pharm. Sci.* **2017**, *96*, 263–272. [CrossRef]

71. Saipriya, D.; Prakash, A.C.; Kini, S.G.; Bhatt, V.G.; Pai, S.R.; Biswas, S.; Mohammed, S.K. Design, synthesis, antioxidant and anticancer activity of novel Schiff's bases of 2-amino benzothiazole. *Ind. J. Pharm. Educat. Res.* **2018**, *52*, S333–S342. [CrossRef]

72. Uddin, N.; Rashid, F.; Ali, S.; Tirmizi, S.A.; Ahmad, I.; Zaib, S.; Zubir, M.; Diaconescu, P.L.; Tahir, M.N.; Iqbal, J.; et al. Synthesis, characterization, and anticancer activity of Schiff bases. *J. Biomol. Struct. Dynam.* **2020**, *38*, 3246–3259. [CrossRef] [PubMed]

73. Mahal, A.; Wu, P.; Jiang, Z.H.; Wei, X. Schiff bases of tetrahydrocurcumin as potential anticancer agents. *Chem. Sel.* **2019**, *4*, 366–369. [CrossRef]

74. Erturk, A.G. Synthesis, structural identifications of bioactive two novel Schiff bases. *J. Mol. Struct.* **2020**, *1202*, 127299. [CrossRef]

75. Suyambulingam, J.K.; Karvembu, R.; Bhuvanesh, N.S.; Enoch, I.V.M.V.; Selvakumar, P.M.; Premnath, D.; Subramanian, C.; Mayakrishnan, P.; Kim, S.H.; Chung, I.M. Synthesis, structure, biological/chemosensor evaluation and molecular docking studies of aminobenzothiazole Schiff bases. *J. Ades. Sci. Technol.* **2020**, *34*, 2590–2612. [CrossRef]

76. Kitdumrongthum, S.; Reabroi, S.; Suksen, K.; Tuchinda, P.; Munyoo, B.; Mahalapbutr, P.; Rungrotmongkol, T.; Ounjai, P.; Chairoungdua, A. Inhibition of topoisomerase IIα and induction of DNA damage in cholangiocarcinoma cells by altholactone and its halogenated benzoate derivatives. *Biomed. Pharmacother.* **2020**, *127*, 110149. [CrossRef]

77. Pozzi, C.; Ferrari, S.; Cortesi, D.; Luciani, R.; Stroud, R.M.; Catalano, A.; Costi, M.P.; Mangani, S. The structure of Enterococcus faecalis thymidylate synthase provides clues about folate bacterial metabolism. *Acta Crystallogr. D Biol. Crystallogr.* **2012**, *68*, 1232–1241. [CrossRef] [PubMed]

78. Mishra, V.R.; Ghanavatkar, C.W.; Mali, S.N.; Chaudhari, H.K.; Sekar, N. Schiff base clubbed benzothiazole: Synthesis, potent antimicrobial and MCF-7 anticancer activity, DNA cleavage and computational study. *J. Biomol. Struct. Dynam.* **2020**, *38*, 1772–1785. [CrossRef] [PubMed]

79. Bhat, S.S.; Shivalingegowda, N.; Revankar, V.K.; Lokanath, N.K.; Kugaji, M.S.; Kumbar, V.; Bhat, K. Synthesis, structural characterization and biological properties of phosphorescent iridium (III) complexes. *J. Inorg. Biochem.* **2017**, *177*, 127–137. [CrossRef]

80. Desai, S.B.; Desai, P.B.; Desai, K.R. Synthesis of some Schiff bases, thiazolidones, and azetidinones derived from 2,6-diaminobenzo[1,2-d:4,5-d']bisthiazole and their anticancer activities. *Heterocycl. Commun.* **2001**, *7*, 83–90. [CrossRef]

81. Padhye, S.; Yang, H.; Jamadar, A.; Cui, Q.C.; Chavan, D.; Dominiak, K.; McKinney, J.; Banerjee, S.; Dou, Q.P.; Sarkar, F.H. New difluoro knoevenagel condensates of curcumin, their schiff bases and copper complexes as proteasome inhibitors and apoptosis inducers in cancer cells. *Pharm. Res.* **2009**, *26*, 1874–1880. [CrossRef] [PubMed]

82. Shokrollahi, S.; Amiri, A.; Fadaei-Tirani, F.; Schenk-Joß, K. Promising anti-cancer potency of 4,5,6,7-tetrahydrobenzo[*d*]thiazole-based Schiff-bases. *J. Mol. Liq.* **2020**, *300*, 112262. [CrossRef]

83. Morsy, N.M.; Hassan, A.S.; Hafez, T.S.; Mahran, M.R.; Sadawe, I.A.; Gbaj, A.M. Synthesis, antitumor activity, enzyme assay, DNA binding and molecular docking of Bis-Schiff bases of pyrazoles. *J. Iran. Chem. Soc.* **2021**, *18*, 47–59. [CrossRef]

applied sciences

Review

Gold Derivatives Development as Prospective Anticancer Drugs for Breast Cancer Treatment

Ileana Ielo [1], Domenico Iacopetta [2], Carmela Saturnino [3], Pasquale Longo [4], Maurilio Galletta [5], Dario Drommi [5], Giuseppe Rosace [6,*], Maria Stefania Sinicropi [2,*] and Maria Rosaria Plutino [1,*]

[1] Institute for the Study of Nanostructured Materials, ISMN—CNR, Palermo, c/o Department of ChiBioFarAm, University of Messina, Viale F. Stagno d'Alcontres 31, Vill. S. Agata, 98166 Messina, Italy; ileana.ielo@ismn.cnr.it

[2] Department of Pharmacy, Health and Nutritional Sciences, University of Calabria, Via Pietro Bucci, 87036 Arcavacata di Rende, Italy; domenico.iacopetta@unical.it

[3] Department of Science, University of Basilicata, Viale dell'Ateneo Lucano 10, 85100 Potenza, Italy; carmela.saturnino@unibas.it

[4] Department of Chemistry and Biology, University of Salerno, 84084 Fisciano, Italy; plongo@unisa.it

[5] Department of ChiBioFarAm, University of Messina, Viale F. Stagno d'Alcontres 31, Vill. S. Agata, 98166 Messina, Italy; mgalletta@unime.it (M.G.); ddrommi@unime.it (D.D.)

[6] Department of Engineering and Applied Sciences, University of Bergamo, Viale Marconi 5, 24044 Dalmine, Italy

* Correspondence: giuseppe.rosace@unibg.it (G.R.); s.sinicropi@unical.it (M.S.S.); mariarosaria.plutino@cnr.it (M.R.P.); Tel.: +39-03-5205-2021 (G.R.); +39-09-8449-3200 (M.S.S.); +39-09-0676-5713 (M.R.P.)

Citation: Ielo, I.; Iacopetta, D.; Saturnino, C.; Longo, P.; Galletta, M.; Drommi, D.; Rosace, G.; Sinicropi, M.S.; Plutino, M.R. Gold Derivatives Development as Prospective Anticancer Drugs for Breast Cancer Treatment. *Appl. Sci.* **2021**, *11*, 2089. https://doi.org/10.3390/app11052089

Academic Editors: Giuliana Muzio and Francesca Silvagno

Received: 30 December 2020
Accepted: 20 February 2021
Published: 26 February 2021

Publisher's Note: MDPI stays neutral with regard to jurisdictional claims in published maps and institutional affiliations.

Copyright: © 2021 by the authors. Licensee MDPI, Basel, Switzerland. This article is an open access article distributed under the terms and conditions of the Creative Commons Attribution (CC BY) license (https://creativecommons.org/licenses/by/4.0/).

Abstract: Commonly used anticancer drugs are *cis*platin and other platinum-based drugs. However, the use of these drugs in chemotherapy causes numerous side effects and the onset of frequent drug resistance phenomena. This review summarizes the most recent results on the gold derivatives used for their significant inhibitory effects on the in vitro proliferation of breast cancer cell models and for the consequences deriving from morphological changes in the same cells. In particular, the study discusses the antitumor activity of gold nanoparticles, gold (I) and (III) compounds, gold complexes and carbene-based gold complexes, compared with *cis*platin. The results of screening studies of cytotoxicity and antitumor activity for the gold derivatives show that the death of cancer cells can occur intrinsically by apoptosis. Recent research has shown that gold (III) compounds with square planar geometries, such as that of *cis*platin, can intercalate the DNA and provide novel anticancer agents. The gold derivatives described can make an important contribution to expanding the knowledge of medicinal bioorganometallic chemistry and broadening the range of anticancer agents available, offering improved characteristics, such as increased activity and/or selectivity, and paving the way for further discoveries and applications.

Keywords: gold derivatives; cancer treatment; breast cancer; cytotoxicity; antitumor activity

1. Introduction

Approximately 9 million people around the world fall ill and die from cancer diseases every year, many of who do not receive adequate treatment due to the high cost. It is esteemed that the incidence of cancer will double by 2035 [1]. In particular, breast cancer is the most frequent type diagnosed, about 25% of all cancers [2,3]. The anticancer therapy currently in use consists of three main approaches: surgical removal of the tumor mass, chemotherapy and radiotherapy. Unfortunately, several types of metastatic tumors have been found to be chemoresistant [4]. The most common form of breast cancer, which makes up 60% of all diagnoses, is hormone receptor (HR) positive and human epidermal growth factor receptor 2 (HER2) negative [5]. Several studies have shown that HR positive patients have a better response to hormone therapy. However, the lack of HER2

113

proteins causes a disposal of drugs specially designed to target these proteins as therapeutic options. The retinoblastoma (Rb) protein, a tumor suppressor, regulates RNA transcription during the G1 cell growth phase to stop the proliferation of malignant cells [6]. Cyclin D1 proteins bind to cyclin-dependent kinase (CDK) enzymes 4 and 6 thereby inhibiting the regulatory function of the Rb protein. The National Comprehensive Cancer Network (NCCN) recognized CDK 4/6 inhibitors are palbociclib, abemaciclib and ribociclib in combination with aromatase inhibitors. Combination therapy provides treatment for HR positive/HER2 negative advanced or metastatic breast cancer by lowering estrogen levels to inhibit cell growth and cyclin-dependent kinase to block malignant cell division and proliferation [7,8]. Abemaciclib and ribociclib showed a relative reduction in the risk of death of 25–30% [9]. By combining an aromatase inhibitor with palbociclib, progression free survival (PFS) is increased by 10 months, compared with hormone monotherapy [10]. Likewise, adding palbociclib to fulvestrant resulted in a double increase in PFS compared to those taking fulvestrant alone [11]. There is currently limited evidence to support the use of CDK 4/6 inhibitors as monotherapy. Increased data on the safety and tolerability of CDK 4/6 inhibitors in patients may help clinicians in the selection of initial therapy for patients with HR positive/HER2 negative breast cancer [12]. Platinum-based anticancer drugs such as *cis*platin, carboplatin and oxaliplatin have been widely employed in the treatment of several types of cancer, such as lung, colorectal, ovarian and breast cancers [13]. However, the efficacy and, therefore, the use of such platinum-based drugs have decreased due to the high risk of serious toxicity, such as neurotoxicity [14,15]. Major advances have recently been made in drug delivery systems, due to miniaturization technologies that have improved the performance of existing drugs to provide new, more effective therapies [16]. The use of biomolecules such as peptides, nucleic acids and others reduced the amount of drug required by improving its targeted action [17]. The reduction in size at the nanometer level (<100 nm) has led to dramatic improvements in the way drugs are delivered [18]. All particles <100 nm in size could be formed via nanocrystals, drug–polymer complexation or using nanoscale shells that could trap drugs [19]. The fine size of nanoparticles or metal complexes allows for a loading of small molecules, peptides, proteins and nucleic acids that escape immunological detection unlike larger particles, which are easily excluded from the body. Recently, innovative research has developed towards metal-based anticancer drugs, such as gold derivatives, with the aim of improving the effectiveness, expanding the activity field and, above all, reducing the general toxicity [20,21]. The pharmacological activity of gold compounds has been tested since ancient times; these compounds have been used in a series of treatments, including that for the rheumatoid arthritis and as an antibacterial and antitumor. Different studies have shown that gold derivatives act differently from platinum anticancer drugs, since their primary target is the proteasome; in this regards future approaches will bring to the development gold complexes selective for specific cancer cells and tumor targets in order to increase their effectiveness and better control of undesired side effects [22,23]. The purpose of this review is to summarize the state of the art of the antitumor activity of gold compounds, complexes and nanoconjugates, providing a brief overview of their use against breast cancer. In particular, the following aspects will be treated:

1. Drug delivery systems;
2. Gold nanosystems;
3. Gold complexes.

The antiproliferative activity of these systems has been studied in several human breast cancer cell lines. In vitro and ex vivo experiments have shown that this class of compounds shows significant antiproliferative activity against ovarian, prostate, lung and particularly breast cancers [24].

2. Drug Delivery Systems

Conventional cancer treatments, such as chemotherapy and radiotherapy, act in biological systems in a non-specific way, affecting both malignant and healthy cells. This affects the optimal therapeutic and implementation of gold derivatives.

Targeted delivery through gold derivatives can take place through two types: passive and active effect, reducing unwanted effects and the development of drug resistance [25]. Passive targeting allows the accumulation of a drug or drug transport system within a specific site due to the variation of physicochemical or pharmacological factors. This type of method exploits the size of the nanoparticles and the properties of the tumor vascular system, effectively improving the bioavailability and efficacy of the drug. The vascularity of the tumor is very different from normal tissue, in fact the blood vessels of the tumor tissues, unlike those in normal tissues, have spaces between the adjacent endothelial cells up to 600–800 nm. These pathophysiological features of tumor vessels induce the Enhanced permeability and retention EPR (Enhanced Permeability and Retention) effect, which allows macromolecules, including nanoparticles, to extravasate through these extravascular spaces and accumulate within tumor tissues [26]. The accumulation of tumor drugs is ten times greater when the drug is administered from a nanoparticle rather than as a free drug. Another contributor to passive targeting is the unique microenvironment that surrounds cancer cells, which is different from that of normal cells. Fast-growing hyperproliferative cancer cells use glycolysis for extra energy, resulting in an acidic environment. The pH-sensitive liposomes are designed to be stable at a physiological pH but degraded to release the active drug into target tissues where the pH is lower, such as in the acidic environment of cancer cells [27]. Active targeting involves the attachment of a fraction, such as a monoclonal antibody or a ligand, to deliver a drug to pathological sites or to cross biological barriers based on molecular recognition processes [28–30]. When designing the synthesis of nanoparticles, it is necessary to consider some factors: for example, the antigen or receptor should be expressed exclusively and homogeneously on tumor cells and not expressed on healthy ones. The internalization of conjugates occurs through receptor-mediated endocytosis. Indeed, when a conjugate binds to its receptor on the cell surface, the plasma membrane envelops the receptor and ligand complex to form an endosome, this is transferred to target organelles.

When the pH value inside the endosome becomes acidic and lysozymes are activated, the drug is released from the conjugate and enters the cytoplasm. The receptor released by the conjugate returns to the cell membrane to begin a second transport cycle by binding with new conjugates. Ligands targeting cell surface receptors can be natural substances that have the advantages of a molar mass and lower immunogenicity than antibodies. Molecular targeted therapy is a potential solution to overcome these challenges, it can be achieved through smart design (Figure 1). Both methods allow increasing the concentration of the anticancer drug directly inside the tumor cell, causing the decrease of toxicity for healthy cells [30]. Gold derivatives (gold compounds, complexes and nanoparticles) can be conjugated to a wide range of biologically active organic molecules, designed to cross the blood–brain barrier, interact with specific receptors entering the cell through an alternative path. In particular, passive targeting of gold nanoparticles is based on the effect of enhanced permeability and retention (EPR) and tumor angiogenesis, while active targeting is based on direct binding from the ligand to receptors expressed by tumor cells [26]. Antitumor agents can be released as a function of pH or temperature [31,32].

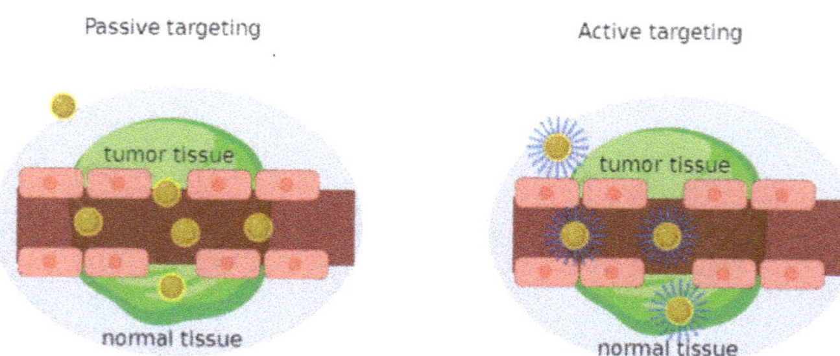

Figure 1. Schematic illustration of targeted strategies for cancer therapy by functional Gold Nanoparticles (Au-NPs, NPs = NanoParticles). Malignant tissue is distributed unevenly concerning healthy cells, hence the gold nanoparticles that cross the empty spaces due to the increase in permeability and the retention effect of passive drug release. This figure also illustrates active targeting with direct binding of gold nanoparticles to receptors with specific ligands.

3. Gold Nanosystems

Medical research on gold nanoparticles has been directed towards the study of drug delivery systems, chemotherapeutic agents, detection and diagnostics [33,34]. Gold nanosystems turned out to be attractive due to their unique properties, mainly dependent on their size and shape [35]. The resulting physical properties of nanoparticles strongly depend on the particle size, density, nature of the protective organic shell and their shape [36]. The quantum size effect occurs when the de Broglie wavelength of the valence electrons has the same dimensional order as the particle itself. Thus, the particles electronically behave like zero-dimensional quantum boxes. Therefore, the electrons are free to move inside these metal boxes with a collective oscillation frequency characteristic of plasma resonance, called the plasmon resonance band (PRB) observed near 530 nm in the 5–20 nm diameter range. The plasmon resonance of nanoparticles is closely related to the size, shape and dielectric properties of the medium surrounding the nanoparticles [37]. By varying the shape of metal nanoparticles, such as nanospheres, nanotubes, nanoprisms or core–shell nanoparticles, their optical properties vary in a quantitative and dependent manner [38]. Gold nanoparticles are used as sensors for the early diagnosis of many diseases [39]. Alzheimer's disease and breast cancer have been targeted for essential early diagnosis [40]. With early diagnosis, current drugs have the ability to postpone the onset of symptoms typical of the disease [41] and are therefore essential for greater treatment efficacy and a higher survival rate [42]. Gold nanoshells have been used as theranostics for the diagnosis and photothermal therapy of breast cancer cells in vitro [40]. Gold nanoshells induce an important photothermal response under illumination of near infrared radiation, showing good potential for cancer therapy, with 100% efficacy in tumor remission [42,43]. The surfaces of gold nanoshells can link targeting, diagnostic and therapeutic functionalities, forming a multifunctional nanocomplex. This system has also been used in vivo, enriching the near infrared fluorescence and, at the same time, the magnetic resonance imaging capability [44]. A gold nanoparticle delivery system conjugated with gemcitabine and cetuximab as a target agent has been tested in vitro and in vivo for the treatment of pancreatic cancer cells [45]. The results of these tests showed a greater inhibition of tumor growth through a targeted system. The targeted release of multifunctional nanoparticles [46], obtained by conjugating three different peptides, was investigated: an epidermal growth factor receptor (EGFR)-recognizing peptide, an aminoterminal peptide that recognizes the urokinase plasminogen activator receptor and a peptide cyclic that recognizes the integrin receptor, to study the accumulation of gold in tumor models. These experiments did not demonstrate a substantial improvement in tumor uptake compared to control particles in vivo. Instead, gold nanoparticles with a thiolate derivative of *cis*platin have been produced and tested against ovarian cancer cells [47]. The results showed that the gold conjugate with *cis*platin had comparable efficacy to *cis*platin alone, but toxicity to healthy cells was almost nothing, unlike the high toxicity of *cis*platin used alone [48]. The decrease in toxicity towards healthy cells is one of the many reasons why therapies with gold nanoparticles can prove superior to the use of drugs alone [49]. In vivo studies have shown how multifunctional fluorescent magnetic nanocomplexes are used to trace the distribution of the nanocomplexes on tumor tissues. Nanocomplexes conjugated with specific antibodies targeting human epidermal growth factor receptor 2 (HER2) that overexpress breast cancer tumors could then be identified using magnetic resonance imaging (MRI) of the nanocomplex. As antibody-conjugated nanocomplexes are tracked throughout the body, we observe clear differences in the amounts of tumor uptake between over-expressed HER2 and low-expression HER2 tumors. This study demonstrated that it is possible to visualize tumors in vivo and that MRI could reveal a detailed picture of the distribution of nanoparticles in tumors and internal organs [50]. The diagnostic capabilities of the nanocomplexes have been visualized in vivo on HER2-expressing breast cancer tumors in animal models. Molecular targeting is achieved by combining anti-HER2 antibodies on the surface of the nanoparticle via the streptavidin-biotin binding procedure. In addition, poly ethylene glycol (PEG) conjugated to nanocomplexes is used to weaken nonspecific binding in vivo, to sterically stabilize the complexes, to implement circulation time, to lower immunogenicity and, in combination with antibodies, to increase accumulation of nanoparticles in the tumor [51]. Advantageous biological systems were investigated that exploit polyvalent interactions, allowing an organism to take advantage of a set of monovalent ligands with lower affinity, rather than using new and higher affinity monovalent ligands for each function. Ligand binding to a gold nanoparticle in the multivalent mode is an effective way to generate a high local concentration of ligands. The binding equilibria between the surface-bound ligand and the receptor can be shifted towards the formation of more ligand–receptor pairs in the presence of a high local ligand concentration according to the Le Chatelier principle [52].

One type of ligands, conjugated to gold nanoparticles, are carbazoles, extensively studied for their antioxidant and antimicrobial properties. Carbazole derivatives have also become important for their efficient inhibition of topoisomerase, tubulin, telomerase, kinase and integrase [53,54]. These compounds induce antiproliferative activity and a significant apoptotic response in a selective manner towards tumor cells [55–57]. The gold nanoparticles functionalized with *N*-thiocarbazole derivatives (Figure 2) have been employed as antiproliferative agents against breast and uterine cancer cell lines without affecting non-tumor cells [58].

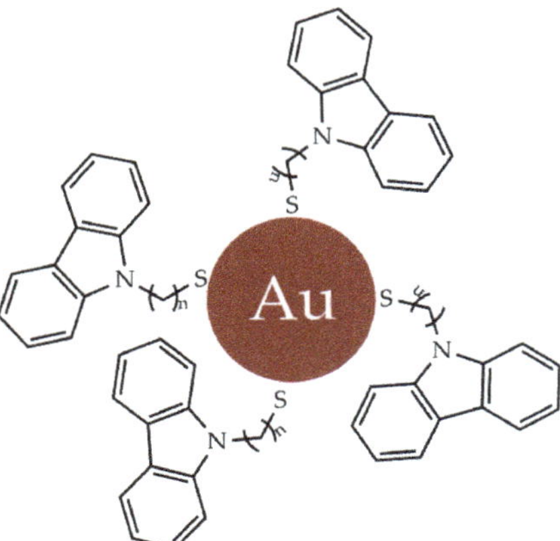

Figure 2. Gold nanoparticles functionalized with *N*-thioalkylcarbazole derivatives.

The unique properties of gold nanoparticles were exploited for detection assays, employing the nanoparticles as cell-permeable multivalent systems, and further investigating promising and targeted therapies for known receptors [59].

4. Gold Complexes

In recent decades, new gold complexes have been developed with antibacterial, antiviral, antiparasitic and antitumor activity [60]. Several intracellular protein targets, such as kinases, reductases, proteases and topoisomerases, interact with gold (III) complexes; in particular, the blocking of the latter leads to programmed cell death [61]. The ability of [Au(C^N^C)(IMe)]CF$_3$SO$_3$ (**1**) (Figure 3) to inhibit the relaxation reaction of supercoiled DNA was analyzed. By comparing the different electrophoretic mobility of both supercoiled and relaxed forms, it was shown that compound **1** was able to inhibit the relaxation activity of human IB topoisomerase.

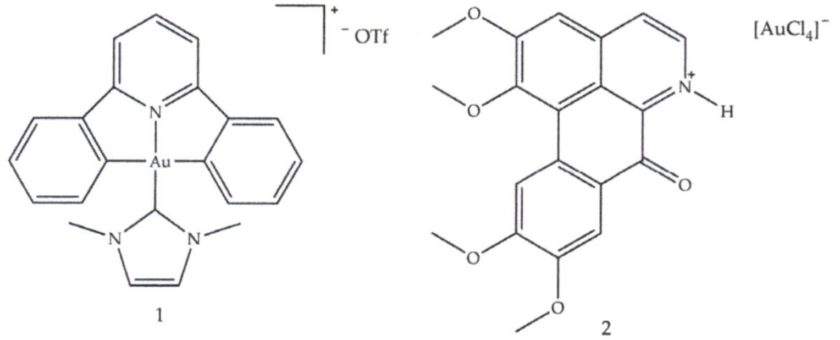

Figure 3. Molecular structure of [Au(C^N^C)(IMe)]CF$_3$SO$_3$] (**1**) and [OGH][AuCl$_4$] (**2**) [60]. Abbreviations: C^N^C is the bi-cyclometallated di-anionderived from 2,6-diphenylpyridine, IMe is *N,N'*-dimethylimidazolium and [OGH](+) is the charged alkaloid oxoglaucine (OG).

A study conducted in 2012 described the use of oxoglaucine (OG) as a ligand, an alkaloid extract of oxoaporphine obtained from the overground parts of different plants, such as Annonaceae, Magnoliaceae and Papaveraceae [62]. The reaction was carried out between OG and a gold (III) salt, to obtain the complex of gold (III) (**2**), an ionic compound made up of the oxoglaucine cation and the anion [AuCl$_4$]$^-$ having a square planar structure (Figure 3). In this compound, the oxoglaucine is protonated on the nitrogen atom while the gold (III) tetrachloride acts as a counter ion, balancing the charge. Crystallographic data confirmed that oxoglaucine organizes itself into a planar structure, capable of intercalating DNA [63]. The ability of compound **2** to block tumor growth was evaluated in vitro against different cell lines where the best results were obtained for human papillomavirus-related endocervical adenocarcinoma BEL7404 cells, with an inhibitory concentration IC$_{50}$ (i.e., the half maximal inhibitory concentration) of 6.1 ± 0.5 μM and for AS49 human lung carcinoma cells (IC$_{50}$ = 1.4 ± 0.7 μM) [62]. Figure 4 shows the square planar gold (III) chelates (**3–6**), synthesized by Wilson et al. [64], in which the pyridyl- or isoquinolylamide bidentate anionic chelators are used as ligands, in order to stabilize gold (III) through the presence of donor σ atoms. Among these, only compound **6** showed a good cytotoxic profile, towards two ovarian cancer cells (OVCAR-3 and IGROV1, with IC$_{50}$ values of 4.0 and 9.8 μM, respectively) and against one colon tumor cell line (SW-620, IC$_{50}$ = 15 μM). Compound **6** is both a topoisomerase IIa (TopoIIa) inhibitor, with a mechanism of action similar to that of zorubicin, and a TopoIB inhibitor, such as Topotecan and 9-methoxycamptothecin. Furthermore, this compound is structurally similar to *cis*platin, but does not possess the same anticancer properties, probably due to the fact that the chloride ligands are not reactive. This condition could inhibit the replacement of water with the metal ion in vivo and the consequent formation of a gold (III)–DNA complex. Another study described the biological properties of a series of planar cationic Au^{3+} macrocycles containing two pyrrole-imine units linked to a quinoxaline group and an alkyl chain. Among these molecules, shown in Figure 5, compound **7** showed the best antiproliferative activity, against different types of cell lines. Moreover, it exhibits a marked inhibition activity of TopoI at 500 nM and a total inhibitor at the dose of 5 µM. In particular, the presence of gold (III) is crucial for the inhibition of topoisomerase and for DNA intercalation [65].

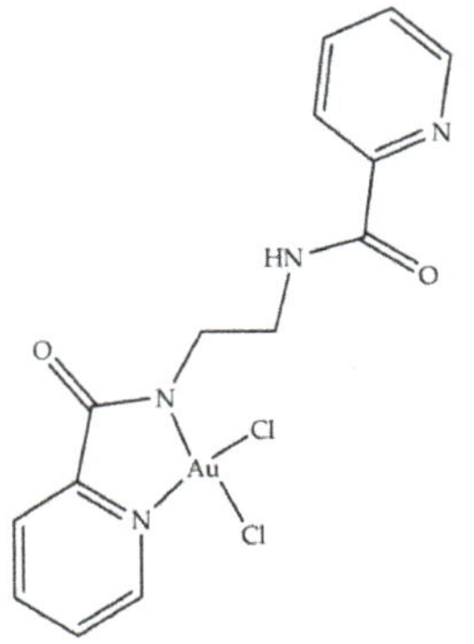

Figure 4. Molecular structures of gold (III) pyridyl and isoquinolylamido chelates (**3–6**) [60].

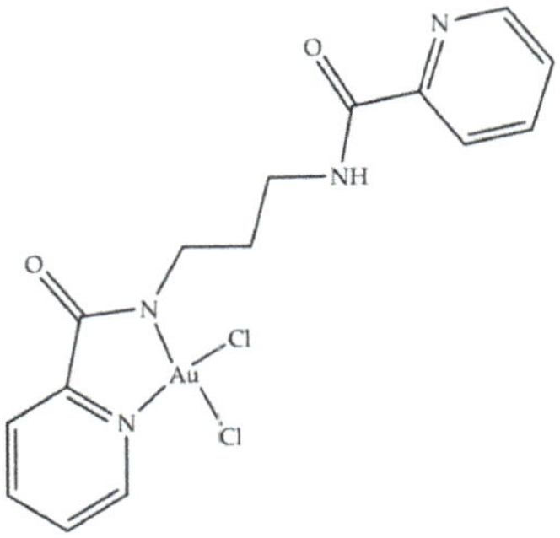

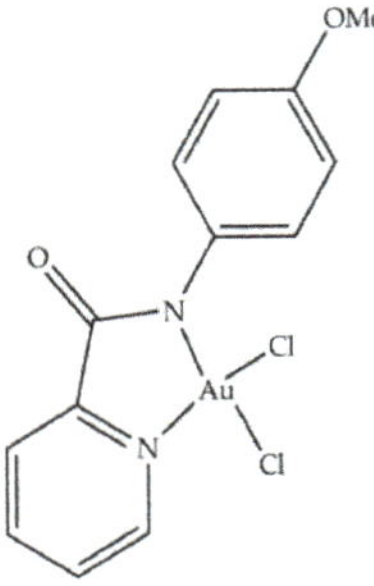

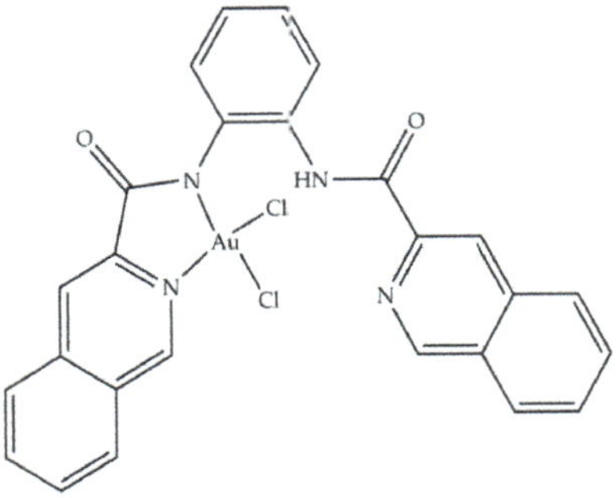

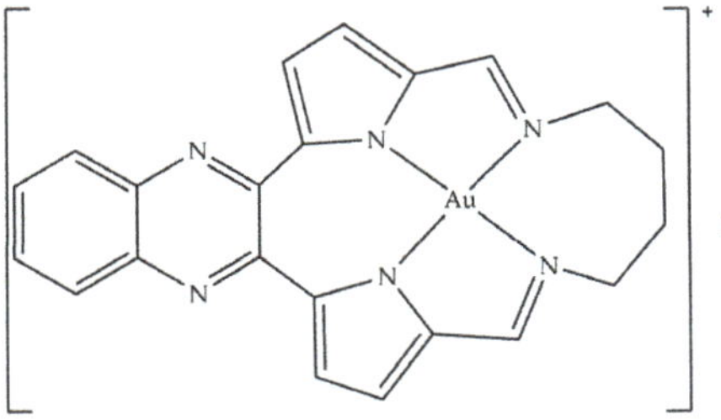

Figure 5. Molecular structure of the gold (III) macrocycle (**7**) [60].

Another study, conducted in 2016, ref. [66] focused on the synthesis of thiosemicarbazones coordinated with different metals (Pt, Pd and Au). Among these, the complex with the best cytotoxic activity against leukemic cells (HL-60 and THP-1, with IC_{50} values of 0.26 ± 0.20 and 0.62 ± 0.49 µM, respectively) and cancer cells breast (MDA-MB-231 and MCF-7, with IC_{50} values of 0.09 ± 0.05 and 0.42 ± 0.01 µM, respectively) resulted in [Au (PyCT$_4$BrPh)Cl]Cl (**8**) (Figure 6). This compound **8** inhibits TopoIB as a function of the administered dose, with a small inhibition at 1.5 µmol L^{-1} and total inhibition starting at 50 µmol L^{-1}. This result could depend on the interaction of gold with the enzyme: in fact, thiosemicarbazone not coordinated with the metal is not able to inhibit TopoIB but could be the vehicle of the metal.

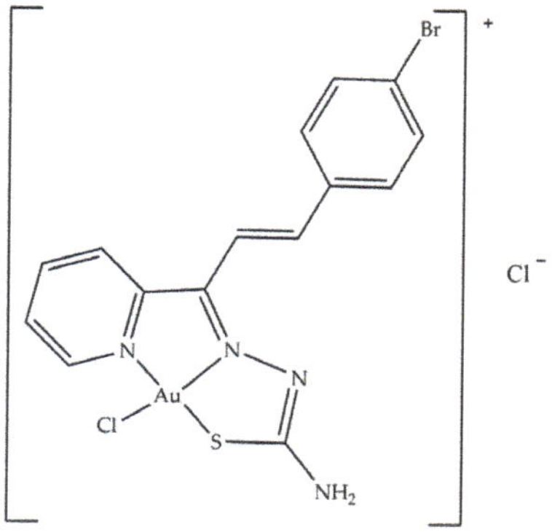

Figure 6. Molecular structure of the gold (III) thiosemicarbazone (**8**).

5. Gold Based Carbene Complexes

Heterocyclic and metallocene complexes have also been investigated for their potential antitumor activity [67–69]. These metal complexes have the general formula L$_n$MX$_m$, where M is the metal that constitutes the reactive center, in its oxidation state equal to 0, L is the carbene, which is the ligand able to influence the electronic properties of the metal and consequently, the possible catalytic activity of the complex and X is a non-carbenic ligand of anionic nature, such as a halide, a carboxylate or an alkoxide [70,71]. The stability and chemical reactivity of these complexes can be influenced by the covalent bonds coordinated with various transition metal centers through σ donation and π-retro-donation, the aromaticity of the NHC ligand and the volume of the side chains [72]. Among these, heterocyclic gold *N*-carbene (NHC) complexes attract particular interest in the pharmaceutical research sector, meeting the requirements for efficient and rapid drug design, and for thence to their chemical stability [73]. A number of gold NHC complexes have been biologically tested and, in particular, some of them have also shown anticancer effects in vitro. Au–NHC complexes offer a wide range of biological activities, such as antiarthritic [74], antimicrobial [75] and especially antitumor ones. The anticancer properties of Au (I/III)–NHC complexes have been studied in different cellular backgrounds, such as melanoma, breast, prostate and hepatocellular carcinoma cell lines. It has been shown that Au–NHC complexes can differently affect cell cycle distribution, expressions of several key regulators of apoptosis, mitochondrial integrity, activation of caspases and generation of reactive intracellular oxygen species (ROS) [76]. The structures of these gold NHC complexes were synthesized to evaluate the influence of increased lipophilicity on their pharmacological effects [77]. The lipophilic cation, whose π-electrons are delocalized, can cross biological membranes more rapidly and concentrate within the mitochondria of cancer cells. The lipophilicity of the complex increases when the nitrogen atoms are functionalized with lipophilic substituents. The effect of different substituents on the N-1 atom on pharmacological activity was tested. In particular, position 1 was functionalized with 2-cyclopentanol (L1), 2-cyclohexanol (L2) and 2-hydroxy-2-phenylethyl (L3) while in position 3 a methyl group was always present (Figure 7). The obtained complexes were studied for their anticancer properties against human breast cancer cells and the underlying molecular mechanism was studied in detail by biological assays and macromolecular docking studies, in order to understand the probable ligand–protein binding modes. One of the tested compounds, AuL3, showed good growth inhibitory activity by inducing apoptosis of breast cancer cells, without exerting any effect on healthy breast epithelial cells.

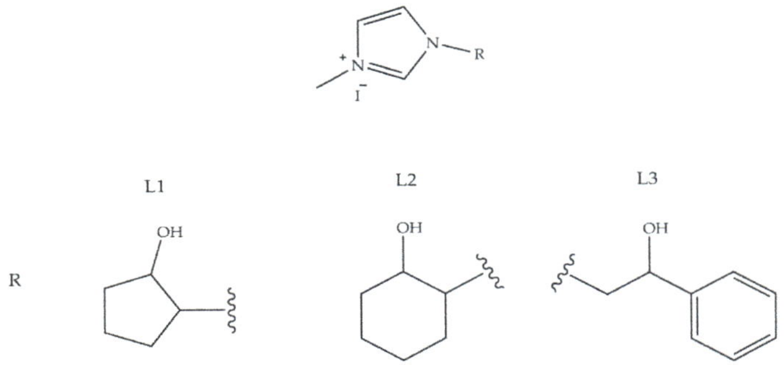

Figure 7. NH-Carbene proligands L1, L2 and L3, employed in AuL_n (n = 1–3) complexes.

The antitumor activity of these complexes was evaluated on two breast cancer cell lines, MCF-7 and MDA-MB-231. The results obtained highlighted an interesting antitumor activity of the compounds shown in Figure 8, in particular on the highly aggressive and metastatic MDA-MB-231 cell line. In vitro studies have shown that these compounds are able to interfere with tubulin polymerization by inhibiting the activity of hTopo I and II, already at the minimum concentration of 1 M. Therefore, these results have shown that these compounds may be useful for further future investigations enabling the development of new multitarget agents in the treatment of breast cancer.

Antiarthritic Gold Compounds in Oncology

The treatment of different tumors with *cis*platin, $(NH_3)_2PtCl_2$, has initiated and focused the pharmacological research towards metal-based drugs, such as gold (I, III) compounds, in particular, those that exhibit antiarthritic, antitumor and cytotoxic activity. The study of the effectiveness of anticancer drugs, e.g., 6-mercaptopurine and cyclophosphamide, in the treatment of rheumatoid arthritis, derived from their known immunosuppressive and anti-inflammatory actions [78,79]. This study established a connection between the two therapies, antiarthritic and antitumor. One of the reasons why scientific research has become interested in gold compounds is their square planar geometry, which unites them to platinum in *cis*platin. Gold (III) is isoelectronic with platinum (II) in *cis*platin and also forming similar planar square complexes. Furthermore, the biological environment is generally reducing, so gold (III) compounds can be reduced in vivo to gold (I) and metallic gold. By coordinating bioactive molecules with gold, it is possible to obtain compounds having greater activity and therapeutic effect than uncoordinated molecules. Promising in vitro inhibitory effects of auranofin (Figure 8b) against HeLa cells [80] and subsequent research on its antitumor activity were investigated. The in vitro cytotoxic activity of auranofin, tested on various cell lines, was similar or even greater than *cis*platin (Figure 8a).

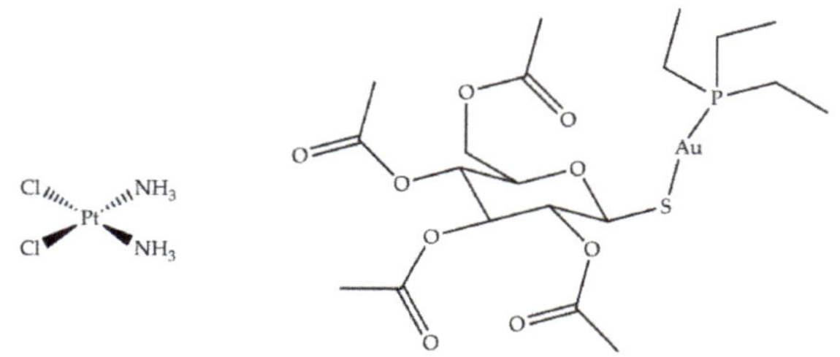

*cis*platin

a)

Auranofin

b)

Figure 8. Chemical structure of (**a**) *cis*platin and (**b**) auranofin.

A good cytotoxic activity was also evaluated in vivo against P388 lymphocytic leukemia implanted in mice [81]. Several gold (I) thiolates have been tested for their anticancer activity. Aurothioglucose (Figure 9a) and aurothiomalate (Figure 9b) inhibited primary tumor growth in mice with Lewis carcinoma and reduced lung metastases [81]. Unlike *cis*platin (**6**), aurothiomalate did not exhibit acute toxicity in vivo.

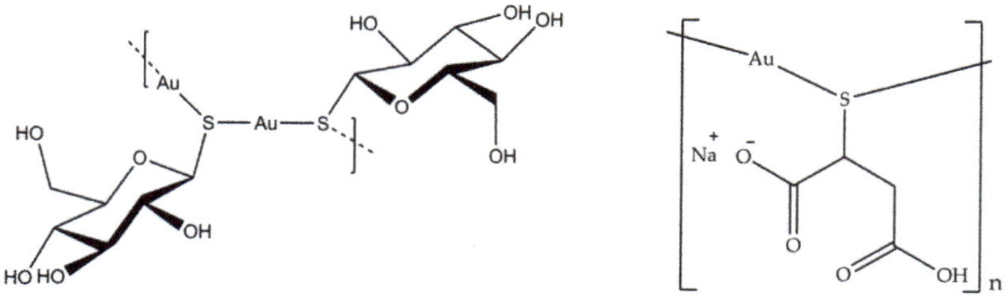

Aurothioglucose

Sodium aurothiomalate

a)

b)

Figure 9. Chemical structure of (**a**) aurothioglucose and (**b**) sodium aurothiomalate.

The in vitro cytotoxicity of several gold compounds against B16 melanoma and P388 leukemic cells was examined in a study of auranofin and similar compounds [81]. The result shows that a wide variety of gold (I) phosphine thiolates possesses a significant cytotoxic activity. The presence of a phosphine or arsine as a binder has resulted in an increase in cytotoxicity and in vivo antitumor activity against P388 leukemia, compared to other species. The importance of the type of phosphine, used in the binuclear gold (I) compounds containing bidentate phosphines, was tested in vivo against leukemia P388, by systematically varying the nature of the phosphine and keeping the thiolate, $R1S^-$, constant. The most active compound was the one containing $(CH_3CH_2)[(CH_3)_2CH]$ P (Figure 10a), while the least active one had $(C_6H_5)_2P$ (Figure 10b).

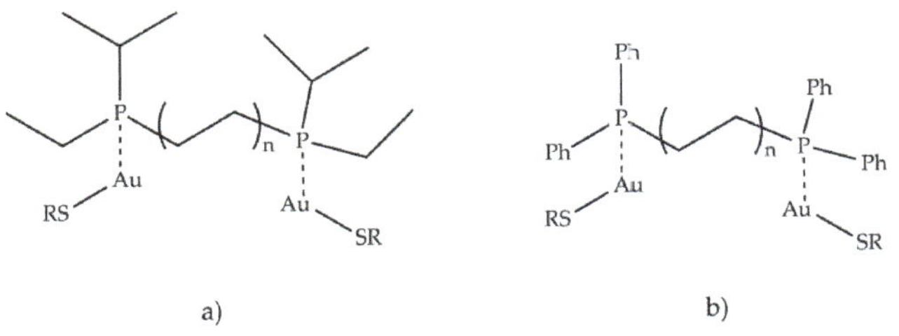

Figure 10. Structures of binuclear gold (I) compounds containing bidentate phosphines (**a**) $(CH_3CH_2)[(CH_3)_2CH]$ P and (**b**) $(C_6H_5)_2P$.

In summary, it can be stated that the most active class, among the compounds examined, was the one containing both phosphine and thioglucose ligands. The cytotoxicity and antitumor activity of several gold compounds with phosphine binders was tested and proved cytotoxic against B16 melancma, subcutaneous breast adenocarcinoma, M5076 reticulum cell sarcoma leukemia and P388 and L1210 leukemia [82]. In addition, two new 1-acridin-9-yl-3-methylthiourea complexes Au(I), [Au(ACRTU)$_2$]Cl (**9**) and [Au (ACRTU)(PPh$_3$)]PF$_6$ (**10**) (Figure 11), showed submicromolar cytotoxic activity against A2780 ovarian cancer cells and antiproliferative activity, superior to *cis*platin, against some breast cancer cell lines, with a potent antiangiogenic effect. Compound **9** showed the best inhibitory effect, completely relaxing the supercoiled plasmid DNA at the concentration of 12 μM.

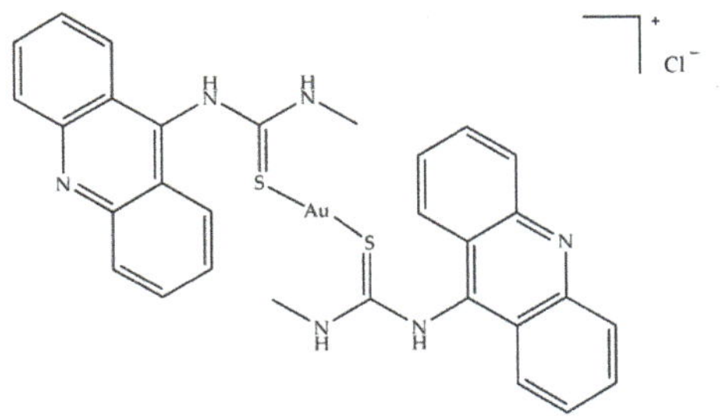

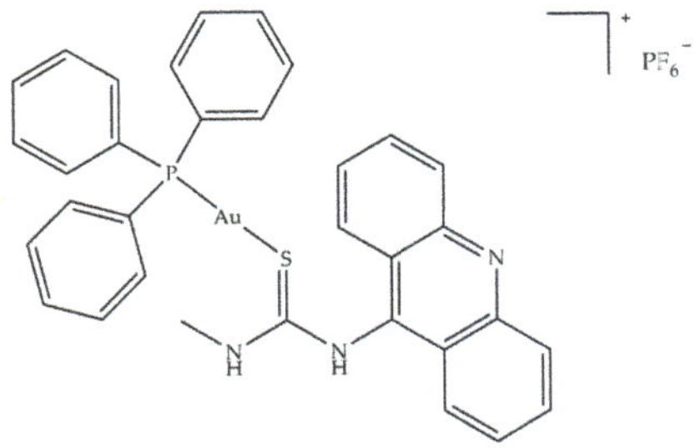

Figure 11. Molecular structure of acridine thiourea gold (I) (**9** and **10**).

6. Conclusions

The fight against cancer is a problem that requires a combination of different therapies. Scientific research has focused on the study of gold compounds that show selectivity and cytotoxicity towards *cis*platin-resistant tumors, and affinity for biological targets such as mitochondria and DNA. This study moved towards the personalization of therapy, to obtain targeted treatments, with greater selectivity and fewer toxic effects. This review summarized the cytotoxic and antitumor activity of some gold derivatives, from nanoparticles to complexes, highlighting their action in blocking human topoisomerases, vital enzymes involved in the proliferation of cancer cells, validating their potential as antitumor agents. All the compounds discussed in this review showed interesting anticancer properties against breast and uterine cancer and were very active against the proliferation of HeLa cells. Further investigations revealed that these derivatives caused the destruction of mitochondria and the release of cytochrome c from its natural site, inducing the activation of enzymes belonging to the intrinsic pathway of apoptosis in a ROS-dependent manner. These characteristics, together with the absence of cytotoxic effects on two different non-tumor cells, make these derivatives very promising and valid candidates for the development of gold-based antitumor drugs and for further preclinical investigations.

Author Contributions: Conceptualization was done by M.S.S., G.R. and M.R.P. Investigation was performed by I.I., D.I., M.G., D.D. Project administration was directed by P.L., C.S., M.S.S., G.R. and M.R.P. Resources were provided by I.I., D.I., C.S., P.L., M.G., D.D., G.R., M.S.S., M.R.P. Supervision was taken care of by M.S.S., G.R. and M.R.P. Validation was carried out by D.I., I.I. Visualization and original draft writing was done by I.I., D.I., M.S.S., G.R. and M.R.P. All authors have read and agreed to the published version of the manuscript.

Funding: This research received no external funding.

Institutional Review Board Statement: Not applicable.

Informed Consent Statement: Not applicable.

Data Availability Statement: Not applicable.

Acknowledgments: MURST: CNR and MIUR are gratefully acknowledged for financial support.

Conflicts of Interest: The authors declare no conflict of interest.

References

1. Prager, G.W.; Braga, S.; Bystricky, B.; Qvortrup, C.; Criscitiello, C.; Esin, E.; Sonke, G.S.; Martínez, G.A.; Frenel, J.-S.; Karamouzis, M.; et al. Global cancer control: Responding to the growing burden, rising costs and inequalities in access. *ESMO Open* **2018**, *3*, e000285. [CrossRef] [PubMed]
2. Ferlay, J.; Soerjomataram, I.; Dikshit, R.; Eser, S.; Mathers, C.; Rebelo, M.; Parkin, D.M.; Forman, D.; Bray, F. Cancer incidence and mortality worldwide: Sources, methods and major patterns in GLOBOCAN 2012. *Int. J. Cancer* **2014**, *136*, E359–E386. [CrossRef] [PubMed]
3. Rizza, P.; Pellegrino, M.; Caruso, A.; Iacopetta, D.; Sinicropi, M.S.; Rault, S.; Lancelot, J.C.; El-Kashef, H.; Lesnard, A.; Rochais, C.; et al. 3-(Dipropylamino)-5-hydroxybenzofuro[2,3-f]quinazolin-1(2H)-one (DPA-HBFQ-1) plays an inhibitory role on breast cancer cell growth and progression. *Eur. J. Med. Chem.* **2016**, *107*, 275–287. [CrossRef]
4. Nagajyothi, P.; Muthuraman, P.; Sreekanth, T.; Kim, D.H.; Shim, J. Green synthesis: In-vitro anticancer activity of copper oxide nanoparticles against human cervical carcinoma cells. *Arab. J. Chem.* **2017**, *10*, 215–225. [CrossRef]
5. Husinka, L.; Koerner, P.H.; Miller, R.T.; Trombatt, W. Review of cyclin-dependent kinase 4/6 inhibitors in the treatment of advanced or metastatic breast cancer. *J. Drug Assess.* **2021**, *10*, 27–34. [CrossRef] [PubMed]
6. Spring, L.M.; Wander, S.A.; Zangardi, M.; Bardia, A. CDK 4/6 Inhibitors in Breast Cancer: Current Controversies and Future Directions. *Curr. Oncol. Rep.* **2019**, *21*, 1–9. [CrossRef] [PubMed]
7. Marra, A.; Curigliano, G. Are all cyclin-dependent kinases 4/6 inhibitors created equal? *NPJ Breast Cancer* **2019**, *5*, 1–9. [CrossRef]
8. Miller, T.W.; Balko, J.M.; Fox, E.M.; Ghazoui, Z.; Dunbier, A.; Anderson, H.; Dowsett, M.; Jiang, A.; Smith, R.A.; Maira, S.-M.; et al. ERα-Dependent E2F Transcription Can Mediate Resistance to Estrogen Deprivation in Human Breast Cancer. *Cancer Discov.* **2011**, *1*, 338–351. [CrossRef] [PubMed]
9. Im, S.-A.; Lu, Y.-S.; Bardia, A.; Harbeck, N.; Colleoni, M.; Franke, F.; Chow, L.; Sohn, J.; Lee, K.-S.; Campos-Gomez, S.; et al. Overall Survival with Ribociclib plus Endocrine Therapy in Breast Cancer. *N. Engl. J. Med.* **2019**, *381*, 307–316. [CrossRef] [PubMed]
10. Finn, R.S.; Martin, M.; Rugo, H.S.; Jones, S.E.; Im, S.-A.; Gelmon, K.A.; Harbeck, N.; Lipatov, O.N.; Walshe, J.M.; Moulder, S.L.; et al. PALOMA-2: Primary results from a phase III trial of palbociclib (P) with letrozole (L) compared with letrozole alone in postmenopausal women with ER+/HER2– advanced breast cancer (ABC). *J. Clin. Oncol.* **2016**, *34*, 507. [CrossRef]

11. Verma, S.; Bartlett, C.H.; Schnell, P.; DeMichele, A.M.; Loi, S.; Ro, J.; Colleoni, M.; Iwata, H.; Harbeck, N.; Cristofanilli, M.; et al. Palbociclib in Combination with Fulvestrant in Women with Hormone Receptor-Positive/HER2-Negative Advanced Metastatic Breast Cancer: Detailed Safety Analysis From a Multicenter, Randomized, Placebo-Controlled, Phase III Study (PALOMA-3). *Oncologist* **2016**, *21*, 1165–1175. [CrossRef] [PubMed]

12. Paranjpe, R.; John, G.; Trivedi, M.; Abughosh, S. Identifying adherence barriers to oral endocrine therapy among breast cancer survivors. *Breast Cancer Res. Treat.* **2018**, *174*, 297–305. [CrossRef]

13. Cafeo, G.; Carbotti, G.; Cuzzola, A.; Fabbi, M.; Ferrini, S.; Kohnke, F.H.; Papanikolaou, G.; Plutino, M.R.; Rosano, C.; White, A.J.P. Drug Delivery with a Calixpyrrole–trans-Pt(II) Complex. *J. Am. Chem. Soc.* **2013**, *135*, 2544–2551. [CrossRef]

14. Saturnino, C.; Napoli, M.; Paolucci, G.; Bortoluzzi, M.; Popolo, A.; Pinto, A.; Longo, P. Synthesis and cytotoxic activities of group 3 metal complexes having monoanionic tridentate ligands. *Eur. J. Med. Chem.* **2010**, *45*, 4169–4174. [CrossRef] [PubMed]

15. Romeo, R.; Scolaro, L.M.; Plutino, M.R.; Albinati, A. Structural properties of the metallointercalator cationic complex (2,2′:6′,2″-terpyridine)methylplatinum(II) ion. *J. Organomet. Chem.* **2000**, *593*, 403–408. [CrossRef]

16. Malik, P.; Mukherjee, T.K. Recent advances in gold and silver nanoparticle based therapies for lung and breast cancers. *Int. J. Pharm.* **2018**, *553*, 483–509. [CrossRef]

17. Langer, C.J.; Gadgeel, S.M.; Borghaei, H.; Papadimitrakopoulou, V.A.; Patnaik, A.; Powell, S.F.; Gentzler, R.D.; Martins, R.G.; Stevenson, J.P.; I Jalal, S.; et al. Carboplatin and pemetrexed with or without pembrolizumab for advanced, non-squamous non-small-cell lung cancer: A randomised, phase 2 cohort of the open-label KEYNOTE-021 study. *Lancet Oncol.* **2016**, *17*, 1497–1508. [CrossRef]

18. Cohen, H.; Levy, R.J.; Gao, J.; Fishbein, I.; Kousaev, V.; Sosnowski, S.; Slomkowski, S.; Golomb, G. Sustained delivery and expression of DNA encapsulated in polymeric nanoparticles. *Gene Ther.* **2000**, *7*, 1896–1905. [CrossRef]

19. Gref, R.; Minamitake, Y.; Peracchia, M.T.; Trubetskoy, V.; Torchilin, V.; Langer, R.; Kamb, A.; Gruis, N.; Weaver-Feldhaus, J.; Liu, Q.; et al. Biodegradable long-circulating polymeric nanospheres. *Science* **1994**, *263*, 1600–1603. [CrossRef] [PubMed]

20. Zanella, A.; Gandin, V.; Porchia, M.; Refosco, F.; Tisato, F.; Sorrentino, F.; Scutari, G.; Rigobello, M.P.; Marzano, C. Cytotoxicity in human cancer cells and mitochondrial dysfunction induced by a series of new copper(I) complexes containing tris(2-cyanoethyl)phosphines. *Investig. New Drugs* **2010**, *29*, 1213–1223. [CrossRef]

21. Dhanyalayam, D.; Scrivano, L.; Parisi, O.I.; Sinicropi, M.S.; Fazio, A.; Saturnino, C.; Plutino, M.R.; Di Cristo, F.; Puoci, F.; Cappello, A.R.; et al. Biopolymeric self-assembled nanoparticles for enhanced antibacterial activity of Ag-based compounds. *Int. J. Pharm.* **2017**, *517*, 395–402. [CrossRef] [PubMed]

22. Tiekink, E.R. Gold derivatives for the treatment of cancer. *Crit. Rev. Oncol.* **2002**, *42*, 225–248. [CrossRef]

23. Kumar, A.; Boruah, B.M.; Liang, X.-J. Gold Complexes as Prospective Metal-Based Anticancer Drugs. *Encycl. Met.* **2013**, *23*, 867–875. [CrossRef]

24. Claffey, J.; Gleeson, B.; Hogan, M.; Müller-Bunz, H.; Wallis, D.; Tacke, M. Fluorinated Derivatives of Titanocene Y: Synthesis and Cytotoxicity Studies. *Eur. J. Inorg. Chem.* **2008**, *2008*, 4074–4082. [CrossRef]

25. Kudgus, R.A.; Bhattacharya, R.; Mukherjee, P. Cancer nanotechnology: Emerging role of gold nanoconjugates. *Anti-Cancer Agents Med. Chem.* **2011**, *11*, 965–973. [CrossRef] [PubMed]

26. Maeda, H.; Wu, J.; Sawa, T.; Matsumura, Y.; Hori, K. Tumor vascular permeability and the EPR effect in macromolecular therapeutics: A review. *J. Control. Release* **2000**, *65*, 271–284. [CrossRef]

27. Brigger, I.; Dubernet, C.; Couvreur, P. Nanoparticles in cancer therapy and diagnosis. *Adv. Drug Deliv. Rev.* **2012**, *64*, 24–36. [CrossRef]

28. Fenart, L.; Casanova, A.; Dehouck, B.; Duhem, C.; Slupek, S.; Cecchelli, R.; Betbeder, D. Evaluation of effect of charge and lipid coating on ability of 60-nm nanoparticles to cross an in vitro model of the blood-brain barrier. *J. Pharmacol. Exp. Ther.* **1999**, *291*, 1017–1022. [PubMed]

29. Huynh, N.; Passirani, C.; Saulnier, P.; Benoit, J. Lipid nanocapsules: A new platform for nanomedicine. *Int. J. Pharm.* **2009**, *379*, 201–209. [CrossRef]

30. Maeda, H. Tumor-Selective Delivery of Macromolecular Drugs via the EPR Effect: Background and Future Prospects. *Bioconjugate Chem.* **2010**, *21*, 797–802. [CrossRef]

31. Cho, K.; Wang, X.; Nie, S.; Chen, Z.; Shin, D.M. Therapeutic Nanoparticles for Drug Delivery in Cancer. *Clin. Cancer Res.* **2008**, *14*, 1310–1316. [CrossRef]

32. Iacopetta, D.; Grande, F.; Caruso, A.; Mordocco, R.A.; Plutino, M.R.; Scrivano, L.; Ceramella, J.; Muià, N.; Saturnino, C.; Puoci, F.; et al. New insights for the use of quercetin analogs in cancer treatment. *Future Med. Chem.* **2017**, *9*, 2011–2028. [CrossRef]

33. Misra, R.; Acharya, S.; Sahoo, S.K. Cancer nanotechnology: Application of nanotechnology in cancer therapy. *Drug Discov. Today* **2010**, *15*, 842–850. [CrossRef]

34. Murphy, C.J.; Gole, A.M.; Stone, J.W.; Sisco, P.N.; Alkilany, A.M.; Goldsmith, E.C.; Baxter, S.C. Gold Nanoparticles in Biology: Beyond Toxicity to Cellular Imaging. *Accounts Chem. Res.* **2008**, *41*, 1721–1730. [CrossRef]

35. Bhattacharya, R.; Mukherjee, P.; Xiong, Z.; Atala, A.; Soker, A.S.; Mukhopadhyay, D. Gold Nanoparticles Inhibit VEGF165-Induced Proliferation of HUVEC Cells. *Nano Lett.* **2004**, *4*, 2479–2481. [CrossRef]

36. Daniel, M.-C.; Astruc, D. Gold Nanoparticles: Assembly, Supramolecular Chemistry, Quantum-Size-Related Properties, and Applications toward Biology, Catalysis, and Nanotechnology. *Chem. Rev.* **2004**, *104*, 293–346. [CrossRef] [PubMed]

37. Brust, M.; Kiely, C.J. Some recent advances in nanostructure preparation from gold and silver particles: A short topical review. *Colloids Surf. A Physicochem. Eng. Asp.* **2002**, *202*, 175–186. [CrossRef]

38. McConnell, W.P.; Novak, J.P.; Brousseau, L.C.; Fuierer, R.R.; Tenent, R.C.; Feldheim, D.L. Electronic and Optical Properties of Chemically Modified Metal Nanoparticles and Molecularly Bridged Nanoparticle Arrays. *J. Phys. Chem. B* **2000**, *104*, 8925–8930. [CrossRef]

39. Chen, W.; Bardhan, R.; Bartels, M.; Perez-Torres, C.; Pautler, R.G.; Halas, N.J.; Joshi, A. A Molecularly Targeted Theranostic Probe for Ovarian Cancer. *Mol. Cancer Ther.* **2010**, *9*, 1028–1038. [CrossRef]

40. Bardhan, R.; Chen, W.; Bartels, M.; Perez-Torres, C.; Botero, M.F.; McAninch, R.W.; Contreras, A.; Schiff, R.; Pautler, R.G.; Halas, N.J.; et al. Tracking of Multimodal Therapeutic Nanocomplexes Targeting Breast Cancer in Vivo. *Nano Lett.* **2010**, *10*, 4920–4928. [CrossRef]

41. Quinn, J.F.; Raman, R.; Thomas, R.G.; Yurko-Mauro, K.; Nelson, E.B.; Van Dyck, C.; Galvin, J.E.; Emond, J.; Jack, C.R.; Weiner, M.; et al. Docosahexaenoic Acid Supplementation and Cognitive Decline in Alzheimer Disease. *JAMA* **2010**, *304*, 1903–1911. [CrossRef]

42. Lu, W.; Arumugam, S.R.; Senapati, D.; Singh, A.K.; Arbneshi, T.; Khan, S.A.; Yu, H.; Ray, P.C.; Yu, S.A.K.H. Multifunctional Oval-Shaped Gold-Nanoparticle-Based Selective Detection of Breast Cancer Cells Using Simple Colorimetric and Highly Sensitive Two-Photon Scattering Assay. *ACS Nano* **2010**, *4*, 1739–1749. [CrossRef]

43. Choi, M.-R.; Stanton-Maxey, K.J.; Stanley, J.K.; Levin, C.S.; Bardhan, R.; Akin, D.; Badve, S.; Sturgis, J.; Robinson, J.P.; Bashir, R.; et al. A Cellular Trojan Horse for Delivery of Therapeutic Nanoparticles into Tumors. *Nano Lett.* **2007**, *7*, 3759–3765. [CrossRef]

44. Kim, D.; Jeong, Y.Y.; Jon, S. A Drug-Loaded Aptamer–Gold Nanoparticle Bioconjugate for Combined CT Imaging and Therapy of Prostate Cancer. *ACS Nano* **2010**, *4*, 3689–3696. [CrossRef]

45. Patra, C.R.; Bhattacharya, R.; Wang, E.; Katarya, A.; Lau, J.S.; Dutta, S.; Muders, M.H.; Wang, S.; Buhrow, S.A.; Safgren, S.L.; et al. Targeted Delivery of Gemcitabine to Pancreatic Adenocarcinoma Using Cetuximab as a Targeting Agent. *Cancer Res.* **2008**, *68*, 1970–1978. [CrossRef]

46. El-Sayed, I.H.; Huang, X.; El-Sayed, M.A. Selective laser photo-thermal therapy of epithelial carcinoma using anti-EGFR antibody conjugated gold nanoparticles. *Cancer Lett.* **2006**, *239*, 129–135. [CrossRef] [PubMed]

47. Patra, C.R.; Bhattacharya, R.; Mukherjee, P. Fabrication and functional characterization of goldnanoconjugates for potential application in ovarian cancer. *J. Mater. Chem.* **2009**, *20*, 547–554. [CrossRef] [PubMed]

48. De Luca, G.; Bonaccorsi, P.; Trovato, V.; Mancuso, A.; Papalia, T.; Pistone, A.; Casaletto, M.P.; Mezzi, A.; Brunetti, B.; Minuti, L.; et al. Tripodal tris-disulfides as capping agents for a controlled mixed functionalization of gold nanoparticles. *New J. Chem.* **2018**, *42*, 16436–16440. [CrossRef]

49. Priya, M.R.K.; Iyer, P.R. Antiproliferative effects on tumor cells of the synthesized gold nanoparticles against Hep2 liver cancer cell line. *Egypt. Liver J.* **2020**, *10*, 1–12. [CrossRef]

50. Bardhan, R.; Chen, W.; Perez-Torres, C.; Bartels, M.; Huschka, R.M.; Zhao, L.L.; Morosan, E.; Pautler, R.G.; Joshi, A.; Halas, N.J. Nanoshells with Targeted Simultaneous Enhancement of Magnetic and Optical Imaging and Photothermal Therapeutic Response. *Adv. Funct. Mater.* **2009**, *19*, 3901–3909. [CrossRef]

51. Akiyama, Y.; Mori, T.; Katayama, Y.; Niidome, T. The effects of PEG grafting level and injection dose on gold nanorod biodistribution in the tumor-bearing mice. *J. Control. Release* **2009**, *139*, 81–84. [CrossRef] [PubMed]

52. Mammen, M.; Choi, S.K.; Whitesides, G.M. Polyvalent interactions in biological systems: Implications for design and use of multivalent ligands and inhibitors. *Angew. Chem.* **1998**, *37*, 2754–2794. [CrossRef]

53. Sinicropi, M.S.; Iacopetta, D.; Rosano, C.; Randino, R.; Caruso, A.; Saturnino, C.; Muià, N.; Ceramella, J.; Puoci, F.; Rodriquez, M.; et al. N-thioalkylcarbazoles derivatives as new anti-proliferative agents: Synthesis, characterisation and molecular mechanism evaluation. *J. Enzym. Inhib. Med. Chem.* **2018**, *33*, 434–444. [CrossRef]

54. Iacopetta, D.; Rosano, C.; Puoci, F.; Parisi, O.I.; Saturnino, C.; Caruso, A.; Longo, P.; Ceramella, J.; Malzert-Fréon, A.; Dallemagne, P.; et al. Multifaceted properties of 1,4-dimethylcarbazoles: Focus on trimethoxybenzamide and trimethoxyphenylurea derivatives as novel human topoisomerase II inhibitors. *Eur. J. Pharm. Sci.* **2017**, *96*, 263–272. [CrossRef] [PubMed]

55. Caruso, A.; Voisin-Chiret, A.S.; Lancelot, J.-C.; Sinicropi, M.S.; Garofalo, A.; Raulta, S. Novel and Efficient Synthesis of 5,8-dimethyl-9hcarbazol- 3-ol via a hydroxydeboronation reaction. *Molecules* **2007**, *71*, 2203–2210.

56. Sinicropi, M.S.; Lappano, R.; Caruso, A.; Santolla, M.F.; Pisano, A.; Rosano, C.; Capasso, A.; Panno, A.; Lancelot, J.C.; Rault, S.; et al. (6-Bromo-1,4-dimethyl-9H-carbazol-3-yl-methylene)-hydrazine (Carbhydraz) Acts as a GPER Agonist in Breast Cancer Cells. *Curr. Top. Med. Chem.* **2015**, *15*, 1035–1042. [CrossRef] [PubMed]

57. Saturnino, C.; Caruso, A.; Longo, P.; Capasso, A.; Pingitore, A.; Caroleo, M.; Cione, E.; Perri, M.; Nicolo, F.; Nardo, V.; et al. Crystallographic Study and Biological Evaluation of 1,4-dimethyl-N-alkylcarbazoles†. *Curr. Top. Med. Chem.* **2015**, *15*, 973–979. [CrossRef]

58. Saturnino, C.; Sinicropi, M.S.; Iacopettab, D.; Ceramella, J.; Caruso, A.; Muià, N.; Longo, P.; Rosace, G.; Galletta, M.; Ielo, I.; et al. N-Thiocarbazole-based gold nanoparticles: Synthesis, characterization and anti-proliferative activity evaluation. *IOP Conf. Ser. Mater. Sci. Eng.* **2018**, *459*, 012023. [CrossRef]

59. Mook-Jung, I.; Han, S.-H.; Chang, Y.J.; Jung, E.S.; Kim, J.-W.; Na, D.L. Effective screen for amyloid β aggregation inhibitor using amyloid β-conjugated gold nanoparticles. *Int. J. Nanomed.* **2010**, *6*, 1–12. [CrossRef]

60. Ceramella, J.; Mariconda, A.; Iacopetta, D.; Saturnino, C.; Barbarossa, A.; Caruso, A.; Rosano, C.; Sinicropi, M.S.; Longo, P. From coins to cancer therapy: Gold, silver and copper complexes targeting human topoisomerases. *Bioorganic Med. Chem. Lett.* **2020**, *30*, 126905. [CrossRef]

61. Kilpin, K.J.; Dyson, P.J. Enzyme inhibition by metal complexes: Concepts, strategies and applications. *Chem. Sci.* **2013**, *4*, 1410–1419. [CrossRef]

62. Chen, Z.-F.; Shi, Y.-F.; Liu, Y.-C.; Hong, X.; Geng, B.; Peng, Y.; Liang, H. TCM Active Ingredient Oxoglaucine Metal Complexes: Crystal Structure, Cytotoxicity, and Interaction with DNA. *Inorg. Chem.* **2012**, *51*, 1998–2009. [CrossRef] [PubMed]

63. Saturnino, C.; Popolo, A.; Ramunno, A.; Adesso, S.; Pecoraro, M.; Plutino, M.R.; Rizzato, S.; Albinati, A.; Marzocco, S.; Sala, M.; et al. Anti-Inflammatory, Antioxidant and Crystallographic Studies of N-Palmitoyl-ethanol Amine (PEA) Derivatives. *Molecules* **2017**, *22*, 616. [CrossRef]

64. Wilson, C.R.; Fagenson, A.M.; Ruangpradit, W.; Muller, M.T.; Munro, O.Q. Gold(III) Complexes of Pyridyl- and Isoquinolylamido Ligands: Structural, Spectroscopic, and Biological Studies of a New Class of Dual Topoisomerase I and II Inhibitors. *Inorg. Chem.* **2013**, *52*, 7889–7906. [CrossRef]

65. Chimento, A.; Saturnino, C.; Iacopetta, D.; Mazzotta, R.; Caruso, A.; Plutino, M.R.; Mariconda, A.; Ramunno, A.; Sinicropi, M.S.; Pezzi, V.; et al. Inhibition of human topoisomerase I and II and anti-proliferative effects on MCF-7 cells by new titanocene complexes. *Bioorganic Med. Chem.* **2015**, *23*, 7302–7312. [CrossRef] [PubMed]

66. Sâmia, L.B.P.; Parrilha, G.L.; Da Silva, J.G.; Ramos, J.P.; Souza-Fagundes, E.M.; Castelli, S.; Vutey, V.; Desideri, A.; Beraldo, H. Metal complexes of 3-(4-bromophenyl)-1-pyridin-2-ylprop-2-en-1-one thiosemicarbazone: Cytotoxic activity and investigation on the mode of action of the gold(III) complex. *BioMetals* **2016**, *29*, 515–526. [CrossRef]

67. Saturnino, C.; Barone, I.; Iacopetta, D.; Mariconda, A.; Sinicropi, M.S.; Rosano, C.; Campana, A.; Catalano, S.; Longo, P.; Andò, S. N-heterocyclic carbene complexes of silver and gold as novel tools against breast cancer progression. *Future Med. Chem.* **2016**, *8*, 2213–2229. [CrossRef] [PubMed]

68. Napoli, M.; Saturnino, C.; Sirignano, E.; Popolo, A.; Pinto, A.; Longo, P. Synthesis, characterization and cytotoxicity studies of methoxy alkyl substituted metallocenes. *Eur. J. Med. Chem.* **2011**, *46*, 122–128. [CrossRef]

69. Caporale, A.; Palma, G.; Mariconda, A.; Del Vecchio, V.; Iacopetta, D.; Parisi, O.I.; Sinicropi, M.S.; Puoci, F.; Arra, C.; Longo, P.; et al. Synthesis and Antitumor Activity of New Group 3 Metallocene Complexes. *Molecules* **2017**, *22*, 526. [CrossRef]

70. Sirignano, E.; Saturnino, C.; Botta, A.; Sinicropi, M.S.; Caruso, A.; Pisano, A.; Lappano, R.; Maggiolini, M.; Longo, P. Synthesis, characterization and cytotoxic activity on breast cancer cells of new half-titanocene derivatives. *Bioorganic Med. Chem. Lett.* **2013**, *23*, 3458–3462. [CrossRef] [PubMed]

71. Saturnino, C.; Sirignano, E.; Botta, A.; Sinicropi, M.S.; Caruso, A.; Pisano, A.; Lappano, R.; Maggiolini, M.; Longo, P. New titanocene derivatives with high antiproliferative activity against breast cancer cells. *Bioorganic Med. Chem. Lett.* **2014**, *24*, 136–140. [CrossRef]

72. Iacopetta, D.; Mariconda, A.; Saturnino, C.; Caruso, A.; Palma, G.; Ceramella, J.; Muià, N.; Perri, M.; Sinicropi, M.S.; Caroleo, M.C.; et al. Novel Gold and Silver Carbene Complexes Exert Antitumor Effects Triggering the Reactive Oxygen Species Dependent Intrinsic Apoptotic Pathway. *ChemMedChem* **2017**, *12*, 2054–2065. [CrossRef]

73. Iacopetta, D.; Rosano, C.; Sirignano, M.; Mariconda, A.; Ceramella, J.; Ponassi. M.; Saturnino, C.; Sinicropi, M.S.; Longo, P. Is the Way to Fight Cancer Paved with Gold? Metal-Based Carbene Complexes with Multiple and Fascinating Biological Features. *Pharmaceuticals* **2020**, *13*, 91. [CrossRef]

74. Talib, J.; Beck, J.L.; Ralph, S.F. A mass spectrometric investigation of the binding of gold antiarthritic agents and the metabolite [Au(CN)2]- to human serum albumin. *JBIC J. Biol. Inorg. Chem.* **2006**, *11*, 559–570. [CrossRef] [PubMed]

75. Özdemir, I.; Denizci, A.; Ozturk, H.T.; Çetinkaya, B. Synthetic and antimicrobial studies on new gold(I) complexes of imidazolidin-2-ylidenes. *Appl. Organomet. Chem.* **2004**, *18*, 318–322. [CrossRef]

76. Yan, J.J.; Chow, A.L.-F.; Leung, C.-H.; Sun, R.W.-Y.; Ma, D.-L.; Che, C.-M. Cyclometalated gold(iii) complexes with N-heterocyclic carbene ligands as topoisomerase I poisons. *Chem. Commun.* **2010**, *46*, 3893–3895. [CrossRef] [PubMed]

77. Champness, N.R. The future of metal-organic frameworks. *Dalton Trans.* **2011**, *40*, 10311–10315. [CrossRef]

78. Ward, J.R. Role of disease-modifying antirheumatic drugs versus cytotoxic agents in the therapy of rheumatoid arthritis. *Am. J. Med.* **1988**, *85*, 39–44. [CrossRef]

79. Fries, J.F.; Bloch, D.; Spitz, P.; Mitchell, D.M. Cancer in rheumatoid arthritis: A prospective long-term study of mortality. *Am. J. Med.* **1985**, *78*, 56–59. [CrossRef]

80. Simon, T.M.; Kunishima, D.H.; Vibert, G.J.; Lorber, A. Inhibitory effects of a new oral gold compound on hela cells. *Cancer* **1979**, *44*, 1965–1975. [CrossRef]

81. Simon, T.M.; Kunishima, D.H.; Vibert, G.J.; Lorber, A. Screening trial with the coordinated gold compound auranofin using mouse lymphocyte leukemia P388. *Cancer Res.* **1981**, *41*, 94–97. [PubMed]

82. Mirabelli, C.K.; Jensen, B.D.; Mattern, M.R.; Sung, C.M.; Mong, S.M.; Hill, D.T.; Dean, S.W.; Schein, P.S.; Johnson, R.K.; Crooke, S.T. Cellular pharmacology of mu-[1,2-bis(diphenylphosphino)ethane]bis[(1-thio-beta-D-gluco pyranosato-S)gold(I)]: A novel antitumor agent. *Anti-Cancer Drug Des.* **1986**, *1*, 223–234.

Article

N-Heterocyclic Carbene-Gold(I) Complexes Targeting Actin Polymerization

Domenico Iacopetta [1,*,†], Jessica Ceramella [1,†], Camillo Rosano [2], Annaluisa Mariconda [3], Michele Pellegrino [1], Marco Sirignano [4], Carmela Saturnino [3,5], Alessia Catalano [6], Stefano Aquaro [1], Pasquale Longo [4] and Maria Stefania Sinicropi [1]

[1] Department of Pharmacy, Health and Nutritional Sciences, University of Calabria, Via Pietro Bucci, Arcavacata di Rende, 87036 Cosenza, Italy; jessicaceramella@gmail.com (J.C.); michele.pellegrino@unical.it (M.P.); stefano.aquaro@unical.it (S.A.); s.sinicropi@unical.it (M.S.S.)

[2] Proteomics and Mass Spectrometry Unit, IRCCS Policlinico San Martino, L.go Rosanna Benzi 10, 16132 Genova, Italy; camillo.rosano@gmail.com

[3] Department of Science, University of Basilicata, Viale dell'Ateneo Lucano 10, 85100 Potenza, Italy; annaluisa.mariconda@unibas.it (A.M.); carmela.saturnino@unibas.it (C.S.)

[4] Department of Chemistry and Biology, University of Salerno, Via Giovanni Paolo II, 132, 84084 Fisciano, Italy; msirignano@unisa.it (M.S.); plongo@unisa.it (P.L.)

[5] Spinoff TNcKILLERS, Viale dell'Ateneo Lucano 10, 85100 Potenza, Italy

[6] Department of Pharmacy-Drug Sciences, University of Bari "Aldo Moro", 70126 Bari, Italy; alessia.catalano@uniba.it

* Correspondence: domenico.iacopetta@unical.it; Tel.: +39-098-449-3200

† These authors contributed equally to this work.

Citation: Iacopetta, D.; Ceramella, J.; Rosano, C.; Mariconda, A.; Pellegrino, M.; Sirignano, M.; Saturnino, C.; Catalano, A.; Aquaro, S.; Longo, P.; et al. N-Heterocyclic Carbene-Gold(I) Complexes Targeting Actin Polymerization. *Appl. Sci.* **2021**, *11*, 5626. https://doi.org/10.3390/app11125626

Academic Editor: Greta Varchi

Received: 10 May 2021
Accepted: 15 June 2021
Published: 18 June 2021

Publisher's Note: MDPI stays neutral with regard to jurisdictional claims in published maps and institutional affiliations.

Copyright: © 2021 by the authors. Licensee MDPI, Basel, Switzerland. This article is an open access article distributed under the terms and conditions of the Creative Commons Attribution (CC BY) license (https://creativecommons.org/licenses/by/4.0/).

Abstract: Transition metal complexes are attracting attention because of their various chemical and biological properties. In particular, the NHC-gold complexes represent a productive field of research in medicinal chemistry, mostly as anticancer tools, displaying a broad range of targets. In addition to the already known biological targets, recently, an important activity in the organization of the cell cytoskeleton was discovered. In this paper, we demonstrated that two NHC-gold complexes (namely **AuL4** and **AuL7**) possessing good anticancer activity and multi-target properties, as stated in our previous studies, play a major role in regulating the actin polymerization, by the means of in silico and in vitro assays. Using immunofluorescence and direct enzymatic assays, we proved that both the complexes inhibited the actin polymerization reaction without promoting the depolymerization of actin filaments. Our outcomes may contribute toward deepening the knowledge of NHC-gold complexes, with the objective of producing more effective and safer drugs for treating cancer diseases.

Keywords: NHC-gold complexes; actin; anticancer

1. Introduction

Transition metal complexes are considered candidates in medicinal chemistry because of their potential as new, diagnostic, and therapeutic agents. These complexes are modular systems in which the metal centers are bound to different ligands, arranged in a well-defined three-dimensional structure that determines the unique characteristics of the complexes. Transition metal complexes possess different photophysical/photochemical features, and numerous biological applications have been described [1–3]. Amongst the plethora of transition metal complexes, gold-based complexes are attracting the attention of many researchers, because of their possible oxidation states (e.g., Au(I) and Au(III)), stability, easy ligand exchange reactions, effective cytotoxicity towards several in vitro models of cancer, and the lack of negative effects on normal cell viability [4,5]. Gold has historically been used to treat several diseases, albeit the first scientific evidences date back to the 1920s, pertaining to the compound $K[Au(CN)_2]$ (potassium dicyanoaurate, Figure 1), which was clinically tested for its anti-tuberculosis activity, and then dismissed because of its toxicity [6]. Instead, the literature shows many in vitro and in vivo studies conducted on the anticancer activities of another important gold-based complex, i.e., auranofin [(tetra-*O*-acetyl-β-D-glucopyranosyl)-thio] (triethylphosphine)-gold(I) (Figure 1), a second generation drug, which targets, in particular, the mitochondrial enzyme thioredoxin reductase (TrxR) [7]. These studies paved the way to the design, synthesis, and biological evaluation of many gold-based agents, mostly for the treatment of cancers. In this scenario, the gold complexes with N-heterocyclic carbene (NHC) ligands represent a challenging thread because of their ability to potently inhibit TrxR and decrease cancer cell proliferation [8,9]. However, TrxR is not the only target of anticancer gold complexes reported on so far. Indeed, amongst the other important functions, DNA and related enzymatic machinery interference [10], mitochondrial damage (mostly for cationic gold(I) biscarbene complexes) [11,12], and the inhibition of protein tyrosine phosphatases (PTP) [13] should also be noted. More recently, evidence indicated that NHC-gold complexes play an important role in regulating the cytoskeleton dynamics [14–18], interfering with the tubulin and/or actin metabolism. In our previous studies, we individuated two lead molecules, i.e., **AuL4** and **AuL7** (Figure 1) [15,18], demonstrating that they are good anticancer compounds and possess different biological properties, including the ability to regulate the tubulin polymerization reaction in two different ways. Indeed, **AuL4** is able to stabilize the tubulin filaments, behaving as one of the most well-known drug, paclitaxel, whereas **AuL7** hampers the tubulin polymerization reaction, such as the vinblastine, one of the most used vinca alkaloids that induces the assembly of tubulin into non-microtubule polymers, such as para-crystals or spiral proto-filamentous structures [19]. In this paper, another important biological target of the above-mentioned NHC-gold complexes has been studied, namely the actin, using as references the molecules latrunculin A (LA) and cytochalasin B (CB) (Figure 2). By using in silico and in vitro assays, we proved that both leads are able to inhibit the actin polymerization, with an efficacy similar to that of cytochalasin B (Phomin), a well-established cell-permeable fungal inhibitor of the actin polymerization. This evidence highlights, once again, the multi-target potential of the studied compounds, and is seminal for the development of potent and selective anticancer drugs targeting the actin, with negligible cytotoxic effects on the normal cells.

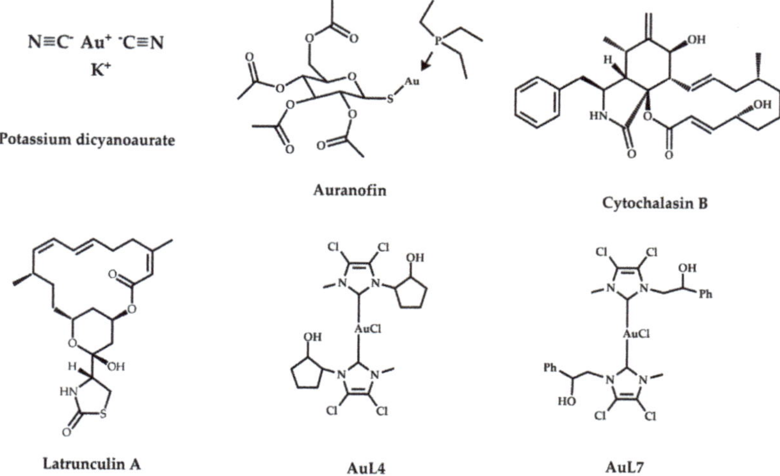

Figure 1. Molecular structure of potassium dicyanoaurate, auranofin, cytochalasin B, latrunculin A, **AuL4** and **AuL7**.

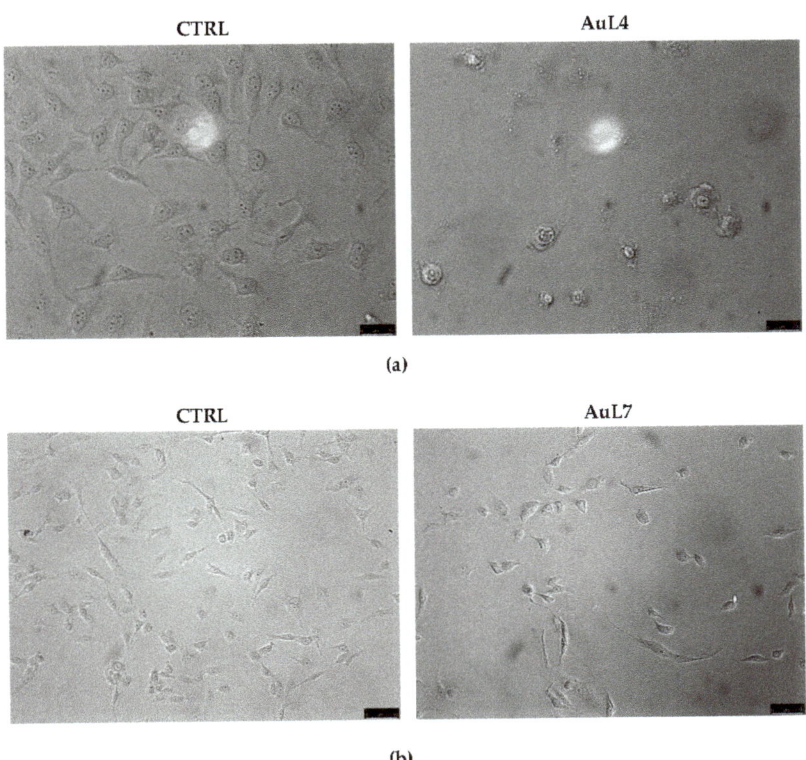

Figure 2. Morphological changes after the exposure of HeLa cells to **AuL4** (a) and MDA-MB-231 cells to **AuL7** (b). Both compounds were used at their IC$_{50}$ values for 72 h. Images were acquired at 20× magnification.

2. Materials and Methods

2.1. Cell Culture

The cell lines used in this work (MDA-MB-231 and HeLa) were purchased from American Type Culture Collection (ATCC, Manassas, VA, USA). Human breast cancer triple negative MDA-MB-231 were grown in DMEM-F12 medium containing 2 mmol/L L-glutamine, 1 mg/mL penicillin-streptomycin, and 5% Fetal Bovine Serum (FBS) [20]. HeLa human epithelial cervix carcinoma cells, estrogen receptor (ER)-negative were maintained in Eagle's minimum essential medium (MEM), supplemented with 10% FBS, 1% l-glutamine, 100 U/mL penicillin/streptomycin, and 1% non-essential amino acids (NEAA). All cell lines were maintained at 37 °C in a humidified atmosphere of 95% air and 5% CO$_2$, and periodically screened for contamination [21].

2.2. Docking Studies

The crystal structure of the complex formed between the beta/gamma-actin with profilin and the acetyltransferase AnCoA-NAA80 [22] [PDB code 6nbw] was used as the protein target for our docking simulations. All structures of the ligands tested in silico were built, and energy was minimized using the program MarvinSketch (ChemAxon ltd, Budapest, Hu). To shed light on the possible binding modes to the protein target and to determine their binding energies, we used the AutoDock v.4.2.2 program suite [23]. We performed a "blind docking" simulation: the docking of the small molecule to the targets was done without a priori knowledge of the location of the binding site by the system. All simulations were performed by adopting the program default values. The protein and the ligands were prepared using the ADT graphical interface [24]. Polar hydrogens were added to each protein, Kollman charges assigned, and finally, the solvation parameters were added. The protein was considered a rigid object and all the ligands as fully flexible. A searching grid was extended all over the protein and affinity maps calculated. The search was carried out with a Lamarckian genetic algorithm: a population of 100 individuals with a mutation rate of 0.02 evolved for 100 generations. Evaluation of the results was performed by listing the different ligand poses according to their predicted binding energy. A cluster analysis based on root-mean-square-deviation (RMSD) values from the starting geometry was performed. The lowest energetic conformation of the most populated cluster was considered the best candidate. When clusters were almost equipopulated, and their energy distribution spread, their corresponding molecules were considered as bad ligands [25,26].

The generated docking poses were ranked in order of increasing binding energy values and clustered on the basis of a RMSD cut-off value of 2.0 Å. From the structural analysis of the lowest energy solutions of each cluster, we could spot the protein binding site. Figure 3 was drawn using the program Chimera [27].

2.3. Immunofluorescence Analysis

Cells were seeded in 48-well culture plates containing glass slides, serum-deprived for 24 h, and incubated with the most active compound for 24 h (concentration equal to its IC_{50} value). Then, the cells fixed with cold methanol were incubated with primary antibody, diluted in bovine serum albumin (BSA) 2% overnight at 4 °C, as previously described [28]. The rabbit anti-β-actin was purchased from Santa Cruz Biotechnology and diluted 1:100 before use. Then the secondary antibody Alexa Fluor® 488 conjugate goat-anti-rabbit (1:500, Thermo Fisher Scientific, MA, USA) was added and incubated for 2 h at 37 °C. Nuclei were stained using DAPI (Sigma Aldrich, Mila, Italy) for 10 min at a concentration of 0.2 µg/mL. Fluorescence was detected using a fluorescence microscope (Leica DM 6000). LAS-X software was used to acquire and process all images.

2.4. Actin Polymerization/Depolymerization Assay

The ability of the tested compounds to interfere with the actin polymerization and depolymerization reaction was measured using an actin Polymerization/Depolymerization Assay Kit purchased from Abcam, following the manufacturer's instructions. Both polymerization and depolymerization reactions occur in a 100 µL final volume, by using Labeled Rabbit Muscle actin reconstituted with the buffer G supplemented with 0.2 mM ATP and 0.5 mM DTT (1,4-dithiothreitol). For the polymerization assay, reconstituted actin was mixed with supplemented buffer G, and samples in a white 96-well plate, and then the polymerization, the reaction, was induced with the addition of the buffer P supplemented with 10 mM ATP. The solution was mixed and the data acquisition started. For the actin depolymerization assay, supplemented buffers P and G were mixed and incubated in a white 96-well plate at room temperature for one hour to polymerize the actin, protected from light. Then samples were added and the data acquisition started. Latrunculin A and cytochalasin B were use as control molecules at a concentration of 5 µM. For both assays, the assembly of actin filaments was determined by measuring the fluorescence (Ex/Em: 365/410 nm) in kinetic mode for 1 h at room temperature using a microplate reader. Images are representative of three independent experiments, each performed in triplicate.

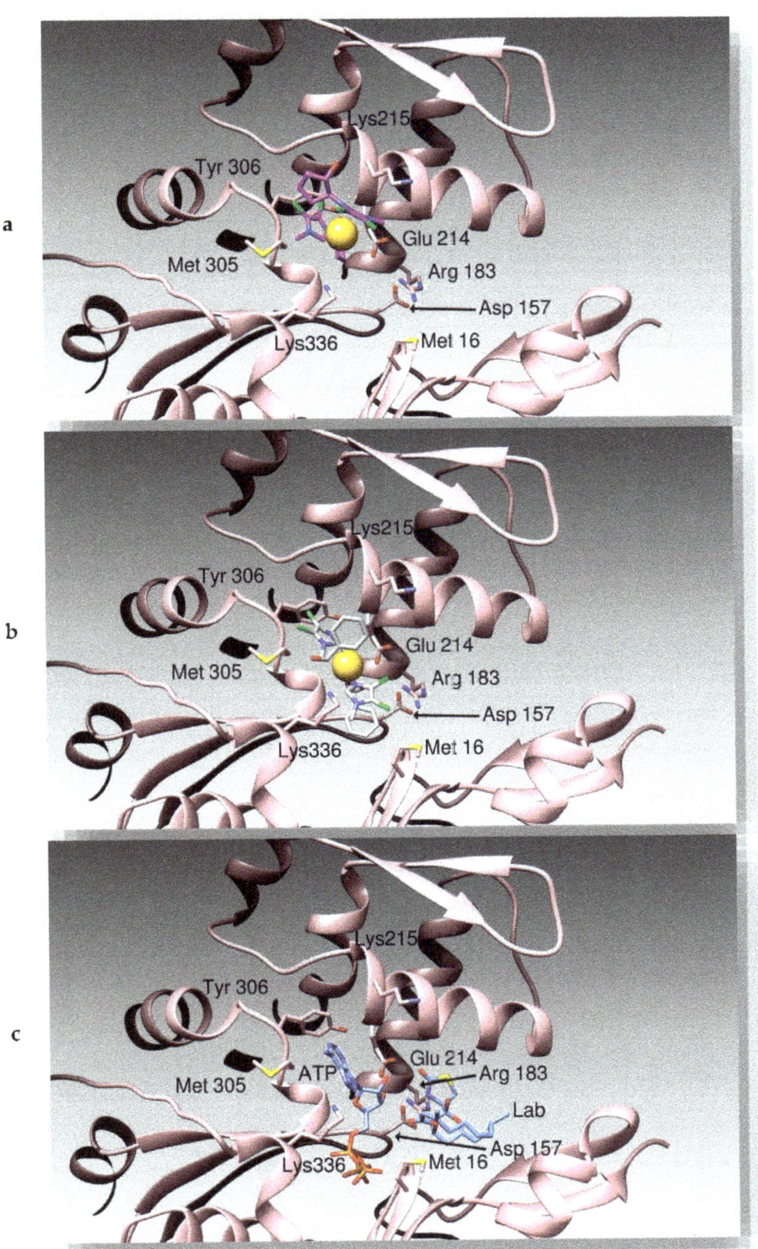

Figure 3. The three-dimensional structure of the nucleotide-binding cleft of actin (pink ribbon) is depicted. (**a**) **AuL4** pose; (**b**) **AuL7** pose; (**c**) binding mode of latrunculin B as determined by X-ray crystallography.

3. Results

3.1. AuL4 and AuL7 Induce Dramatic Cancer Cells Morphology Changes

Our previous studies demonstrated the anticancer activities of **AuL4** and **AuL7** against different in vitro models, being particularly effective against HeLa and MDA-MB-231 cancer cells, respectively, and triggering the ROS-mediated intrinsic apoptotic pathway. Moreover, we proved that these complexes possessed a multi-target action, namely they inhibited human topoisomerases I and II activities and interfered with the microtubules dynamics. However, this latter feature resides in the ability of **AuL7** to inhibit the tubulin polymerization reaction, acting like the vinblastine, whereas **AuL4** behaves as the paclitaxel, thus stabilizing the microtubules, as demonstrated by immunofluorescence and direct enzymatic assays. In both cases, the exposure of HeLa and MDA-MB-231 cancer cells to **AuL4** (at its IC_{50} = 1.63 ± 0.5 µM) and **AuL7** (at its IC_{50} = 2.1 ± 0.7 µM), respectively, for 72 h, produced dramatic morphological changes, as visible by the observation of cancer cells (Figure 2), using an inverted microscope (20× magnification). Indeed, **AuL4**-treated HeLa cells appeared round and shrunk, and **AuL7**-treated MDA-MB-231 cells exhibited both round and tread-like shapes, with respect to the vehicle-treated cells that have a normal morphology. These changes were attributed to the already proved interference with tubulin, but we wondered if our complexes could regulate the actin system, as well. This hypothesis sounded very challenging for us, because this would mean that the complexes possess another unexplored intracellular target, *viz.* the actin. In order to assess this supposition, we performed in silico and in vitro studies.

3.2. Docking Studies

To first evaluate the binding poses and the calculated affinities between **AuL4** and **AuL7** to the protein actin, we performed molecular docking simulations, using as a target the three-dimensional structure of actin [22]. Affinities of the two compounds to the protein were calculated by AutoDock, according to the expression K_i = exp (deltaG/(R*T)). As discussed in our previous works [29–31], clusterization of the outcomes from our simulation runs, together with the visual inspection of the ligand:protein binding sites, were considered as markers of the quality of the interaction.

AuL4 and **AuL7** share the same actin binding site with latrunculin B (Figure 3), previously determined by X-ray crystallography [22]. Among the two compounds, **AuL4** is positioned within the nucleotide binding site of actin, forming hydrogen bonds with protein residues Asp211, Lys215, and Tyr306. Its chloride atoms are involved into halogen bonds with Glu214, while the gold atom is coordinated by Lys336. The binding site is stabilized by hydrophobic interactions with residues Met16, Met305, and Tyr306. On the other hand, **AuL7** is positioned with part of its structure exposed to the solvent. This compound forms hydrogen bonds with actin residues Asp157, Thr303, and the peptidic nitrogen atom of Gly182. Arg201 form a halogen bond with a chlorine atom of the ligand, which is also stabilized by hydrophobic interactions with Met16, Met305, and Tyr306. The Au atom is coordinated to Glu214.

3.3. *AuL4 and AuL7 Interfere with the Normal Intracellular Actin Organization*

The cell cytoskeleton is an important intracellular structure that is involved in several physiological processes, and is implied, as well, in tumor progression, most importantly in the epithelial–mesenchymal transition (EMT) process. The major components are tubulin and actin, which are targets of effective anticancer drugs. Some metal complexes have been proved to interfere with the cytoskeleton dynamics [15,32–36], particularly inhibiting the tubulin polymerization but, more recently, they were discovered to play a different and specific role in regulating the actin system [37]. Since we have already proved that the two metal complexes exerted inhibitory activity against the tubulin polymerization [15], we wondered whether they could also interfere directly with the other major component of the cell cytoskeleton, *viz.* the actin, in the two cancer cell models used for the previous study. With this in mind, HeLa and MDA-MB-231 cells were treated for 24 h with compounds **AuL4** (at its IC_{50} = 1.63 ± 0.5 µM) and **AuL7** (at its IC_{50} = 2.1 ± 0.7 µM), respectively, together with the only vehicle (DMSO, negative control). As a reference molecule, we used latrunculin A (positive control), a well-known natural compound that targets the actin cytoskeleton, at a concentration of 1 µM, under the same experimental conditions.

After that, cancer cells were further processed and imaged under a fluorescent microscope, as described in the Material and Methods section. As visible in Figures 4 and 5, in the negative control (only DMSO), the actin filaments exhibited a normal organization in the cell cytoplasm, with differences relative to the normal HeLa and MDA-MB-231 cell morphology. Conversely, under LA exposure, both cancer cell lines showed a remarkable disorganization of the actin system. Indeed, in Figure 4, the actin bundles are very thick and bright, with respect to the vehicle-treated HeLa cells, and one can notice that the actin system is more compact, not fairly distributed, and some dot-like structures appeared (see white arrows). A similar arrangement was seen in **AuL4**-treated HeLa cells; indeed, the cell size is reduced and the actin is thickened around the cell nuclei, with respect to the vehicle-treated cells (negative control). The MDA-MB-231 cells under **AuL7** treatment lost their normal morphology, as well, but assumed different shapes; indeed, in Figure 5, some cells appear round, others with a thread-like structure or abnormally enlarged cytoplasm (see white arrows). Moreover, the actin filaments look very bright and arranged in dot-like structures or unevenly distributed within the cell cytoplasm. To summarize, we can affirm that both compounds interfere with the regular organization of the actin system in the cancer cells under investigation.

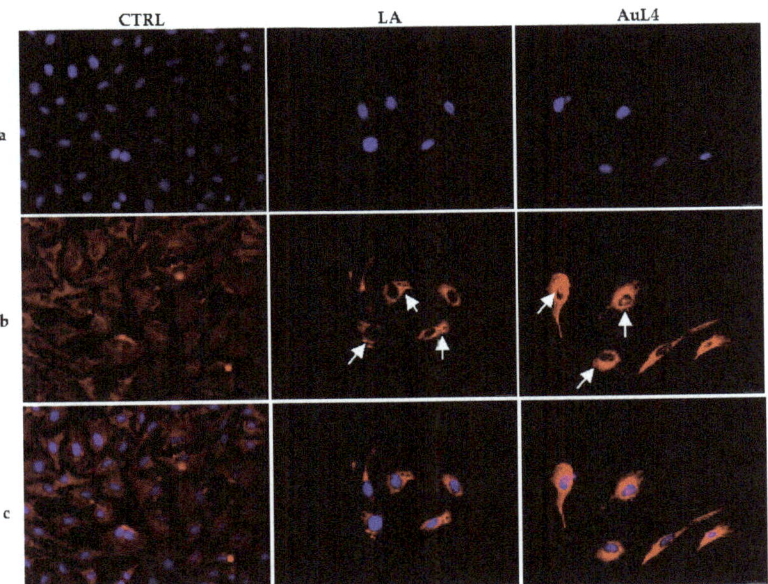

Figure 4. Immunofluorescence analysis. HeLa cells were treated with DMSO (CTRL, negative control) 0.1 µM Latrunculin A (LA, positive control) or **AuL4** (at its IC_{50} value) for 24 h. After treatment, the cells were further processed (see Materials and Methods section) and imaged under inverted fluorescence microscope at 20× magnification. In the CTRL cells, the actin filaments exhibited a normal organization in the cells cytoplasm. HeLa cells treated with LA and **AuL4**, showed an irregular arrangement and organization of the actin cytoskeleton: the actin system appear more compact, not fairly distributed and some dot-like structures appeared (see white arrows). (**Panels a**): nuclear stain with DAPI ($\lambda ex/\lambda em = 350/460$ nm); (**Panels b**): β-actin (Alexa Fluor® 568; $\lambda ex/\lambda em = 644/665$ nm); (**Panels c**): show the merge. Representative fields are shown.

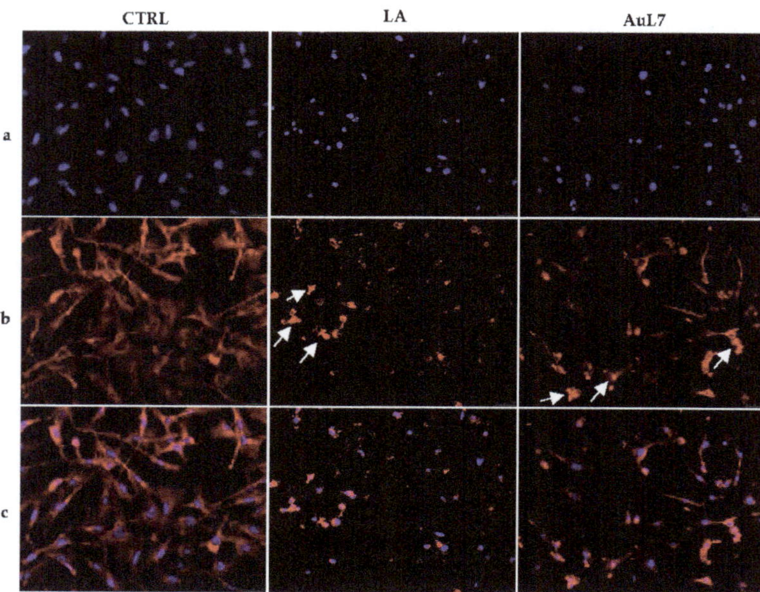

Figure 5. Immunofluorescence analysis. MDA-MB-231 cells were treated with DMSO (CTRL, negative control) 0.1 µM Latrunculin A (LA, positive control) or **AuL7** (at its IC$_{50}$ value) for 24 h. After treatment, the cells were further processed (see Materials and Methods section) and imaged under the inverted fluorescence microscope at 20× magnification (see Materials and Methods section). In the CTRL cells, the actin filaments exhibited a normal organization in the cells cytoplasm. Actin filaments in MDA-MB-231 cells treated with LA and **AuL7**, look very bright and arranged in dot-like structures: cells appear round and others with a thread-like structure (see white arrows). (**Panels a**): nuclear stain with DAPI (λex/λem = 350/460 nm); (**Panels b**): β-actin (Alexa Fluor® 568; λex/λem = 644/665 nm); (**Panels c**): show the merge. Representative fields are shown.

3.4. *AuL4 and AuL7 Block the In Vitro Actin Polymerization Reaction but Do Not Accelerate the Depolymerization Reaction*

Once established that the two metal complexes provoked a dramatic change of the cytoskeleton actin structure, we wondered whether the observed effect was due to a direct action toward the actin or through indirect effects, given that these compounds have already been proved as tubulin polymerization inhibitors. Thus, we adopted an in vitro actin assay that exploits a fluorescent-labeled purified rabbit actin, which polymerization/depolymerization reactions can be easily followed using a fluorimeter, in order to confirm the immunofluorescence results, and establish if **AuL4** and **AuL7** could act as actin polymerization inhibitors and/or accelerate the filamentous actin depolymerization. In this assay, we employed two reference molecules, namely the LA and CB. Particularly, LA can bind and sequester G-actin monomers, hampering the actin polymerization and, at the same time, accelerate the rate of the actin subunits dissociation from the assembled filaments. As a result, LA is able to block the F-actin formation, one way or another. Conversely, CB inhibits the actin polymerization, avoiding the monomers addition to the "barbed" end of the actin filaments, but it cannot increase the depolymerization reaction. The results from this assay were plotted and, as visible in Figure 6, in the control reaction (DMSO), the G-actin polymerizes very fast; indeed, the reaction curve reached a value of about 37,000 RFU (see experimental section for details), and kept the plateau until the end of the experiment. Conversely, the two reference molecules, LA and CB, used at a concentration of 5 µM, blocked the actin polymerization; indeed, the initial rapid curve growth, seen in the control reaction, was absent. In more detail, the LA curve decreased until a value of about 12,000 RFU after 13–15 min and reached a value of about 10,000 RFU at the end of the experiment. The exposure of actin monomers to CB produced a curve with a little increase (about 22,000 RFU) in the first 4–5 min, then decreased, and ended at an RFU value of 22,000, which is equal to the initial value. Finally, **AuL4** and **AuL7**, at a concentration equal to 5 µM, exhibited both an inhibiting activity similar to that of CB, and to a lesser extent, with respect to LA. Specifically, the **AuL4** curve was lower than that of CB in the first 5 min and then rose, maintaining RFU values ranging from 24,000 to 26,000, and ended at a final value of about 27,000 RFU. The **AuL7** curve trend was similar to that of **AuL4**, with the exception of the first 5 min, when it decreased to values lesser than LA, ending with a final value of 24,000 RFU. Overall, the inhibitory activity of **AuL7** seems better than **AuL4** but, in both cases, this assay demonstrates a direct effect on actin polymerization reaction, supporting the immunofluorescence assays.

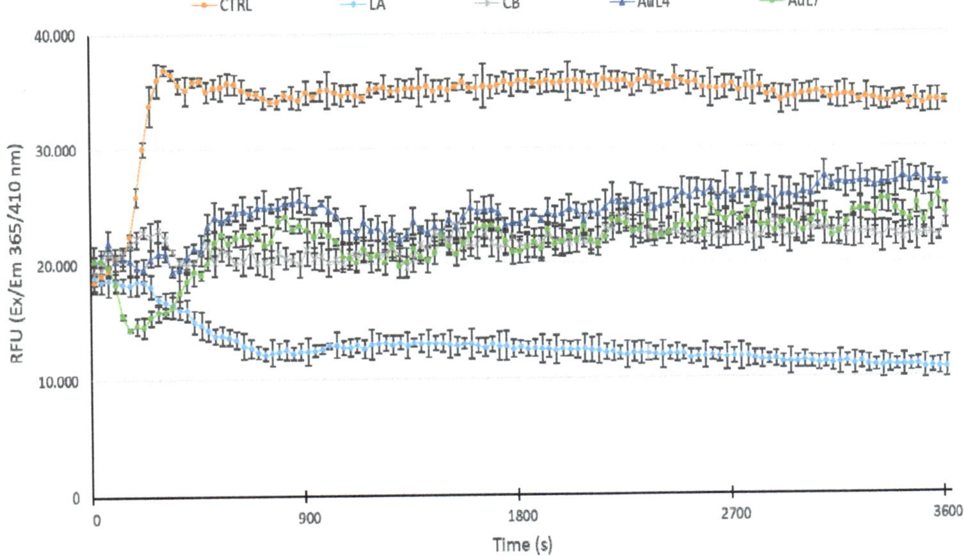

Figure 6. In vitro actin polymerization assay. The effect of compounds **AuL4** and **AuL7** (5 µM) on in vitro actin polymerization was examined. DMSO was used as a negative control (CTRL). Latrunculin A (LA) and cytochalasin B (CB), were employed as reference molecules and used at the concentration of 5 µM. **AuL4** and **AuL7**, incubated with the labeled rabbit muscle actin, demonstrate a direct effect on actin polymerization reaction. The graphic was obtained measuring the fluorescence (Ex/Em: 365/410 nm) given by actin filaments assembly in a kinetic mode for 1 h at room temperature. Data are representative of three independent experiments; standard deviations (SDs) are shown.

Considering that LA can trigger the F-actin disassembly, we performed a depolymerization assay, in order to determine whether our compounds could act as actin depolymerizing agents. Actin polymerization was allowed for one hour, under the same experimental conditions used in the previous assay, then **AuL4** or **AuL7** were added, at the concentration of 5 µM, and the reactions were monitored for one more hour. Negative (only vehicle and CB, 5 µM) and positive (LA, 5 µM) control reactions were performed, as well. Our results showed that the LA curve sensibly decreased until a value of 16,800 RFU in the first 8–9 min, meaning that the actin depolymerization was occurring (Figure 7). The initial value was around 28,000 RFU and after 9 min, the curve decreased slowly until the final value of 14,000 RFU. The negative control reactions (only vehicle and CB, 5 µM) showed no relevant changes, maintaining the same initial value and indicating that the F-actin did not depolymerizes. The curves relative to **AuL4** and **AuL7** showed significant reduction in RFU values with respect to the control reactions, suggesting that they did not accelerate the depolymerization reaction. To summarize, these data proved that both **AuL4** and **AuL7** are actin polymerization inhibitors with an efficacy similar to CB, used at the same concentrations and under the adopted experimental conditions, and do not accelerate actin depolymerization, contrarily to LA.

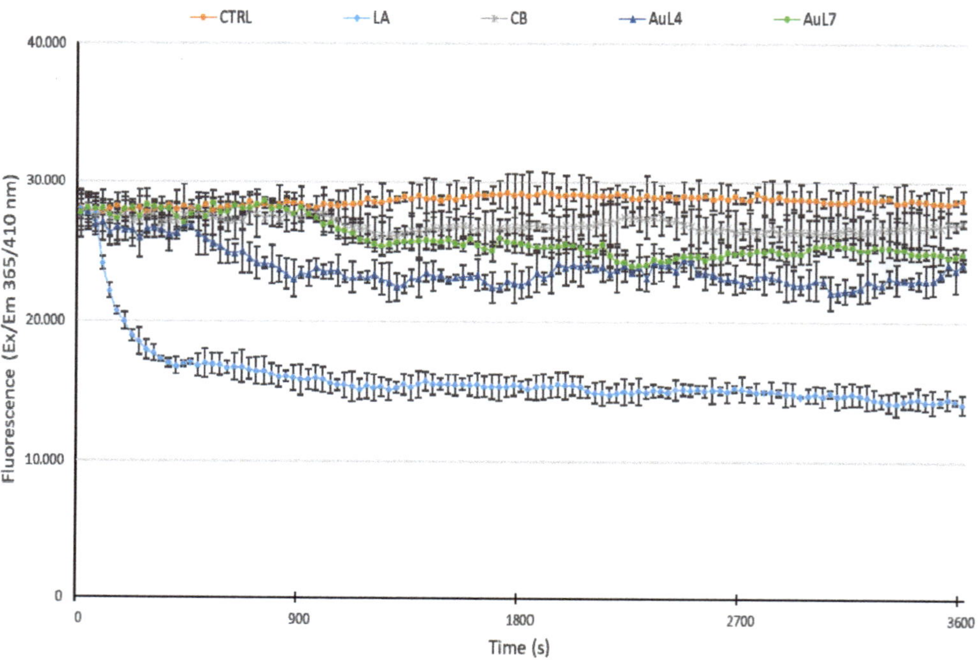

Figure 7. In vitro actin depolymerization assay. The effect of compounds **AuL4** and **AuL7** (5 μM) on in vitro actin depolymerization was examined. DMSO was used as a negative control (CTRL). Latrunculin A (LA) and cytochalasin B (CB), were employed as reference molecules and used at the concentration of 5 μM. **AuL4** and **AuL7**, added to the polymerized actin, do not accelerate the depolymerization reaction. The graphic was obtained measuring the fluorescence (Ex/Em: 365/410 nm) given by actin filaments disassembly in a kinetic mode for 1 h at room temperature. Data are representative of three independent experiments; standard deviations (SDs) are shown.

4. Discussion

Microtubules and actin filaments are cytoskeletal components that play pivotal roles in cell signaling, division, and motility, and regulate tumor relevant processes, for instance, morphological changes or cell migration [38]. For these reasons, they are interesting targets of an even higher number of anticancer drugs, such as the clinically used vinca alkaloids or taxanes that target microtubules, whereas compounds directed toward the actin have relatively lesser therapeutic applications, probably because of poor knowledge on their underlying mechanisms of action [39]. Under physiological conditions, the monomeric form of actin, called G-actin, polymerizes in a head-to-tail manner and constitutes the filamentous F-actin. The turnover of these forms is ensured by a tightly regulated equilibrium between G- and F-actin, but is substantially modified in cancer cells, together with changes in the filament-associated regulatory proteins, and is involved in the uncontrolled growth of tumor cells and, ultimately, in the increased ability to metastasize [40,41]. Generally, the migrating cancer cells undergo to shape rearrangements that depend on the formation of actin-based protrusions, cellular contractility, and new adhesions to surfaces [42]. For example, actin participates in the formation of specialized invasive and adhesive structures with proteolytic activity, called invadosomes [41]. Since we have already proved that the two NHC-gold complexes exerted inhibitory activity against the tubulin polymerization, we wondered whether they could also interfere directly with the other major component of the cell cytoskeleton, *viz.* the actin. Indeed, some literature data reported that the actin system disorganization can be a consequence of the inhibition of the tubulin polymerization and not a direct regulation of the actin system [39,43]. Amongst the most well-known molecules able to interfere directly with the intracellular actin, different types of cytochalasin and latrunculin were employed in many research studies, because of their easy membrane permeation and different types of behavior against the actin polymerization and depolymerization processes. In our assays, we used, as references molecules, cytochalasin B (CB), which inhibits the addition of monomers to the "barbed" end of the actin filaments [44], and latrunculin A (LA), which blocks the actin polymerization, and increases its depolymerization by trapping the actin monomers [45]. In our previous studies, we studied the anticancer properties of **AuL4** and **AuL7**, individuating some intracellular targets, as the human topoisomerases I and II, and the cytoskeleton tubulin, which decreased cancer cell growth by inducing the ROS-depending intrinsic apoptotic pathway. Encouraged by the outcomes, we continued studying other interactions of our compounds with other key-points belonging to the cell cytoskeleton. In particular, we investigated the possible implication of our complexes in regulating the actin dynamics in a direct way, and not as a consequence of the already observed interference with the tubulin polymerization reaction. Indeed, some microtubule poisons may promote the reorganization of the actin system, which is a consequent effect rather than a direct interaction. In other words, the microtubule metabolism can influence the organization, spatial distribution, and function of the nearby actin filaments [46]. With this in mind, we started with in silico studies that allowed us to assume that both complexes may bind the actin, sharing the same binding site of LA, used as the reference molecule. In particular, **AuL4** is positioned inside the nucleotide-binding site of actin, whereas **AuL7** exposes part of its structure to the solvent.

Next, we performed immunofluorescence assays in the two cancer cell models abovementioned. The outcomes suggest that both complexes induced dramatic changes in actin organization; indeed, the filaments appeared thicker and with an abnormal spatial distribution, with respect to the control-treated cells, and in a similar rearrangement produced by LA treatment. In order to fully understand if these changes were due to a direct effect rather than the already studied interference with the tubulin polymerization reaction, we carried out some direct enzymatic assays on the purified rabbit actin, so that any other possible secondary effect could be excluded. Effectively, both complexes inhibited the actin polymerization in a similar extent to the CB, but lesser than LA, used as reference molecules. Conversely to LA, **AuL4** and **AuL7** were not able to promote the F-actin depolymerization. In conclusion, both the complexes inhibited the actin polymerization directly, but not the inverse reaction, and induced dramatic change in the organization and function of the actin system; this is a feature that discloses another—and not foreseen—important anticancer target of our NHC-gold complexes.

5. Conclusions

Herein, we reported evidence that the NHC-gold complexes **AuL4** and **AuL7** are inhibitors of the actin polymerization reaction with similar efficacy of CB, but do not accelerate the rate of F-actin depolymerization, as it happens for the LA. These complexes induce dramatic changes in cell morphology, triggering the formation of abnormal actin structures. We strongly believe that this evidence contributes to better knowledge of the variety of intracellular targets in the cancer fight.

Author Contributions: Conceptualization, M.S.S.; methodology, D.I.; software, C.R. and M.S.; validation, P.L., formal analysis, J.C. and A.M.; investigation, J.C.; data curation, A.C. and M.S.S.; writing—original draft preparation, D.I. and J.C.; writing—review and editing, S.A.; visualization, C.S. and M.P.; supervision, M.S.S.; project administration, P.L.; funding acquisition, C.R. All authors have read and agreed to the published version of the manuscript.

Funding: C.R. acknowledges the support by a grant from the Italian Ministry of Health (Ricerca Corrente 2020).

Institutional Review Board Statement: Not Applicable.

Informed Consent Statement: Not Applicable.

Data Availability Statement: Not Applicable.

Conflicts of Interest: The authors declare no conflict of interest.

References

1. Jakupec, M.A.; Galanski, M.; Arion, V.B.; Hartinger, C.G.; Keppler, B.K. Antitumour metal compounds: More than theme and variations. *Dalton Trans.* **2008**, 183–194. [CrossRef]
2. Wang, D.; Lippard, S.J. Cellular processing of platinum anticancer drugs. *Nat. Rev. Drug Discov.* **2005**, *4*, 307–320. [CrossRef] [PubMed]
3. Zeglis, B.M.; Pierre, V.C.; Barton, J.K. Metallo-intercalators and metallo-insertors. *Chem. Commun.* **2007**, *44*, 4565–4579. [CrossRef] [PubMed]
4. Bertrand, B.; Casini, A. A golden future in medicinal inorganic chemistry: The promise of anticancer gold organometallic compounds. *Dalton Trans.* **2014**, *43*, 4209–4219. [CrossRef] [PubMed]
5. Gaynor, D.; Griffith, D.M. The prevalence of metal-based drugs as therapeutic or diagnostic agents: Beyond platinum. *Dalton Trans.* **2012**, *41*, 13239–13257. [CrossRef]
6. Orvig, C.; Abrams, M.J. Medicinal inorganic chemistry: Introduction. *Chem. Rev.* **1999**, *99*, 2201–2204. [CrossRef] [PubMed]
7. Gottlieb, N.L. Comparative pharmacokinetics of parenteral and oral gold compounds. *J. Rheumatol. Suppl.* **1982**, *8*, 99–109.
8. Hindi, K.M.; Panzner, M.J.; Tessier, C.A.; Cannon, C.L.; Youngs, W.J. The medicinal applications of imidazolium carbene-metal complexes. *Chem. Rev.* **2009**, *109*, 3859–3884. [CrossRef]
9. Liu, W.; Gust, R. Metal N-heterocyclic carbene complexes as potential antitumor metallodrugs. *Chem. Soc. Rev.* **2013**, *42*, 755–773. [CrossRef]
10. Ceramella, J.; Mariconda, A.; Iacopetta, D.; Saturnino, C.; Barbarossa, A.; Caruso, A.; Rosano, C.; Sinicropi, M.S.; Longo, P. From coins to cancer therapy: Gold, silver and copper complexes targeting human topoisomerases. *Bioorg. Med. Chem. Lett.* **2020**, *30*, 126905. [CrossRef]
11. Hickey, J.L.; Ruhayel, R.A.; Barnard, P.J.; Baker, M.V.; Berners-Price, S.J.; Filipovska, A. Mitochondria-targeted chemotherapeutics: The rational design of gold(I) N-heterocyclic carbene complexes that are selectively toxic to cancer cells and target protein selenols in preference to thiols. *J. Am. Chem. Soc.* **2008**, *130*, 12570–12571. [CrossRef] [PubMed]
12. Barnard, P.J.; Berners-Price, S.J. Targeting the mitochondrial cell death pathway with gold compounds. *Coord. Chem. Rev.* **2007**, *251*, 1889–1902. [CrossRef]
13. Krishnamurthy, D.; Karver, M.R.; Fiorillo, E.; Orru, V.; Stanford, S.M.; Bottini, N.; Barrios, A.M. Gold(I)-mediated inhibition of protein tyrosine phosphatases: A detailed in vitro and cellular study. *J. Med. Chem.* **2008**, *51*, 4790–4795. [CrossRef] [PubMed]
14. Fung, S.K.; Zou, T.; Cao, B.; Lee, P.Y.; Fung, Y.M.; Hu, D.; Lok, C.N.; Che, C.M. Cyclometalated Gold(III) Complexes Containing N-Heterocyclic Carbene Ligands Engage Multiple Anti-Cancer Molecular Targets. *Angew. Chem. Int. Ed. Engl.* **2017**, *56*, 3892–3896. [CrossRef] [PubMed]
15. Iacopetta, D.; Rosano, C.; Sirignano, M.; Mariconda, A.; Ceramella, J.; Ponassi, M.; Saturnino, C.; Sinicropi, M.S.; Longo, P. Is the Way to Fight Cancer Paved with Gold? Metal-Based Carbene Complexes with Multiple and Fascinating Biological Features. *Pharmaceuticals* **2020**, *13*, 91. [CrossRef]
16. Magherini, F.; Fiaschi, T.; Valocchia, E.; Becatti, M.; Pratesi, A.; Marzo, T.; Massai, L.; Gabbiani, C.; Landini, I.; Nobili, S.; et al. Antiproliferative effects of two gold(I)-N-heterocyclic carbene complexes in A2780 human ovarian cancer cells: A comparative proteomic study. *Oncotarget* **2018**, *9*, 28042–28068. [CrossRef]
17. Muenzner, J.K.; Biersack, B.; Albrecht, A.; Rehm, T.; Lacher, U.; Milius, W.; Casini, A.; Zhang, J.J.; Ott, I.; Brabec, V.; et al. Ferrocenyl-Coupled N-Heterocyclic Carbene Complexes of Gold(I): A Successful Approach to Multinuclear Anticancer Drugs. *Chemistry* **2016**, *22*, 18953–18962. [CrossRef]
18. Iacopetta, D.; Mariconda, A.; Saturnino, C.; Caruso, A.; Palma, G.; Ceramella, J.; Muia, N.; Perri, M.; Sinicropi, M.S.; Caroleo, M.C.; et al. Novel Gold and Silver Carbene Complexes Exert Antitumor Effects Triggering the Reactive Oxygen Species Dependent Intrinsic Apoptotic Pathway. *ChemMedChem* **2017**, *12*, 2054–2065. [CrossRef]
19. Dhamodharan, R.; Jordan, M.A.; Thrower, D.; Wilson, L.; Wadsworth, P. Vinblastine suppresses dynamics of individual microtubules in living interphase cells. *Mol. Biol. Cell* **1995**, *6*, 1215–1229. [CrossRef]
20. Rizza, P.; Pellegrino, M.; Caruso, A.; Iacopetta, D.; Sinicropi, M.S.; Rault, S.; Lancelot, J.C.; El-Kashef, H.; Lesnard, A.; Rochais, C.; et al. 3-(Dipropylamino)-5-hydroxybenzofuro[2,3-f]quinazolin-1(2H)-one (DPA-HBFQ-1) plays an inhibitory role on breast cancer cell growth and progression. *Eur. J. Med. Chem.* **2016**, *107*, 275–287. [CrossRef]
21. Ceramella, J.; Caruso, A.; Occhiuzzi, M.A.; Iacopetta, D.; Barbarossa, A.; Rizzuti, B.; Dallemagne, P.; Rault, S.; El-Kashef, H.; Saturnino, C.; et al. Benzothienoquinazolinones as new multi-target scaffolds: Dual inhibition of human Topoisomerase I and tubulin polymerization. *Eur. J. Med. Chem.* **2019**, *181*, 111583. [CrossRef] [PubMed]
22. Rebowski, G.; Boczkowska, M.; Drazic, A.; Ree, R.; Goris, M.; Arnesen, T.; Dominguez, R. Mechanism of actin N-terminal acetylation. *Sci. Adv.* **2020**, *6*, eaay8793. [CrossRef]

23. Morris, G.M.; Huey, R.; Lindstrom, W.; Sanner, M.F.; Belew, R.K.; Goodsell, D.S.; Olson, A.J. AutoDock4 and AutoDockTools4: Automated docking with selective receptor flexibility. *J. Comput. Chem.* **2009**, *30*, 2785–2791. [CrossRef] [PubMed]

24. Sanner, M.F.; Duncan, B.S.; Carrillo, C.J.; Olson, A.J. Integrating computation and visualization for biomolecular analysis: An example using python and AVS. *Pac. Symp. Biocomput.* **1999**, 401–412. [CrossRef]

25. Cesarini, S.; Spallarossa, A.; Ranise, A.; Schenone, S.; Rosano, C.; La Colla, P.; Sanna, G.; Busonera, B.; Loddo, R. N-acylated and N,N′-diacylated imidazolidine-2-thione derivatives and N,N′-diacylated tetrahydropyrimidine-2(1H)-thione analogues: Synthesis and antiproliferative activity. *Eur. J. Med. Chem.* **2009**, *44*, 1106–1118. [CrossRef]

26. Rosano, C.; Hunt, M.J.O.; Brach, J.; Newman, A.B.; Studenski, S.; Verghese, J.; Weissfeld, L. Brain Anatomical Correlates of Gait Variability in High Functioning Older Adults: Repeatability across Two Independent Studies (Cns). *Gerontologist* **2012**, *52*, 319–320.

27. Pettersen, E.F.; Goddard, T.D.; Huang, C.C.; Couch, G.S.; Greenblatt, D.M.; Meng, E.C.; Ferrin, T.E. UCSF Chimera–A visualization system for exploratory research and analysis. *J. Comput. Chem.* **2004**, *25*, 1605–1612. [CrossRef]

28. Ceramella, J.; Loizzo, M.R.; Iacopetta, D.; Bonesi, M.; Sicari, V.; Pellicano, T.M.; Saturnino, C.; Malzert-Freon, A.; Tundis, R.; Sinicropi, M.S. Anchusa azurea Mill. (Boraginaceae) aerial parts methanol extract interfering with cytoskeleton organization induces programmed cancer cells death. *Food Funct.* **2019**, *10*, 4280–4290. [CrossRef] [PubMed]

29. Lappano, R.; Rosano, C.; Pisano, A.; Santolla, M.F.; De Francesco, E.M.; De Marco, P.; Dolce, V.; Ponassi, M.; Felli, L.; Cafeo, G.; et al. A calixpyrrole derivative acts as an antagonist to GPER, a G-protein coupled receptor: Mechanisms and models. *Dis. Model Mech.* **2015**, *8*, 1237–1246. [CrossRef] [PubMed]

30. Sinicropi, M.S.; Lappano, R.; Caruso, A.; Santolla, M.F.; Pisano, A.; Rosano, C.; Capasso, A.; Panno, A.; Lancelot, J.C.; Rault, S.; et al. (6-bromo-1,4-dimethyl-9H-carbazol-3-yl-methylene)-hydrazine (carbhydraz) acts as a GPER agonist in breast cancer cells. *Curr. Top. Med. Chem.* **2015**, *15*, 1035–1042. [CrossRef]

31. Stec-Martyna, E.; Ponassi, M.; Miele, M.; Parodi, S.; Felli, L.; Rosano, C. Structural comparison of the interaction of tubulin with various ligands affecting microtubule dynamics. *Curr. Cancer Drug Targets* **2012**, *12*, 658–666. [CrossRef] [PubMed]

32. Acharya, S.; Maji, M.; Ruturaj; Purkait, K.; Gupta, A.; Mukherjee, A. Synthesis, Structure, Stability, and Inhibition of Tubulin Polymerization by Ru(II)-p-Cymene Complexes of Trimethoxyaniline-Based Schiff Bases. *Inorg. Chem.* **2019**, *58*, 9213–9224. [CrossRef] [PubMed]

33. Ceresa, C.; Nicolini, G.; Semperboni, S.; Gandin, V.; Monfrini, M.; Avezza, F.; Alberti, P.; Bravin, A.; Pellei, M.; Santini, C.; et al. Evaluation of the Profile and Mechanism of Neurotoxicity of Water-Soluble [Cu(P)4]PF6 and [Au(P)4]PF6 (P = thp or PTA) Anticancer Complexes. *Neurotox Res.* **2018**, *34*, 93–108. [CrossRef] [PubMed]

34. Kaps, L.; Biersack, B.; Muller-Bunz, H.; Mahal, K.; Munzner, J.; Tacke, M.; Mueller, T.; Schobert, R. Gold(I)-NHC complexes of antitumoral diarylimidazoles: Structures, cellular uptake routes and anticancer activities. *J. Inorg. Biochem.* **2012**, *106*, 52–58. [CrossRef]

35. Khanna, S.; Jana, B.; Saha, A.; Kurkute, P.; Ghosh, S.; Verma, S. Targeting cytotoxicity and tubulin polymerization by metal-carbene complexes on a purine tautomer platform. *Dalton Trans.* **2014**, *43*, 9838–9842. [CrossRef]

36. Muenzner, J.K.; Biersack, B.; Kalie, H.; Andronache, I.C.; Kaps, L.; Schuppan, D.; Sasse, F.; Schobert, R. Gold(I) biscarbene complexes derived from vascular-disrupting combretastatin A-4 address different targets and show antimetastatic potential. *ChemMedChem* **2014**, *9*, 1195–1204. [CrossRef] [PubMed]

37. Cheng, X.; Haeberle, S.; Shytaj, I.L.; Gama-Brambila, R.A.; Theobald, J.; Ghafoory, S.; Wolker, J.; Basu, U.; Schmidt, C.; Timm, A.; et al. NHC-gold compounds mediate immune suppression through induction of AHR-TGFbeta1 signalling in vitro and in scurfy mice. *Commun. Biol.* **2020**, *3*, 10. [CrossRef]

38. Pasquier, E.; Kavallaris, M. Microtubules: A dynamic target in cancer therapy. *IUBMB Life* **2008**, *60*, 165–170. [CrossRef] [PubMed]

39. Fenteany, G.; Zhu, S. Small-molecule inhibitors of actin dynamics and cell motility. *Curr. Top. Med. Chem.* **2003**, *3*, 593–616. [CrossRef]

40. Aseervatham, J. Cytoskeletal Remodeling in Cancer. *Biology* **2020**, *9*, 385. [CrossRef]

41. Yamaguchi, H.; Condeelis, J. Regulation of the actin cytoskeleton in cancer cell migration and invasion. *Biochim. Biophys. Acta* **2007**, *1773*, 642–652. [CrossRef] [PubMed]

42. Olson, M.F.; Sahai, E. The actin cytoskeleton in cancer cell motility. *Clin. Exp. Metastasis* **2009**, *26*, 273–287. [CrossRef] [PubMed]

43. Rosenblum, M.D.; Shivers, R.R. 'Rings' of F-actin form around the nucleus in cultured human MCF7 adenocarcinoma cells upon exposure to both taxol and taxotere. *Comp. Biochem. Physiol. Part C Pharmacol. Toxicol. Endocrinol.* **2000**, *125*, 121–131. [CrossRef]

44. Foissner, I.; Wasteneys, G.O. Wide-ranging effects of eight cytochalasins and latrunculin A and B on intracellular motility and actin filament reorganization in characean internodal cells. *Plant Cell Physiol.* **2007**, *48*, 585–597. [CrossRef]

45. Fujiwara, I.; Zweifel, M.E.; Courtemanche, N.; Pollard, T.D. Latrunculin A Accelerates Actin Filament Depolymerization in Addition to Sequestering Actin Monomers. *Curr. Biol.* **2018**, *28*, 3183–3192. [CrossRef]

46. Danowski, B.A. Fibroblast contractility and actin organization are stimulated by microtubule inhibitors. *J. Cell Sci.* **1989**, *93 Pt 2*, 255–266. [CrossRef]

Article

Simple Thalidomide Analogs in Melanoma: Synthesis and Biological Activity

Alexia Barbarossa [1,†], Alessia Catalano [1,†], Jessica Ceramella [2], Alessia Carocci [1,*], Domenico Iacopetta [2,*], Camillo Rosano [3], Carlo Franchini [1,‡] and Maria Stefania Sinicropi [2,‡]

[1] Department of Pharmacy-Drug Sciences, University of Bari "Aldo Moro", 70126 Bari, Italy; alexia.barbarossa@uniba.it (A.B.); alessia.catalano@uniba.it (A.C.); carlo.franchini@uniba.it (C.F.)

[2] Department of Pharmacy, Health and Nutritional Sciences, University of Calabria, Via Pietro Bucci, Arcavacata di Rende, 87036 Cosenza, Italy; jessicaceramella@gmail.com (J.C.); s.sinicropi@unical.it (M.S.S.)

[3] Proteomics and Mass Spectrometry Unit, IRCCS Policlinico San Martino, L.go Rosanna Benzi 10, 16132 Genova, Italy; camillo.rosano@hsanmartino.it

* Correspondence: alessia.carocci@uniba.it (A.C.); domenico.iacopetta@unical.it (D.I.)

† These authors equally contributed to this work.

‡ Co-senior authors.

Abstract: Thalidomide is an old well-known drug that is still of clinical interest, despite its teratogenic activities, due to its antiangiogenic and immunomodulatory properties. Therefore, efforts to design safer and effective thalidomide analogs are continually ongoing. Research studies on thalidomide analogs have revealed that the phthalimide ring system is an essential pharmacophoric fragment; thus, many phthalimidic compounds have been synthesized and evaluated as anticancer drug candidates. In this study, a panel of selected in vitro assays, performed on a small series of phthalimide derivatives, allowed us to characterize compound **2k** as a good anticancer agent, acting on A2058 melanoma cell line, which causes cell death by apoptosis due to its capability to inhibit tubulin polymerization. The obtained data were confirmed by in silico assays. No cytotoxic effects on normal cells have been detected for this compound that proves to be a valid candidate for further investigations to achieve new insights on possible mechanism of action of this class of compounds as anticancer drugs.

Keywords: anticancer; apoptosis; drugs repositioning; molecular docking studies; thalidomide analogs; tubulin

check for **updates**

Citation: Barbarossa, A.; Catalano, A.; Ceramella, J.; Carocci, A.; Iacopetta, D.; Rosano, C.; Franchini, C.; Sinicropi, M.S. Simple Thalidomide Analogs in Melanoma: Synthesis and Biological Activity. *Appl. Sci.* **2021**, *11*, 5823. https://doi.org/10.3390/app11135823

Received: 13 May 2021
Accepted: 21 June 2021
Published: 23 June 2021

Publisher's Note: MDPI stays neutral with regard to jurisdictional claims in published maps and institutional affiliations.

Copyright: © 2021 by the authors. Licensee MDPI, Basel, Switzerland. This article is an open access article distributed under the terms and conditions of the Creative Commons Attribution (CC BY) license (https://creativecommons.org/licenses/by/4.0/).

1. Introduction

Cancer is still a great health concern, as it is one of the main causes of death worldwide. A total of 19.3 million new cases of cancer were diagnosed last year, of which 10 million were fatal [1]. Among the different types of cancer, malignant melanoma is one of the most aggressive and deadly skin cancers, with an enormously increasing occurrence over the past few decades [2–4]. Melanoma is notoriously chemoresistant; nevertheless, chemotherapy represents the primary treatment for metastatic melanoma. Considering the major impact of chemotherapy in cancer management, and the problems related to chemotherapy that have not yet been solved, such as drug toxicity and resistance, particularly in the case of melanoma, research for newer and potent anticancer agents is of great significance. Over the past few years, numerous biologically active heterocyclic compounds have been synthesized and evaluated as anticancer drugs. Among these, many phthalimidic compounds have turned out to be antitumor drug candidates. The phthalimidic core represents the main pharmacophoric feature of thalidomide (Figure 1), a drug infamous for its teratogenic action, which, nevertheless, represents one of the most successful examples of drug repositioning in the treatment of cancer. Indeed, over the last few decades, the interest in this old drug has been renewed because of its efficacy in several

143

important disorders (such as multiple myeloma, breast cancer, and HIV-related diseases [5]) and its antiangiogenic and immunomodulatory properties. Several studies regarding the effects of thalidomide on metastatic melanoma have been reported. Since angiogenesis, and specifically the increased expression of vascular endothelial growth factor (VEGF), has been reported as an important step in disease progression, the use of thalidomide to enhance the antitumor activity of chemotherapy has been investigated in metastatic melanoma [6]. Furthermore, a number of literature data report the effect of thalidomide and its derivatives on tubulin polymerization, causing a perturbation of microtubule dynamics. Microtubules represent important biological targets in cancer treatment due to their involvement in crucial cellular functions [7,8]. It became clear that thalidomide exerts multifaceted properties; however, many aftereffects, such as deep vein thrombosis, peripheral neuropathy, constipation, somnolence, pyrexia, pain, and teratogenicity, have been reported, showing the need for careful and monitored use [9]. Thus, research efforts are directed toward the synthesis and optimization of new thalidomide analogs, mostly as new anticancer agents, lacking toxic effects and that are able to remove these limits and improve the pharmacological profile. [10,11]. As a part of an ongoing research program, directed toward the synthesis of several antitumor heterocyclic compounds [12–16], and our previous paper [17] concerning the antitumor activity of a thalidomide-correlated compound, we report herein the synthesis of a small series of phthalimide derivatives, obtained by directly connecting the N-terminus of the phthalimide moiety with a differently substituted aromatic ring (**2a–l**, Figure 1). These compounds have been investigated for their cytotoxic effects in a large panel of cancer cells, including two melanoma cell lines, namely A2058 derived from a metastatic site, and malignant melanoma cells Sk-Mel28, two breast cancer cell lines (triple negative MDA-MB-231 and ER(+) MCF-7) and a human cervix carcinoma HeLa cell line. The best cytotoxic activity was recovered for compound **2k** on A2058 melanoma cell line. The ability of compound **2k** to inhibit tubulin polymerization has also been reported and a molecular docking study supported the obtained results.

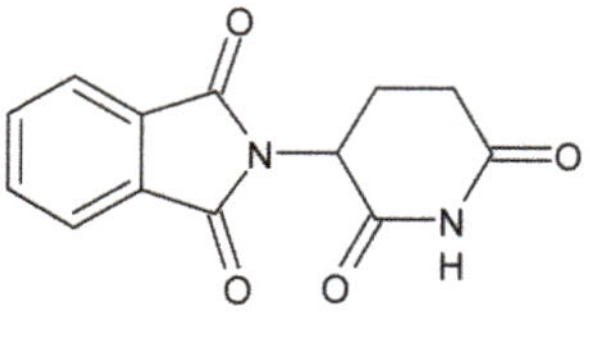

thalidomide

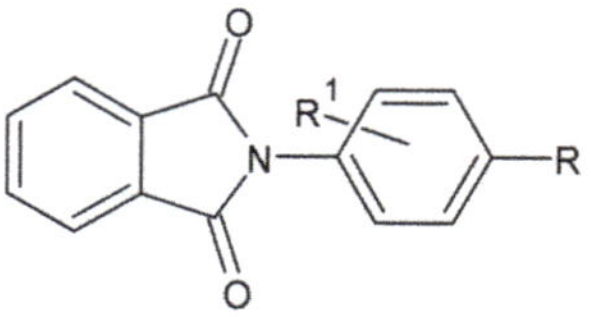

2a: R = R^1 = H	**2g**: R = F; R^1 = 2-Cl
2b: R = H; R^1 = 2-F	**2h**: R = F; R^1 = 3-F
2c: R = 4-MeOPh; R^1 = H	**2i**: R = 2-ClOPh; R^1 = H
2d: R = OPh; R^1 = H	**2j**: R = 2-OMeOPh; R^1 = H
2e: R = Cl; R^1 = 2-F	**2k**: R = 2-MeOPh; R^1 = H
2f: R = Cl; R^1 = 2-Br	**2l**: R = H; R^1 = OPh

2a–l

Figure 1. Structure of thalidomide and correlated compounds (**2a–l**).

2. Materials and Methods

2.1. Chemistry

All chemicals were supplied by Sigma-Aldrich or Alfa Aesar and were the highest-grade purity that were commercially available. Solvents were reagent grade unless otherwise indicated. Yields refer to purified products and were not optimized. Compound structures were confirmed by routine spectrometric analyses. Melting points were determined on a Gallenkamp melting point apparatus in open glass capillary tubes and are uncorrected. ^{1}H NMR and ^{13}C NMR spectra were recorded on a Varian VX Mercury spectrometer operating at 300 and 75 MHz for ^{1}H and ^{13}C, respectively, or an Agilent 500 MHz operating at 500 and 125 MHz for ^{1}H and ^{13}C, respectively. Chemical shifts are reported in parts per million (ppm) relative to solvent (CDCl$_3$) resonance: δ 7.26 (^{1}H NMR) and δ 77.0 (^{13}C NMR). *J* values are given in Hz. The following abbreviations are used: s-singlet, d-doublet, t-triplet, and m-multiplet. Gas chromatography (GC)/mass spectroscopy (MS) was performed on a Hewlett-Packard 6890–5973 MSD at low resolution. Liquid chromatography (LC)/mass spectroscopy (MS) was performed on a spectrometer (Agilent 1100 series LC-MSD Trap System VL). Molecular ion was designed as "M$^+$". Elemental analyses were carried out on a Eurovector Euro EA 3000 analyzer and the data for C, H, N were within $\pm$0.4 of theoretical values.

Synthesis of 2-phenyl-1H-isoindol-1,3(2H)-dione (**2a**). Aniline (0.77 g, 8.3 mmol), phthalic anhydride (1.23 g, 8.3 mmol) and triethylamine (0.12 mL, 0.9 mmol) were refluxed in toluene (10 mL) for 5 h. The solvent was removed in vacuo and the residue, taken up with CHCl$_3$, was washed 3 N HCl and water, and then dried over anhydrous Na$_2$SO$_4$ to give 905 mg (49%) of **2a** as a white solid: mp 211–213 °C; lit. 200 °C [18]; 210–211 °C (acetone) [19]. GC/MS (70 eV) *m/z* (%): 223 (M$^+$, 100); LC/MS: *m/z* (%): 222 [M$^+$–1]; ^{1}H NMR (300 MHz, CDCl$_3$): δ 7.36–7.54 (m, 5H, Ar), 7.54–7.82 (m, 2H, Ar), 7.90–7.98 (m, 2H, Ar); ^{13}C NMR (75 MHz, CDCl$_3$): δ 123.7 (2C), 126.6 (1C), 128.1 (2C), 129.1 (2C), 131.8 (2C), 134.4 (2C), 167.3 (2C). Anal C$_{13}$H$_9$NO$_2$ (223.06): calcd %: C 75.33; H 4.06; N 6.27. Found %: C 74.98; H 4.04; N 6.30.

Synthesis of 2-(2-fluorophenyl)-1H-isoindol-1,3(2H)-dione (**2b**). Prepared as described for 2a starting from 2-fluoroaniline. White solid. Yield: 65%: mp 190–191 °C (CHCl$_3$/hexane); lit. 199 °C [19]; GC/MS (70 eV) *m/z* (%): 241 (M$^+$, 92), 197 (100); ^{1}H NMR (500 MHz, CDCl$_3$): δ 7.23–7.30 (m, 2H, Ar), 7.35–7.38 (m, 1H, Ar), 7.42–7.47 (m, 1H, Ar), 7.79–7.82 (m, 2H, Ar), 7.95–7.98 (m, 2H, Ar); ^{13}C NMR (75 MHz, CDCl$_3$): δ 116.76 (d, *J* = 19.6 Hz, 1C), 119.4 (d, *J*$_{CF}$ = 13.7 Hz, 1C), 123.9 (1C), 124.6 (d, *J*$_{CF}$ = 4.6 Hz, 1C), 129.9 (2C), 130.7 (d, *J*$_{CF}$ = 8.0 Hz, 1C), 131.9 (2C), 134.4 (2C), 157.9 (d, *J*$_{CF}$ = 251.9 Hz, 1C), 166.5 (2C). Anal C$_{14}$H$_8$FNO$_2$ (241.05): calcd %: C 69.71; H 3.34; N 5.81. Found %: C 69.54; H 3.32; N 5.80.

Synthesis of 2-[4-(4-methylphenoxy)phenyl]-1H-isoindol-1,3(2H)-dione (**2c**). Prepared as described for **2a** starting from 4-*p*-tolyloxyaniline. Grey solid. Yield: 55%: mp 201–202 °C (CHCl$_3$/hexane); GC/MS (70 eV) *m/z* (%): 329 (M$^+$, 100); ^{1}H NMR (300 MHz, CDCl$_3$): δ 2.15–2.36 (m, 3H, CH$_3$), 6.97 (d, *J* = 8.2 Hz, 2H, Ar), 7.11 (d, *J* = 21.7 Hz, 2H, Ar), 7.16 (d, *J* = 8.2 Hz, 2H, Ar), 7.34 (d, *J* = 8.8 Hz, 2H, Ar), 7.76–7.79 (m, 2H, Ar), 7.92–7.95 (m, 2H, Ar); ^{13}C NMR (125 MHz, CDCl$_3$): δ 20.7 (1C), 118.3 (2C), 119.7 (2C), 123.7 (2C), 125.9 (1C), 128.0 (2C), 130.4 (2C), 131.8 (2C), 133.6 (1C), 134.4 (2C), 154.0 (1C), 157.8 (1C), 167.4 (2C). Anal C$_{21}$H$_{15}$NO$_3$ (329.11): calcd %: C 76.58; H 4.59; N 4.25. Found %: C 76.43; H 4.53; N 4.35.

Synthesis of 2-(4-phenoxyphenyl)-1H-isoindole-1,3(2H)-dione (**2d**). Prepared as described for **2a** starting from 4-phenoxyaniline. Grey solid. Yield: 65%: mp 161–162 °C (CHCl$_3$/hexane); GC/MS (70 eV) *m/z* (%): 315 (M$^+$, 100); ^{1}H NMR (500 MHz, CDCl$_3$): δ 7.06–7.15 (m, 5H, Ar), 7.35–7.38 (m, 4H, Ar), 7.78–7.79 (m, 2H, Ar), 7.94–7.95 (m, 2H, Ar); ^{13}C NMR (125 MHz, CDCl$_3$): δ 118.8 (2C), 119.5 (2C), 123.7 (2C), 123.9 (1C), 126.3 (1C), 128.0 (2C), 129.9 (2C), 131.7 (2C), 134.4 (2C), 156.5 (1C), 157.2 (1C), 167.4 (2C). Anal C$_{20}$H$_{13}$NO$_3$·0.20 H$_2$O (318.69): calcd %: C 75.32; H 4.23; N 4.39. Found %: C 75.60; H 4.08; N 4.48.

Synthesis of 2-(4-chloro-2-fluorophenyl)-1H-isoindol-1,3(2H)-dione (**2e**). Prepared as described for 2a starting from 4-chloro-2-fluoroaniline. White solid. Yield: 35%: mp 145–146 °C; GC/MS (70 eV) *m/z* (%): 275 (M$^+$, 100); ^{1}H NMR (500 MHz, CDCl$_3$): δ 7.25–7.32 (m, 3H, Ar), 7.78–7.82 (m, 2H, Ar), 7.90–7.98 (m, 2H, Ar); ^{13}C NMR (125 MHz, CDCl$_3$): δ 117.7 (d, *J*$_{CF}$ = 22.9 Hz, 1C), 118.2 (d, *J*$_{CF}$ = 96.4 Hz, 1C), 124.0 (2C), 125.1 (d, *J*$_{CF}$ = 3.8 Hz, 1C), 130.5 (1C), 131.8 (2C), 134.6 (2C), 135.8 (d, *J*$_{CF}$ = 9.6 Hz, 1C), 157.7 (d, *J*$_{CF}$ = 256.5 Hz, 1C), 166.2 (2C). Anal C$_{14}$H$_7$ClFNO$_2$·0.25 H$_2$O (279.51): calcd %: C 60.02; H 2.70; N 5.00. Found %: C 60.42; H 2.50; N 5.07.

Synthesis of 2-(2-bromo-4-chlorophenyl)-1H-isoindol-1,3(2H)-dione (**2f**). Prepared as described for **2a** starting from 2-bromo-4-cloroaniline. White solid. Yield: 18%: mp 140–141 °C (CHCl$_3$/hexane); GC/MS (70 eV) *m/z* (%): 337 (M$^+$, 2), 302 (100); ^{1}H NMR (500 MHz,

CDCl$_3$): δ 7.24–7.30 (m, 1H, Ar), 7.45 (d, *J* = 8.8 Hz, 1H, Ar), 7.75 (s, 1H, Ar), 7.79–7.86 (m, 2H, Ar), 7.94–8.01 (m, 2H, Ar); ^{13}C NMR (125 MHz, CDCl$_3$): δ 123.9 (1C), 124.0 (1C), 128.3 (2C), 130.1 (1C), 131.5 (1C), 131.8 (2C), 133.4 (1C), 134.6 (2C), 136.3 (1C), 166.3 (2C). Anal C$_{14}$H$_7$BrClNO$_2$ (344.93): calcd %: C 49.96; H 2.10; N 4.16. Found %: C 49.96; H 2.05; N 4.15.

Synthesis of 2-(2-cloro-4-fluorophenyl)-1H-isoindol-1,3(2H)-dione (**2g**). Prepared as described for **2a** starting from 2-cloro-4-fluoroaniline. White solid. Yield: 68%: mp 188–189 °C (CHCl$_3$/hexane); GC/MS (70 eV) *m/z* (%): 275 (M$^+$, 7), 240 (100); ^{1}H NMR (500 MHz, CDCl$_3$): δ 7.13 (t, *J* = 8.0 Hz, 1H, Ar), 7.32–7.35 (m, 2H, Ar), 7.80–7.84 (m, 2H, Ar), 7.94–8.00 (m, 2H, Ar); ^{13}C NMR (125 MHz, CDCl$_3$): δ 115.1 (d, *J*$_{CF}$ = 22.9 Hz, 1C), 117.9 (d, *J*$_{CF}$ = 25.8 Hz, 1C), 124.0 (2C), 125.8 (d, *J*$_{CF}$ = 3.9 Hz, 1C), 131.8 (d, *J*$_{CF}$ = 8.7 Hz, 1C + s, 2C), 134.4 (d, *J*$_{CF}$ = 11.4 Hz, 1C), 134.6 (2C), 162.7 (d, *J*$_{CF}$ = 252.8 Hz, 1C), 166.6 (2C). Anal. C$_{14}$H$_7$ClFNO$_2$ (275.01): calcd %: C 61.00; H 2.56; N 5.08. Found %: C 60.71; H 2.49; N 5.06.

Synthesis of 2-(3,4-difluorophenyl)-1H-isoindol-1,3(2H)-dione (**2h**). Prepared as described for **2a** starting from 3,4-difluoroaniline. White solid. Yield: 52%: mp 199–200 °C; GC/MS (70 eV) *m/z* (%): 259 (M$^+$, 100); ^{1}H NMR (300 MHz, CDCl$_3$): δ 7.22–7.39 (m, 3H, Ar), 7.78–7.84 (m, 2H, Ar), 7.93–7.99 (m, 2H, Ar); ^{13}C NMR (75 MHz, CDCl$_3$): δ 116.0 (d, *J*$_{CF}$ = 19.4 Hz, 1C), 117.5 (d, *J* = 18.6 Hz, 1C), 112.6 (d, *J*$_{CF}$ = 4.6 Hz, 1C), 122.7 (d, *J*$_{CF}$ = 3.4 Hz, 1C), 123.9 (2C), 127.9 (1C), 131.4 (2C), 134.6 (2C), 148.3 (dd, *J*$_{CF}$ = 29.8, 12.6 Hz, 1C), 151.6 (dd, *J* = 28.0, 13.2 Hz, 1C), 166.8 (2C). Anal C$_{14}$H$_7$FNO$_2$·0.20 H$_2$O (262.64): calcd %: C 63.98; H 2.84; N 5.33. Found %: C 64.32; H 2.69; N 5.38.

Synthesis of 2-[4-(2-chlorophenoxy)phenyl]-1H-isoindol-1,3(2H)-dione (**2i**). Prepared as described for **2a** starting from 4-(2-chlorophenoxy)aniline. Grey solid. solid. Yield: 55%: mp 135–136 °C (CHCl$_3$/hexane); GC/MS (70 eV) *m/z* (%): 349 (M$^+$, 100); ^{1}H NMR (500 MHz, CDCl$_3$): δ 7.11–7.15 (m, 4H, Ar), 7.24–7.28 (m, 1H, Ar), 7.37–7.40 (m, 2H, Ar), 7.47–7.48 (m, 1H, Ar), 7.78–7.80 (m, 2H, Ar), 7.93–7.96 (m, 2H, Ar); ^{13}C NMR (125 MHz, CDCl$_3$): δ 117.9 (2C), 121.7 (1C), 123.8 (2C), 125.4 (1C), 126.4 (1C), 126.5 (1C), 128.0 (1C), 128.1 (1C), 130.9 (2C), 131.7 (2C), 134.4 (2C), 151.8 (1C), 156.7 (1C), 167.4 (2C). Anal. C$_{20}$H$_{12}$ClNO$_3$·0.50 H$_2$O (257.02): calcd %: C 66.95; H 3.65; N 3.90. Found %: C 67.32; H 3.36; N 4.02.

Synthesis of 2-[4-(2-methoxyphenoxy)phenyl]-1H-isoindol-1,3(2H)-dione (**2j**). Prepared as described for **2a** starting from 4-(2-methoxyphenoxy)aniline. Light brown solid. Yield: 75%: mp 148–150 °C (AcOEt/hexane). GC/MS (70 eV) *m/z* (%) 345 (M$^+$, 100), 133 (100); ^{1}H NMR (300 MHz, CDCl$_3$): δ 3.84 (s, 3H, CH$_3$), 6.93–7.09 (m, 5H, Ar), 7.15–7.22 (m, 1H, Ar), 7.31–7.36 (m, 2H, Ar), 7.76–7.81 (m, 2H, Ar), 7.92–7.97 (m, 2H, Ar); ^{13}C NMR (125 MHz, CDCl$_3$): δ 55.9 (1C), 113.0 (1C), 117.1 (2C), 121.1 (1C), 122.0 (1C) 123.7 (2C), 125.5 (1C), 125.6 (1C), 127.9 (2C) 131.8 (2C), 134.4 (2C), 144.0 (1C), 151.7 (1C), 157.9 (1C), 167.5 (2C). Anal C$_{21}$H$_{15}$NO$_4$·H$_2$O (363.35): calcd %: C 69.63; H 4.91; N 3.54. Found: C 69.41; H 4.72; N 3.85.

Synthesis of 2-[4-(2-methylphenoxy)phenyl]-1H-isoindol-1,3(2H)-dione (**2k**). Prepared as described for **2a** starting from 4-(2-methylphenoxy)aniline. White solid. Yield: 33%: mp 123–124 °C (AcOEt/hexane); GC/MS (70 eV) *m/z* (%) 329 (M$^+$, 100); ^{1}H NMR (500 MHz, CDCl$_3$): δ 2.25 (s, 3H, CH$_3$), 6.97–7.02 (m, 3H, Ar), 7.11 (t, *J* = 7.4 Hz, 1H, Ar), 7.21 (t, *J* = 7.4 Hz, 1H, Ar), 7.25–7.29 (m, 1H, Ar), 7.32–7.36 (m, 2H, Ar), 7.77–7.81 (m, 2H, Ar), 7.92–7.97 (m, 2H, Ar); ^{13}C NMR (75 MHz, CDCl$_3$): δ 16.1 (1C), 117.3 (2C), 120.5 (1C), 123.7 (2C), 124.6 (1C), 125.6 (1C), 127.3 (1C), 127.9 (2C), 130.3 (1C), 131.6 (2C), 131.8 (2C), 134.3 (1C), 153.8 (1C), 157.7 (1C), 167.4 (2C). Anal C$_{21}$H$_{15}$NO$_3$·0.50 H$_2$O (338.11): calcd %: C 74.79; H 4.96; N 4.45. Found: 74.54; H 4.77; N 4.14.

Synthesis of 2-(2-phenoxyphenyl)-1H-isoindol-1,3(2H)-dione (**2l**). Prepared as described for **2a** starting from 2-phenoxyaniline. Light pink solid. Yield: 70%: mp 146–148 °C; GC/MS (70 eV) *m/z* (%) 315 (M$^+$, 100); ^{1}H NMR (300 MHz, CDCl$_3$): δ 6.96–7.10 (m, 4H, Ar), 7.18–7.31 (m, 3H, Ar), 7.33–7.42 (m, 2H, Ar), 7.71–7.79 (m, 2H, Ar), 7.87–7.95 (m, 2H, Ar). ^{13}C NMR (125 MHz, CDCl$_3$): δ 119.0 (1C), 119.6 (2C), 122.4 (1C), 123.4 (1C), 123.7 (2C), 123.9 (1C), 129.7 (2C), 130.2 (1C), 130.5 (1C), 132.1 (2C), 134.1 (2C), 154.0 (1C), 156.3 (1C), 167.0 (2C). Anal C$_{20}$H$_{13}$NO$_3$ (315.33): calcd %: C 76.06; H 4.20; N 4.43. Found: C 76.18; H 4.16; N 4.44.

2.2. Biological Methods

2.2.1. Cell Culture

The cell lines used in the present work (A2058, Sk-Mel28, MDA-MB-231, MCF-7, HeLa and HEK-293) were purchased from the American Type Culture Collection (ATCC, Manassas, VA, USA). All cell lines were maintained at 37 °C in a humidified atmosphere of 95% air and 5% CO_2 and periodically screened for contamination [20]. Human melanoma Sk-Mel-28 cells and human embryonic kidney HEK-293 cells, were cultured in Dulbecco's modified Eagle's medium (DMEM) high glucose (4.5 g/L) supplemented with 1% L-glutamine, 100 U/mL penicillin/streptomycin and 10% fetal bovine serum (FBS). The human melanoma cells A2058 were grown in DMEM low glucose (1 g/L) supplemented with 1% L-glutamine and 100 U/mL penicillin/streptomycin and 20% FBS. MCF-7 and MDA-MB-231 breast cancer cells were maintained in Dulbecco's modified Eagle's medium/nutrient mixture Ham F-12 (DMEM/F12), supplemented with 5% FBS and 100 U/mL penicillin/streptomycin. HeLa human epithelial cervix carcinoma cells were maintained in minimum essential Eagle's medium (MEM), supplemented with 10% FBS, 1% l-glutamine, 100 U mL 1 penicillin/streptomycin and 1% non-essential amino acids (NEAA).

2.2.2. MTT Assay

Cell viability was evaluated using the 3-(4,5-dimethylthiazol-2-y1)-2,5-diphenyltetrazolium (MTT, Sigma–Aldrich, Milan, Italy) assay as already reported [21]. Results are represented as percent (%) of basal and the IC_{50} values were calculated using GraphPad Prism 9 (GraphPad Software, La Jolla, CA, USA).

2.2.3. Immunofluorescence Analysis

Cells were seeded in 48-well culture plates containing glass slides and incubated with the most active compound for 24 h (concentration equal to its IC_{50} value), following a previously described protocol [22]. Rabbit anti β-tubulin was purchased from Santa Cruz Biotechnology and diluted to 1:100 before the use. The secondary antibody, Alexa Fluor® 568 conjugate goat-anti-mouse (1:500, Thermo Fisher Scientific, Waltham, MA, USA), was added and incubated for 2 h at 37 °C. Images were acquired using a fluorescence microscope (Leica DM 6000; 40× magnification). LAS-X software was used to process all images.

2.2.4. Tubulin Polymerization Assay

Tubulin polymerization inhibition was measured using an in vitro Tubulin Polymerization Assay Kit purchased from EMD Millipore Corporation [23], as already described [22]. The turbidity variation was measured every 30 s at 350 nm for 90 min, stirring the mixture for 10 s before each measurement.

2.2.5. TUNEL Assay

Apoptosis was detected using a TUNEL assay, according to the guidelines of the manufacturer (CFTM488A TUNEL Assay Apoptosis Detection Kit, Biotium, Hayward, CA, USA), as already reported [24]. Cells were observed and imaged under a fluorescence microscope (Leica DM6000; 20× magnification) with excitation/emission wavelength maxima of 490/515 nm (CF 488A) or 350/460 nm (DAPI). Images are representative of three independent experiments.

2.2.6. Western Blotting Analysis

Protein lysates were subjected to Western blot analysis, as already reported [25] Poly(ADP-ribose) polymerase 1 (Parp-1) and glyceraldehyde 3-phosphate dehydrogenase (GAPDH, used to verify the equal loading of proteins) were purchased from Santa Cruz Biotechnology (Santa Cruz, CA, USA). The antigen–antibody complex was detected by incubation of the membrane with a secondary antibody (peroxidase-coupled anti-rabbit

IgG, VWR International PBI, Milano, Italy). Immunoreactive bands were revealed using an ECL Western blotting detection system (Amersham Pharmacia Biotech, Piscataway, NJ, USA).

2.2.7. Docking Studies

The crystal structure of the complex formed between a tetrameric assembly of tubulin and vinblastine [26]; PDB code 5j2t, was used as the protein target for our docking simulations. All the structures of the ligands tested in silico were designed, their 3D structures were created, charges were assigned and further energy was minimized using the program MarvinSketch (ChemAxon ltd, Budapest, Hungary). To better characterize the binding mode and the mechanisms of action of the tested compounds to the protein target and to determine their binding energies, we used Autodock v.4.2.2. [27] and its graphical interface, ADT [28]. We performed all simulations using a searching grid with a cubic shape, composed of 126 points for each dimension, centered on the protein center. Nodes were spaced 0.375 Å from each other. Each of our simulation runs was performed adopting the default program values and the docking procedures were carried out as described in our previous works [20,29–32]. Briefly, the protein and the ligands were prepared using the ADT graphical interface by adding polar hydrogens to the protein and solvation parameters. During all the simulations, we considered the protein as a rigid object and the ligands as fully flexible. The Lamarckian genetic algorithm was used in our search with a population of 100 individuals that were evolved for 100 generations. The docking poses as resulting for our simulation runs were ranked in order of binding energy and collected into clusters on the basis of RMSD.

The generated docking poses were ranked in order of increasing binding energy values and clustered on the basis of a RMSD cut-off value of 2.0 Å. From the structural analyses of the lowest energy solutions of each cluster, we could identify the ligand binding mode. All the figures concerning the ligand binding modes to the protein target were drawn using the program Chimera [33].

3. Results and Discussion

3.1. Chemistry

The synthesis of the tested compounds was obtained by mixing equimolar quantities of phthalic anhydride and the suitable aryl amine in the presence of triethylamine (Scheme 1) [19].

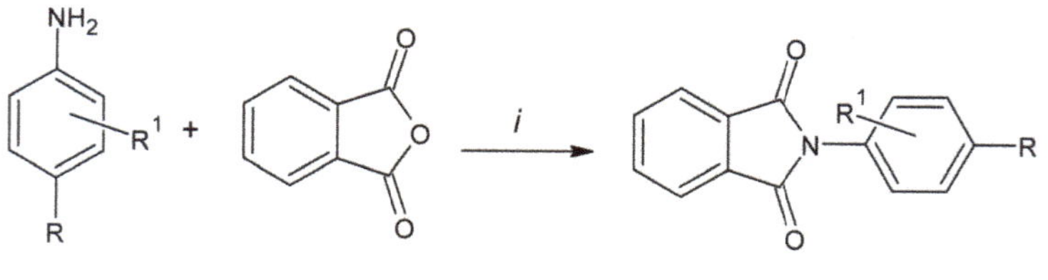

1a-l   **2a-l**

Scheme 1. Synthesis of tested compounds. Reagents and conditions: (i) Et$_3$N, toluene, Δ, 5 h.

3.2. Biological Results

3.2.1. Cell Viability Assay

The cytotoxic activities of synthesized compounds (**2a–l**) were examined against two human melanoma cell lines, A2058, derived from a metastatic site, and malignant melanoma cells, Sk-Mel28, two breast cancer cells, ER(+) MCF-7 and triple negative MDA-MB-231 and the human epithelial cervix carcinoma HeLa cells. The calculated IC$_{50}$ values of studied compounds in comparison with the reference drug, thalidomide, are listed in Table 1. As shown, thalidomide did not exhibit any activity on both the cell lines at the higher concentration tested. None of the compounds were shown to possess anticancer activity against the Sk-Mel28 cell line. Regarding the A2058 cell line, a slight activity was found for compound **2d**, which had no substituents on the phenoxy moiety. On the contrary, the presence of a chlorine substituent at the ortho position of the phenoxy nucleus seemed to enhance the activity since an IC_{50} of 23.26 ± 1.1 µM was calculated for compound **2i** on A2058 cells. A notable inhibitory effect on the viability of the A2058 cell line, with an IC_{50} of 15.37 ± 0.7 µM, was obtained for compound **2k**, bearing a methyl group at the ortho position of the phenoxy moiety, which performed best out of the compound in the series. Similarly, compound **2k** exerted a good antitumor activity against both breast cancer cells used in this assay (IC_{50} = 20.99 ± 0.8 and 22.72 ± 0.9 µM on MDA-MB-231 and MCF-7, respectively), while no activity was recorded on HeLa cells. Compound **2i** was able to reduce the breast cancer cells growth with IC_{50} values amounting to 25.34 ± 1.0 and 31.2 ± 1.2 µM on MDA-MB-231 and MCF-7 respectively, and, to a lesser extent, against cervix carcinoma HeLa cells growth (IC_{50} = 54.5 ± 0.8 µM). A slight anticancer activity was also recovered for compound **2d** towards the breast cancer cells MDA-MB-231 (IC_{50} = 76.5 ± 1.2 µM) and MCF-7 (IC_{50} = 177.0 ± 0.7 µM) and towards HeLa cells (IC_{50} = 42.1 ± 1.1 µM). No interesting activity was detected for the compounds lacking the phenoxy group against all cell lines used. To assess potential cytotoxicity towards normal cells, a screening of the compounds was performed on the non-tumoral human embryonic renal cell line Hek-293. As can be seen in Table 1, most of the compounds and thalidomide did not interfere with the normal cell viability, at least until the concentration of 200 µM and under the conditions used for this assay. A slight cytotoxicity was found for compound **2i**, though at a concentration 6-fold higher than the active one. A certain cytotoxicity against Hek-293 cell lines was found for compound **2c**, which, however, did not exhibit antitumor activity on the melanoma cells.

Table 1. IC_{50} values of compounds **2a–l** and thalidomide expressed in µM.

| | IC$_{50}$ (µM) | | | | | |
Compound	A2058	Sk-Mel28	MDA-MB-231	MCF-7	HeLa	Hek-293
2a	>200	>200	>200	>200	>200	>200
2b	>200	>200	>200	>200	>200	>200
2c	>200	>200	>200	>200	>200	65.63 ± 0.8
2d	124.7 ± 0.9	>200	76.5 ± 1.2	177.0 ± 0.7	42.1 ± 1.1	>200
2e	>200	>200	>200	>200	>200	>200
2f	>200	>200	>200	>200	>200	>200
2g	>200	>200	>200	>200	>200	>200
2h	>200	>200	>200	>200	>200	>200
2i	23.26 ± 1.1	>200	25.34 ± 1.0	31.2 ± 1.2	54.5 ± 0.8	149.2 ± 0.9
2j	>200	>200	>200	>200	>200	>200
2k	15.37 ± 0.7	>200	20.99 ± 0.8	22.72 ± 0.9	>200	>200
2l	>200	>200	>200	>200	>200	>200
thalidomide	>200	>200	>200	>200	>200	>200

3.2.2. Effect of Compound **2k** on Microtubule Dynamics and Cell Death by Apoptosis

Microtubules represent vital cell components shaped by polymerization of α and β tubulin heterodimers and play crucial roles in several cellular functions, including cell division [34]. Several classes of naturally occurring, as well as synthetic and semi-synthetic, compounds disrupt microtubule organization, altering the tubulin dynamics. Among them, vinca alkaloids and taxanes are two families of microtubule-binding agents currently used for the treatment of different solid and hematological tumors, and are characterized by strong side effects [35]. Generally, microtubule-binding agents are divided into two main groups: microtubule-destabilizing and microtubule-stabilizing agents, able to suppress microtubule dynamics with different mechanisms. Particularly, microtubule-destabilizing agents, including vinca alkaloids, inhibit the tubulin polymerization and thus destabilize microtubules; instead, taxanes, including paclitaxel, bind to polymerized microtubules, stabilizing tubulin filaments [36]. Recently, agents able to regulate microtubule dynamics, hampering tubulin polymerization or blocking microtubule disassembly, and free of adverse effects, have received great interest in anticancer therapy [37–39]. Given that our most active compound, **2k**, did not exert any cytotoxicity against the normal cells, we evaluated its ability to interfere with tubulin by using both an immunofluorescence assay and an in vitro tubulin-polymerization inhibition assay. Regarding the immunofluorescence assay, melanoma A2058 cells were exposed to compound **2k** or vinblastine, used as a reference molecule, for 24 h and then processed as described in the materials and methods. Figure 2 shows that, in the CTRL cells (vehicle-treated), tubulin resulted widely distributed in the cytoplasm, forming a regular arrangement and organization of the microtubules with normal and well-assembled filaments (Panel B, CTRL). On the contrary, A2058 cells exposed to vinblastine, as well as those treated with compound **2k**, reveled a dramatic disorganization of microtubules in which tubulin accumulated in the form of crystals within the cell cytoplasm (see white arrows, Figure 2, Panel B, vinblastine and **2k**). Thus, compound **2k** is able to perturb tubulin assembly, interfering with normal microtubule organization.

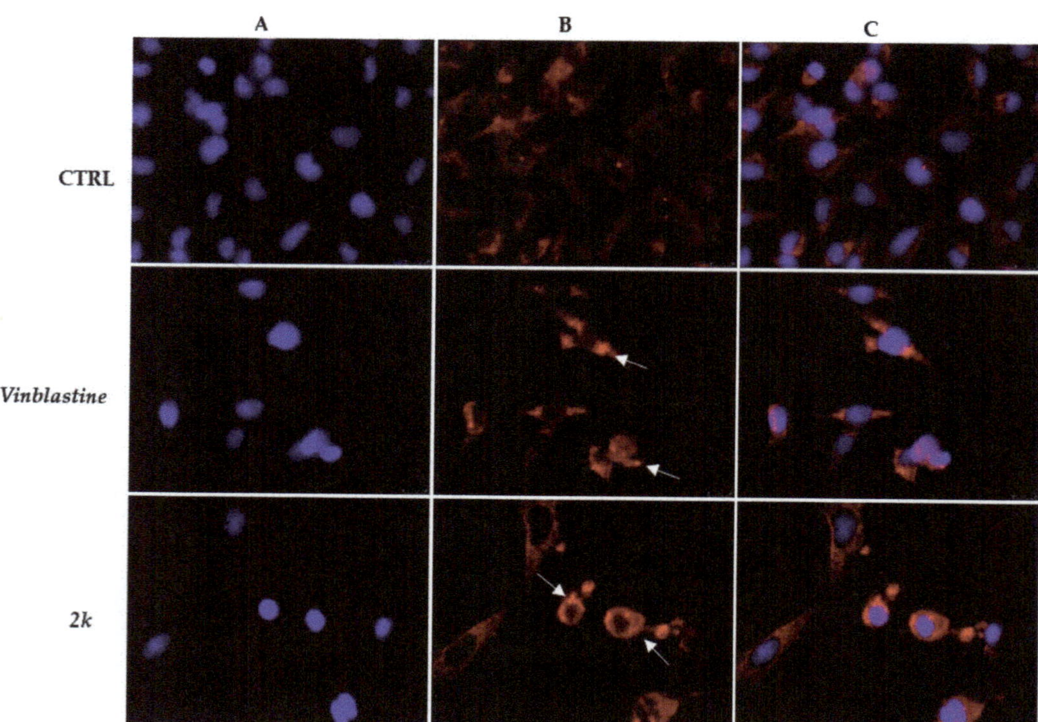

Figure 2. Immunofluorescence studies. Human melanoma A2058 cells were incubated with compound **2k**, with vinblastine (both used at their IC_{50} values), or with DMSO (CTRL) for 24 h. Cells were observed under the inverted fluorescence microscope at 40× magnification. CTRL cells displayed a normal arrangement and organization of the tubulin cytoskeleton. Instead, A2058 cells treated with compound **2k** and vinblastine exhibited an irregular microtubules organization (see white arrows). (**A**) nuclear stain with DAPI ($\lambda_{ex} = 350$ nm/$\lambda_{em} = 460$ nm); (**B**) β-tubulin (Alexa Fluor® 568; $\lambda_{ex} = 644$ nm/$\lambda_{em} = 665$ nm); (**C**) overlay channel. Representative fields are shown.

Next, we evaluated if the detected microtubule disorder could be linked to the ability of compound **2k** to act as a destabilizing agent, inhibiting tubulin polymerization, or as a stabilizing agent. Thus, we performed an in vitro tubulin-polymerization inhibition assay, using paclitaxel and vinblastine as reference molecules, and compound **2k**, all used at a concentration of 10 µM. In this assay, we measured the turbidity at 350 nm for 3600 s at 37 °C to evaluate the tubulin assembly. The obtained results, shown in Figure 3, demonstrated that, in the CTRL reaction, tubulin heterodimers polymerize in a time-dependent manner, reaching a plateau after about 15 min with a final optical density value at 350 nm (OD350) of about 0.48. Instead, the exposure of tubulin to the microtubule-stabilizing agent, paclitaxel, caused an increase in tubulin polymerization, which reached a plateau after about 10 min with a final OD350 higher than CTRL (OD350 paclitaxel = 0.56). Contrarily, the microtubule-destabilizing agent vinblastine strongly blocked the tubulin polymerization and its curve reached the steady state only after 70 min with a very low final OD350 (0.19). Similarly, the exposure of tubulin to the compound **2k** exerted an important inhibitory activity on the protein polymerization, even if in a lesser extent if compared with vinblastine. Indeed, the **2k** curve reaches the plateau in about 25 min with a turbidity value 1.5-fold less than the control reaction (OD350 **2k** = 0.32). Taken together, these results suggest that compound **2k** perturbs microtubule dynamics due to its capability to inhibit tubulin polymerization.

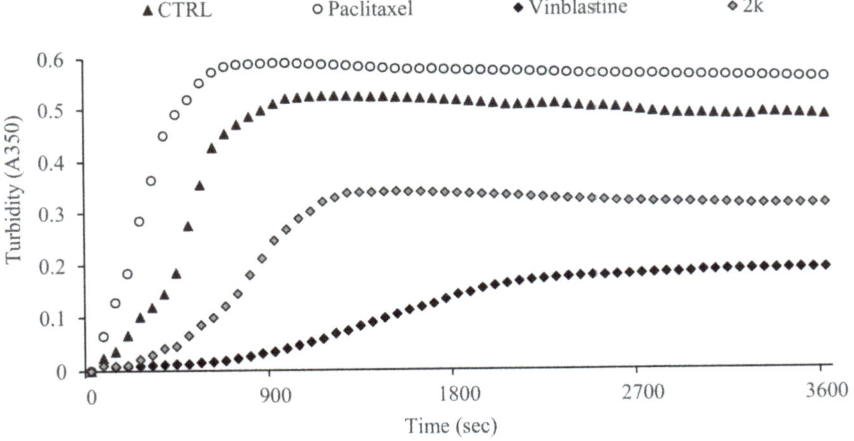

Figure 3. Tubulin polymerization assay. The in vitro tubulin polymerization was evaluated after tubulin incubation with compound **2k** used at the concentration of 10 µM. Vinblastine and paclitaxel (10 µM) were used as positive controls, while vehicle (DMSO) was used as a negative control. Tubulin polymerization was measured by determining the turbidity at 350 nm for 3600 s at 37 °C.

Due to the great involvement of microtubules in cellular growth, they represent a good target for cancer treatment. Indeed, the disorganization of microtubules can be responsible of cell cycle arrest, subsequently triggering the programmed cell death of tumoral cells [22,23,40–42]. In order to investigate if compound **2k** induced apoptosis, we performed a TUNEL assay on A2058 cells, exposed for 24 h to this compound, used at its IC$_{50}$ value. The obtained results indicate that compound **2k** induced apoptosis in these cells. Indeed, in the treated cells is notable a clear green nuclear fluorescence due to the formation of fragmented DNA, substrate of the terminal deoxynucleotidyl transferase (TdT) enzyme (Figure 4, Panel B, compound **2k**). Contrarily, green fluorescence is totally absent in the nuclei of the CTRL cells (vehicle-treated cells), indicating no DNA breakage (Figure 4, Panel B, CTRL).

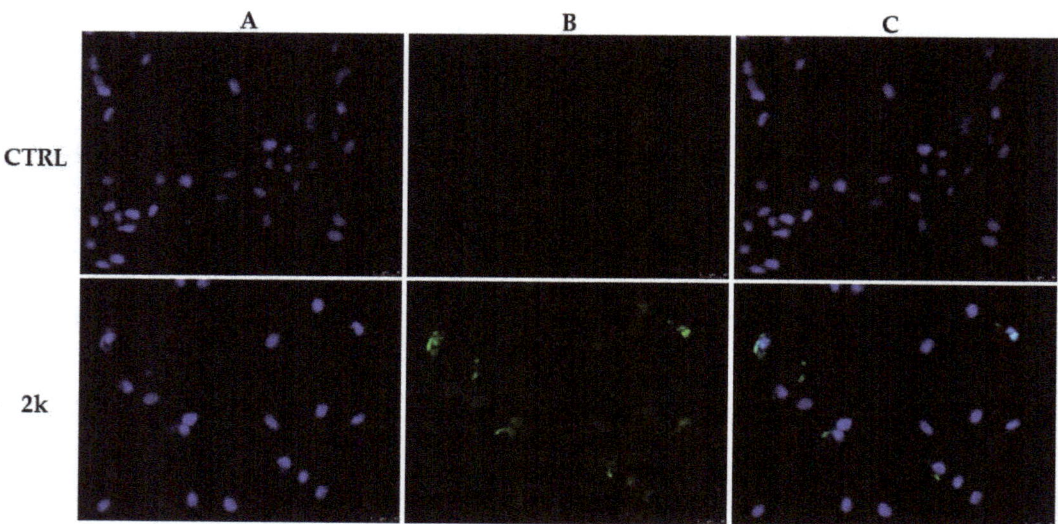

Figure 4. TUNEL assay. A2058 cells were treated with compound **2k** used at its IC$_{50}$ value or with the vehicle (CTRL) for 24 h. Apoptotic death is clearly indicated by the green nuclear fluorescence, present in the A2058 cells treated with compound **2k** and absent in the CTRL cells. Images were acquired under an inverted fluorescence microscope at 20× magnification. (**A**) DAPI (CTRL and compound **2k**) λ_{ex} = 350 nm/λ_{em} = 460 nm. (**B**) CF™488A (CTRL and compound **2k**) λ_{ex} = 490 nm/λ_{em} = 515 nm. (**C**) Overlay channel. Fields are representative of three separate experiments.

The **2k**-induced apoptosis of melanoma A2058 cells was accompanied by Poly(ADP-ribose) polymerase-1 (PARP-1) cleavage. During the apoptosis process, cells exhibit several biochemical modifications, including the degradation of many proteins by caspases [43]. Among them, PARP-1 is converted from the 116-kDa form to a fragment of 89 kDa. PARP-1 is an eukaryotic protein that plays a vital role in DNA repair, replication, and differentiation, and is involved in the activation of cellular defense mechanisms against DNA damage [44]. During apoptotic death, caspases cause PARP-1 cleavage and inactivation, making it no longer able to repair DNA damage [45]. Thus, considering the importance of PARP-1 cleavage as a hallmark of apoptosis, we analyzed the status of this protein in A2058 cells treated with compound **2k** (used at its IC$_{50}$ value) at different times (48 h and 72 h), performing a Western blot analysis. Treatment with compound **2k** induced proteolytic cleavage of PARP-1. Indeed, an accumulation of the 89-kDa cleavage fragment was observed (Figure 5, 48 h and 72 h) in a time dependent manner: the cleaved form of the protein was already visible 48 h after **2k** treatment and it gradually increased until 72 h. Contrarily, the native form of PARP-1 (116-kDa) decreased in a time-dependent manner in cells treated with compound **2k**, becoming hardly visible at 72 h. Instead, at time 0 h (Figure 5, 0 h), the uncleaved form is highly represented. Thus, these data indicate that apoptosis cell death induced by compound **2k** involves PARP-1 degradation.

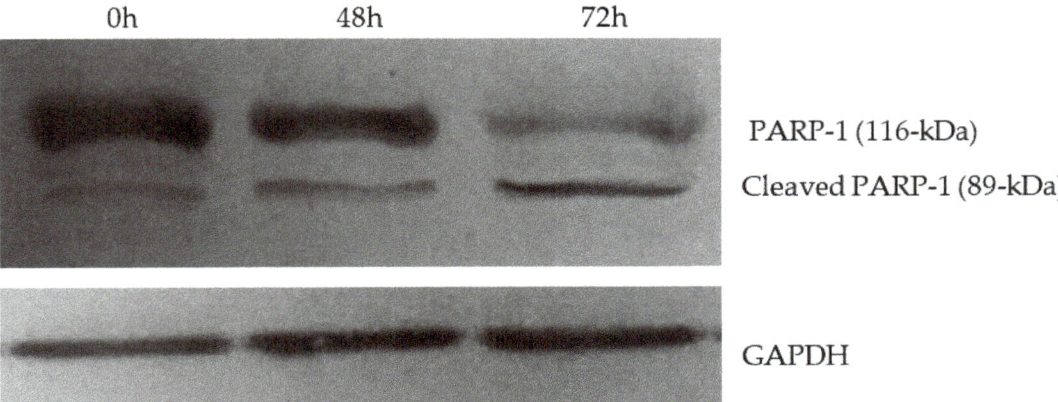

Figure 5. PARP-1 cleavage. A2058 melanoma were treated with compound **2k** (used at its IC$_{50}$ value) for 48 and 72 h and then the total protein extracts were analyzed by Western blot. PARP-1 native form (116-kDa) and the cleaved form (89 kDa) were visualized. Time points 0 h, 48 h and 72 h for protein extraction. GAPDH was used for loading normalization. Blots are representative of three independent experiments.

3.2.3. Docking Studies

To first evaluate the binding poses and the calculated affinities between our more active compounds (**2d**, **2i** and **2k**) and tubulin and in order to screen the molecules for the best lead candidate, we performed molecular docking simulations using, as a target, the three-dimensional structure of tubulin in complex with its well-known ligand vinblastin, a vinca alkaloid that inhibits the formation of microtubules within cells [26]. Affinities of the three compounds to the protein were calculated using Autodock according to the expression $K_i = \exp(\mathrm{deltaG}/(R^*T))$. As discussed in our previous works [46–48], to identify the most promising compound, we adopted a strategy based on the clusterization of the results from our simulation runs, together with the visual inspection of the ligand-protein binding mode. All our compounds share the same protein binding site with vinblastine, at the interface between the β- and α-tubulin subunits (Figure 6) as determined by X-ray crystallography [26]. All the tested compounds seem to be good ligands of the protein. They are almost superimposed over each other and share most of the interactions with the protein amino acids. Particularly, molecule **2d** is positioned within the vinblastine binding site of tubulin, forming hydrogen bonds with protein residues Asp β179 and Tyr β224. This binding mode is stabilized by hydrophobic interactions with residues Tyr β210, Phe β214, Val β217, Pro β222, Ile α332, and Phe α351. Ligands **2k** and **2i** share the same interactions described above with **2d**, but form an additional hydrophobic interaction with Val α328. The slightly different activity of **2i** in comparison with **2k** might be due to the hydrophobic environment surrounding the CH$_3$ group of **2k**, unlike the more hydrophilic chlorine atom present in ortho position of the phenoxy portion of **2i**.

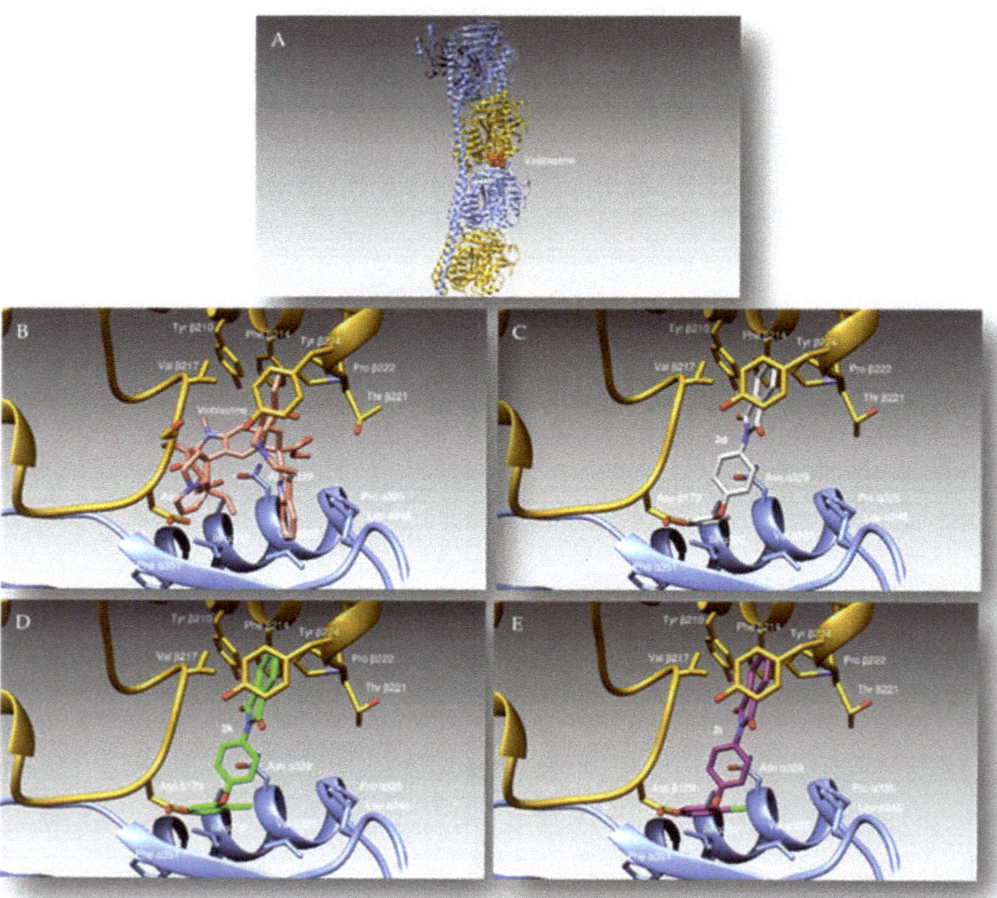

Figure 6. A schematic representation of the tetrameric 2α2β tubulin structure. α chain is reported as cyan ribbons, β as Golden ribbons (**A**). Compounds **2d**, **2k** and **2i** superpose to vinblastine binding site (**B**). (**C–E**) report the compounds **2d**, **2k** and **2i**, respectively, drawn as white, green and dark magenta sticks, bound to the vinblastine binding task of the protein.

4. Conclusions

Thalidomide, a drug well-known in the pharmaceutical history for its teratogenicity has been re-evaluated due to its antiangiogenic and immunomodulatory actions, which have made it effective against several malignancies. As a result, in the last decade, many researchers have concentrated their efforts on the design and development of new thalidomide analogs with improved efficacy and no or reduced toxicity. The focal point of our work was the synthesis and the biological evaluation of a series of thalidomide analogs (**2a–l**) for their potential anticancer activity toward a series of cell lines, including melanoma cells. Interestingly, compound **2k**, bearing a methyl group at the ortho position of the phenoxy moiety, exhibited a noteworthy cytotoxic effect against A2058 cells, thus becoming the most efficient compound of the series. A weak or no activity was detected for the other compounds. All the compounds under investigation have proven to be safe on the non-tumoral human embryonic renal cell line Hek-293. Furthermore, several studies agree on the importance of microtubule dynamics in governing the main cell functions, such as mitosis, maintaining of cell shape, cell division, intracellular transport, and chromosomes segregation. Following this, we tested the ability of compound **2k** to target tubulin polymerization performing both an immunofluorescence assay in A2058 cells and an in vitro tubulin-polymerization assay. Data extrapolated from both the experiments evidenced that the compound **2k** interferes with the microtubule dynamics because of its capability of preventing tubulin polymerization. These interesting findings were also supported by docking studies that shown the ability of **2k** to superpose to vinblastine binding site in tubulin structure. Next, we performed a TUNEL assay proving that compound **2k** can induce the apoptotic process in A2058 melanoma cell line. To conclude, although most of the examined compounds did not display an encouraging activity, we aimed to focus our attention on a single compound (**2k**) from which we obtained auspicious results that make it a promising ally in the fight against cancer. However, further studies must be carried out to optimize this candidate and validate its efficacy in vivo.

Author Contributions: Conceptualization, M.S.S.; methodology, D.I. and A.C. (Alessia Carocci); software, C.R.; validation, C.F., formal analysis, A.B. and A.C. (Alessia Catalano); investigation, A.B. and A.C. (Alessia Catalano); data curation, D.I. and J.C.; writing—original draft preparation, A.C. (Alessia Carocci) and A.B.; supervision, M.S.S. and C.F. All authors have read and agreed to the published version of the manuscript.

Funding: The work of C.R. has been partially supported by a grant from the Italian Ministry of Heatlth (Ricerca Corrente).

Institutional Review Board Statement: Not applicable.

Informed Consent Statement: Not applicable.

Conflicts of Interest: The authors declare no conflict of interest.

References

1. Sung, H.; Ferlay, J.; Siegel, R.L.; Laversanne, M.; Soerjomataram, I.; Jemal, A.; Bray, F. Global cancer statistics 2020: GLOBOCAN estimates of incidence and mortality worldwide for 36 cancers in 185 countries. *CA Cancer J. Clin.* **2021**, *71*, 209–249. [CrossRef]
2. Paolino, G.; Bekkenk, M.W.; Didona, D.; Eibenschutz, L.; Richetta, A.G.; Cantisani, C.; Viti, G.P.; Carbone, A.; Buccini, P.; De Simone, P.; et al. Is the prognosis and course of acral melanoma related to site-specific clinicopathological features? *Eur. Rev. Med. Pharmacol. Sci.* **2016**, *20*, 842–848. [PubMed]
3. Scali, E.; Mignogna, C.; Di Vito, A.; Presta, I.; Camastra, C.; Donato, G.; Bottoni, U. Inflammation and macrophage polarization in cutaneous melanoma: Histopathological and immunohistochemical study. *Int. J. Immunopathol. Pharmacol.* **2016**, *29*, 715–719. [CrossRef] [PubMed]
4. Chang, Q.; Long, J.; Hu, L.; Chen, Z.; Li, Q.; Hu, G. Drug repurposing and rediscovery: Design, synthesis, and preliminary biological evaluation of 1-arylamino-3-aryloxypropan-2-ols as anti-melanoma agents. *Bioorg. Med. Chem.* **2020**, *28*, 115404. [CrossRef] [PubMed]
5. Mercurio, A.; Adriani, G.; Catalano, A.; Carocci, A.; Rao, L.; Lentini, G.; Cavalluzzi, M.M.; Franchini, C.; Vacca, A.; Corbo, F. A mini-review on thalidomide: Chemistry, mechanisms of action, therapeutic potential and anti-angiogenic properties in multiple myeloma. *Curr. Med. Chem.* **2017**, *24*, 2736–2744. [CrossRef] [PubMed]
6. Zhang, S.; Li, M.; Gu, Y.; Liu, Z.; Xu, S.; Cui, Y.; Sun, B. Thalidomide influences growth and vasculogenic mimicry channel formation in melanoma. *J. Exp. Clin. Cancer Res.* **2008**, *27*, 1–9. [CrossRef]
7. Rashid, A.; Kuppa, A.; Kunwar, A.; Panda, D. Thalidomide (5HPP-33) suppresses microtubule dynamics and depolymerizes the microtubule network by binding at the vinblastine binding site on tubulin. *Biochemistry* **2015**, *54*, 2149–2159. [CrossRef]
8. Inatsuki, S.; Noguchi, T.; Miyachi, H.; Oda, S.; Iguchi, T.; Kizaki, M.; Hashimoto, Y.; Kobayashi, H. Tubulin-polymerization inhibitors derived from thalidomide. *Bioorg. Med. Chem. Lett.* **2005**, *15*, 321–325. [CrossRef]
9. Ghobrial, I.M.; Rajkumar, S.V. Management of thalidomide toxicity. *J. Support. Oncol.* **2003**, *1*, 194–205.
10. Pessoa, C.; Ferreira, P.M.P.; Lotufo, L.V.C.; de Moraes, M.O.; Cavalcanti, S.M.; Coêlho, L.C.D.; Hernandes, M.Z.; Leite, A.C.L.; De Simone, C.A.; Costa, V.M.; et al. Discovery of phthalimides as immunomodulatory and antitumor drug prototypes. *ChemMedChem Chem. Enabling Drug Discov.* **2010**, *5*, 523–528. [CrossRef]
11. Da Costa, P.M.; da Costa, M.P.; Carvalho, A.A.; Cavalcanti, S.M.T.; de Oliveira Cardoso, M.V.; de Oliveira Filho, G.B.; Viana, D.D.A.; Fechine-Jamacaru, F.V.; Leite, A.L.; De Moraes, M.O.; et al. Improvement of in vivo anticancer and antiangiogenic potential of thalidomide derivatives. *Chem. Biol. Interact.* **2015**, *239*, 174–183. [CrossRef]
12. Saturnino, C.; Caruso, A.; Longo, P.; Capasso, A.; Pingitore, A.; Cristina Caroleo, M.; Cione, E.; Perri, M.; Nicolo, F.; Mollica Nardo, V.; et al. Crystallographic study and biological evaluation of 1, 4-dimethyl-N-alkylcarbazoles. *Curr. Top. Med. Chem.* **2015**, *15*, 973–979. [CrossRef] [PubMed]

13. Ceramella, J.; Caruso, A.; Occhiuzzi, M.A.; Iacopetta, D.; Barbarossa, A.; Rizzuti, B.; Dallemagne, P.; Rault, S.; El-Kashef, H.; Saturnino, C.; et al. Benzothienoquinazolinones as new multi-target scaffolds: Dual inhibition of human Topoisomerase I and tubulin polymerization. *Eur. J. Med. Chem.* **2019**, *181*, 111583. [CrossRef] [PubMed]

14. Sinicropi, M.S.; Iacopetta, D.; Rosano, C.; Randino, R.; Caruso, A.; Saturnino, C.; Muià, N.; Ceramella, J.; Puoci, F.; Rodriquez, M. et al. N-thioalkylcarbazoles derivatives as new anti-proliferative agents: Synthesis, characterisation and molecular mechanism evaluation. *J. Enzym. Inhib. Med. Chem.* **2018**, *33*, 434–444. [CrossRef] [PubMed]

15. Ceramella, J.; Mariconda, A.; Rosano, C.; Iacopetta, D.; Caruso, A.; Longo, P.; Sinicropi, M.S.; Saturnino, C. α–ω Alkenyl-bis-S-Guanidine Thiourea Dihydrobromide Affects HeLa Cell Growth Hampering Tubulin Polymerization. *ChemMedChem* **2020**, *15*, 2306–2316. [CrossRef] [PubMed]

16. Iacopetta, D.; Lappano, R.; Mariconda, A.; Ceramella, J.; Sinicropi, M.S.; Saturnino, C.; Talia, M.; Cirillo, F.; Martinelli, F.; Puoci, F.; et al. Newly Synthesized Imino-Derivatives Analogues of Resveratrol Exert Inhibitory Effects in Breast Tumor Cells. *Int. J. Mol. Sci.* **2020**, *21*, 7797. [CrossRef]

17. Iacopetta, D.; Carocci, A.; Sinicropi, M.S.; Catalano, A.; Lentini, G.; Ceramella, J.; Curcio, R.; Caroleo, M.C. Old drug scaffold, new activity: Thalidomide-correlated compounds exert different effects on breast cancer cell growth and progression. *ChemMedChem* **2017**, *12*, 381–389. [CrossRef]

18. Aliabadi, A.; Gholamine, B.; Karimi, T. Synthesis and antiseizure evaluation of isoindoline-1,3-dione derivatives in mice. *Med. Chem. Res.* **2014**, *23*, 2736–2743. [CrossRef]

19. Assis, S.P.O.; Araújo, T.G.; Sena, V.L.; Catanho, M.T.J.; Ramos, M.N.; Srivastava, R.M.; Lima, V.L. Synthesis, hypolipidemic, and anti-inflammatory activities of arylphthalimides. *Med. Chem. Res.* **2014**, *23*, 708–716. [CrossRef]

20. Iacopetta, D.; Grande, F.; Caruso, A.; Mordocco, R.A.; Plutino, M.R.; Scrivano, L.; Ceramella, J.; Muià, N.; Saturnino, C.; Puoci, F.; et al. New insights for the use of quercetin analogs in cancer treatment. *Future Med. Chem.* **2017**, *9*, 2011–2028. [CrossRef]

21. Fazio, A.; Iacopetta, D.; La Torre, C.; Ceramella, J.; Muià, N.; Catalano, A.; Carocci, A.; Sinicropi, M.S. Finding solutions for agricultural wastes: Antioxidant and antitumor properties of pomegranate Akko peel extracts and β-glucan recovery. *Food Funct.* **2018**, *9*, 6618–6631. [CrossRef]

22. Ceramella, J.; Loizzo, M.R.; Iacopetta, D.; Bonesi, M.; Sicari, V.; Pellicanò, T.M.; Saturnino, C.; Malzert-Fréon, A.; Tundis, R.; Sinicropi, M.S. Anchusa azurea Mill. (Boraginaceae) aerial parts methanol extract interfering with cytoskeleton organization induces programmed cancer cells death. *Food Funct.* **2019**, *10*, 4280–4290. [CrossRef]

23. Iacopetta, D.; Rosano, C.; Sirignano, M.; Mariconda, A.; Ceramella, J.; Ponassi, M.; Saturnino, C.; Sinicropi, M.S.; Longo, P. Is the way to fight cancer paved with gold? Metal-based carbene complexes with multiple and fascinating biological features. *Pharmaceuticals* **2020**, *13*, 91. [CrossRef]

24. Tundis, R.; Iacopetta, D.; Sinicropi, M.S.; Bonesi, M.; Leporini, M.; Passalacqua, N.G.; Ceramella, J.; Menichini, F.; Loizzo, M. Assessment of antioxidant, antitumor and pro-apoptotic effects of Salvia fruticosa Mill. subsp. thomasii (Lacaita) Brullo, Guglielmo, Pavone & Terrasi (Lamiaceae). *Food Chem. Toxicol.* **2017**, *106*, 155–164.

25. Rechoum, Y.; Rovito, D.; Iacopetta, D.; Barone, I.; Andò, S.; Weigel, N.L.; O'Malley, B.W.; Brown, P.H.; Fuqua, S.A.W. AR collaborates with ERα in aromatase inhibitor-resistant breast cancer. *Breast Cancer Res. Treat.* **2014**, *147*, 473–485. [CrossRef]

26. Waight, A.B.; Bargsten, K.; Doronina, S.; Steinmetz, M.O.; Sussman, D.; Prota, A.E. Structural Basis of Microtubule Destabilization by Potent Auristatin Anti-Mitotics. *PLoS ONE* **2016**, *11*, e0160890. [CrossRef]

27. Morris, G.M.; Huey, R.; Lindstrom, W.; Sanner, M.F.; Belew, R.K.; Goodsell, D.S.; Olson, A.J. Autodock4 and AutoDockTools4: Automated docking with selective receptor flexiblity. *J. Comput. Chem.* **2009**, *30*, 2785–2791. [CrossRef]

28. Sanner, M.F.; Duncan, B.S.; Carrillo, C.J.; Olson, A.J. Integrating computation and visualization for biomolecular analysis: An example using python and AVS. *Pac. Symp. Biocomput.* **1999**, *8*, 401–412.

29. Cesarini, S.; Spallarossa, A.; Ranise, A.; Schenone, S.; Rosano, C.; La Colla, P.; Sanna, G.; Busonera, B.; Loddo, R. N-acylated and N,N'-diacylated imidazolidine-2-thione derivatives and N,N'-diacylated tetrahydropyrimidine-2(1H)-thione analogues: Synthesis and antiproliferative activity. *Eur. J. Med. Chem.* **2009**, *44*, 1106–1118. [CrossRef]

30. Viale, M.; Cordazzo, C.; de Totero, D.; Budriesi, R.; Rosano, C.; Leoni, A.; Ioan, P.; Aiello, C.; Croce, M.; Andreani, A.; et al. Inhibition of MDR1 activity and induction of apoptosis by analogues of nifedipine and diltiazem: An in vitro analysis. *Investig. New Drugs* **2011**, *29*, 98–109. [CrossRef]

31. Rosano, C.; Lappano, R.; Santolla, M.F.; Ponassi, M.; Donadini, A.; Maggiolini, M. Recent advances in the rationale design of GPER ligands. *Curr. Med. Chem.* **2012**, *19*, 6199–6206. [CrossRef]

32. Saturnino, C.; Iacopetta, D.; Sinicropi, M.S.; Rosano, C.; Caruso, A.; Caporale, A.; Marra, N.; Marengo, B.; Pronzato, M.A.; Parisi, O.I.; et al. N-alkyl carbazole derivatives as new tools for Alzheimer's disease: Preliminary studies. *Molecules* **2014**, *19*, 9307–9317. [CrossRef]

33. Pettersen, E.F.; Goddard, T.D.; Huang, C.C.; Couch, G.S.; Greenblatt, D.M.; Meng, E.C.; Ferrin, T.E. UCSF Chimera- A visualization system for exploratory research and analysis. *J. Comput. Chem.* **2004**, *25*, 1605–1612. [CrossRef]

34. Brouhard, G.J.; Rice, L.M. Microtubule dynamics: An interplay of biochemistry and mechanics. *Nat. Rev. Mol. Cell Biol.* **2018**, *19*, 451–463. [CrossRef]

35. Zhang, D.; Kanakkanthara, A. Beyond the paclitaxel and Vinca alkaloids: Next generation of plant-derived microtubule-targeting agents with potential anticancer activity. *Cancers* **2020**, *12*, 1721. [CrossRef] [PubMed]

36. Dumontet, C.; Jordan, M.A. Microtubule-binding agents: A dynamic field of cancer therapeutics. *Nat. Rev. Drug Discov.* **2010**, *9*, 790–803. [CrossRef]
37. Yele, V.; Pindiprolu, S.K.S.; Sana, S.; Ramamurty, D.S.V.N.M.; Madasi, J.R.; Vadlamani, S. Synthesis and preclinical evaluation of indole triazole conjugates as microtubule targeting agents that are effective against MCF-7 breast cancer cell lines. *Anti-Cancer Agents Med. Chem.* **2021**, *21*, 1047–1055. [CrossRef]
38. Karahalil, B.; Yardım-Akaydin, S.; Nacak Baytas, S. An overview of microtubule targeting agents for cancer therapy. *Arch. Hyg. Rada Toksikol.* **2019**, *70*, 160–172. [CrossRef]
39. Kaur, R.; Kaur, G.; Gill, R.K.; Soni, R.; Bariwal, J. Recent developments in tubulin polymerization inhibitors: An overview. *Eur. J. Med. Chem.* **2014**, *87*, 89–124. [CrossRef]
40. Liu, J.; Xue, D.; Zhu, X.; Yu, L.; Mao, M.; Liu, Y. Anticancer evaluation of a novel dithiocarbamate hybrid as the tubulin polymerization inhibitor. *Investig. New Drugs* **2020**, *38*, 525–532. [CrossRef]
41. Donthiboina, K.; Anchi, P.; Ramya, P.S.; Karri, S.; Srinivasulu, G.; Godugu, C.; Shankaraiah, N.; Kamal, A. Synthesis of substituted biphenyl methylene indolinones as apoptosis inducers and tubulin polymerization inhibitors. *Bioorg. Chem.* **2019**, *86*, 210–223. [CrossRef] [PubMed]
42. Iacopetta, D.; Catalano, A.; Ceramella, J.; Barbarossa, A.; Carocci, A.; Fazio, A.; La Torre, C.; Caruso, A.; Ponassi, M.; Rosano, C.; et al. Synthesis, anticancer and antioxidant properties of new indole and pyranoindole derivatives. *Bioorg. Chem.* **2020**, *105*, 104440. [CrossRef]
43. Elmore, S. Apoptosis: A review of programmed cell death. *Toxicol. Pathol.* **2007**, *35*. 495–516. [CrossRef] [PubMed]
44. Almahli, H.; Hadchity, E.; Jaballah, M.Y.; Daher, R.; Ghabbour, H.A.; Kabil, M.M.; Al-Shakliah, N.S.; Eldehna, W.M. Development of novel synthesized phthalazinone-based PARP-1 inhibitors with apoptosis inducing mechanism in lung cancer. *Bioorg. Chem.* **2018**, *77*, 443–456. [CrossRef]
45. Chaitanya, G.V.; Alexander, J.S.; Babu, P.P. PARP-1 cleavage fragments: Signatures of cell-death proteases in neurodegeneration. *Cell. Commun. Signal* **2010**, *8*, 1–11. [CrossRef] [PubMed]
46. Lappano, R.; Rosano, C.; Pisano, A.; Santolla, M.F.; De Francesco, E.M.; De Marco, P.; Dolce, V.; Ponassi, M.; Felli, L.; Cafeo, G.; et al. Calixpyrrole derivative acts as an antagonist to GPER, a G-protein coupled receptor: Mechanisms and models. *Dis. Model. Mech.* **2015**, *8*, 1237–1246. [CrossRef]
47. Sinicropi, M.S.; Lappano, R.; Caruso, A.; Santolla, M.F.; Pisano, A.; Rosano, C.; Capasso, A.; Panno, A.; Lancelot, J.C.; Rault, S.; et al. (6-bromo-1,4-dimethyl-9H-carbazol-3-yl-methylene)-hydrazine (carbhydraz) acts as a GPER agonist in breast cancer cells. *Curr. Top. Med. Chem.* **2015**, *15*, 1035–1042. [CrossRef] [PubMed]
48. Stec-Martyna, E.; Ponassi, M.; Miele, M.; Parodi, S.; Felli, L.; Rosano, C. Structural comparison of the interaction of tubulin with various ligands affecting microtubule dynamics. *Curr. Cancer Drug Targets* **2012**, *12*, 658–666. [CrossRef]

MDPI

St. Alban-Anlage 66

4052 Basel

Switzerland

Tel. +41 61 683 77 34

Fax +41 61 302 89 18

www.mdpi.com

Applied Sciences Editorial Office

E-mail: applsci@mdpi.com

www.mdpi.com/journal/applsci

www.ingramcontent.com/pod-product-compliance
Lightning Source LLC
LaVergne TN
LVHW070852170726
843515LV00005B/627